BUILDING THE CASE
FOR BIOTECHNOLOGY

BUILDING THE CASE FOR BIOTECHNOLOGY

Management Case Studies in Science, Laws, Regulations, Politics, and Business

Mark J. Ahn, Michael A. Alvarez, Arlen D. Meyers, Anne S. York

LOGOS PRESS

BUILDING THE CASE FOR BIOTECHNOLOGY
Management Case Studies in Science, Laws, Regulations, Politics, and Business
Mark J. Ahn, Michael A. Alvarez, Arlen D. Meyers, Anne S. York

Published in The United States of America
by
Logos Press, Washington DC
www.Logos-Press.com
info@Logos-Press.com

10 9 8 7 6 5 4 3 2 1

ISBN-13
Softcover: 978-1-934899-15-1
Harcover: 978-1-934899-16-8

Cover images:
Courtesy of NASA, wikipedia user user Coolcaesar, and Flickr user thenails1

Library of Congress Cataloging-in-Publication Data

Building the case for biotechnology : management case studies in science, laws, regulations, politics, and business / Mark J. Ahn ... [et al.].
p. cm.
Includes bibliographical references.
ISBN 978-1-934899-15-1
1. Biotechnology industries--Management--Case studies. I. Ahn, Mark J.
HD9999.B442B85 2010
660.6068--dc22
2010020307

Contents

Foreword iii

Preface v

Introduction 1

How to Teach Bioscience Using the Case Discussion Method 3

PART I: SCIENCE   15

Medarex: Realizing its Potential? 17

FoxHollow Technologies: The SilverHawk® Cuts Open a New Market 37

OraPharma: Reformulation of an Existing Product 51

DesignMedix: Maintain Product Focus or Leverage Technology Platform? 73

Oxigene: Realizing Value from Multiple Technology Platforms 89

PART II: LAWS, REGULATIONS, POLITICS   105

Tysabri Re-launch Decision: Promise and Perils of Addressing Unmet Needs 107

XDx: Navigating Regulatory & Reimbursement Challenges 121

Dyadic International: From Doom to Dawn—What's Next? 143

The Prince Edward Island Bioscience Cluster: Creating a Knowledge-Based Economy 167

The Founding and Growth of On-Q-ity: Developing Advances in Personalized Medicine 189

Airway Tools Company: Changing Medical Device Standards of Care 201

PART III: BUSINESS   217

Genentech Acquisition by Roche: Will Innovation Wither? 219

Myogen: Are We There Yet? 245

Compression Dynamics: In Search of Sales 269

iKaryos Diagnostics: The Rocky Road from Concept to Startup 277

Biocon: From Local to Global 303

Adnexus: Strategic and Resource Considerations When Developing Novel Biotechnology Medicines 317

Gardasil: From University Discovery to Global Blockbuster Drug 331

Guru Instruments: Bootstrapping a Medical Device Startup 349

Growing Pains at Camelot Biopharmaceuticals 367

Sandhill Scientific: Where to Manufacture? 387

Lumina Life Sciences: The Challenges of Raising Capital to take to Market a Promising Technology Innovation 395

RESOURCES   409

	Technology Type			Geographic Focus		Company Status	
	Biopharmaceuticals	Molecular Diagnostics	Medical Devices	US	Int'l	Private	Public
PART I: SCIENCE							
Medarex-Bristol Myers Squibb Company: Realizing Its Potential?	X			X			X
FoxHollow Technologies: The SilverHawk Cuts Open a New Market			X	X		X	
OraPharma: Reformulation of an Existing Product	X			X			X
DesignMedix: Maintain Product Focus or Leverage Technology Platform?	X			X		X	
Oxigene: Realizing Value from Multiple Technology Platforms	X			X			X
PART II: LAWS, REGULATIONS AND POLITICS							
Biogen-Idec and Elan: Promise and Perils of Addressing Unmet Needs—Tysabri Re-launch Decision	X			X			X
XDx: Navigating Regulatory & Reimbursement Challenges		X		X		X	
Dyadic International: From Doom to Dawn—What's Next?				X	X		X
The Prince Edward Island Bioscience Cluster: Creating a Knowledge-Based Economy	X	X	X		X		
Developing Advances in Personalized Medicine: The Founding and Growth of On-Q-Ity		X		X		X	
Changing Medical Device Standards of Care: Airway Tools Company			X		X	X	
PART III: THE BUSINESS OF BIOTECHNOLOGY							
Genentech-Roche: Will Innovation Whither?	X			X	X		X
Myogen-Gilead: Are We There Yet?	X			X			X
Compression Dynamics: In Search of Sales			X	X		X	
iKaryos Diagnostics: The Rocky Road from Concept to Startup		X		X		X	
Biocon: From Local to Global	X				X		X
Adnexus: Strategic and Resource Considerations When Developing Novel Biotechnology Medicines	X			X		X	
Gardasil: From University Discovery to Global Blockbuster Drug	X				X		X
Guru Instruments: Bootstrapping a Bioscience Device Startup			X	X		X	
Growing Pains at Camelot Biopharmaceuticals	X			X		X	
Sandhill Scientific: Where to Manufacture?			X	X		X	
Lumina Life Sciences: The Challenges of Raising Capital to take to Market a Promising Technology Innovation		X			X	X	

Foreword

G. STEVEN BURRILL
CEO, Burrill & Company

The management and boards of directors of life science companies face many critical decisions in their long journey from idea to the marketplace. To move beyond the point from which good science becomes good business, companies will be faced with extremely tough choices. In their formative years these will mainly revolve around financings. When will be the best time to finance? What valuation and how much ownership of the company must be ceded in order to obtain the necessary funding required to advance lead products along the pipeline?

If all goes well and as companies mature they will face another difficult choice about whether to "roll the dice" and persevere toward becoming vertically integrated and building development capabilities, manufacturing facilities and marketing and sales teams, or use partners and share success, or simply be acquired. The excellently written case of Myogen, a development-stage biotechnology company and its reaction to an acquisition offer from Gilead Sciences is typical of the types of decision making that this exciting industry of ours faces on a regular basis.

You cannot but marvel at the fact that the biotechnology industry has been something of an economic miracle. The expectations universally held by biotechnology company management - that capital would always be there to help fulfill development milestones – have by and large been met. However, the collapse of the global financial markets and financial systems is a major force in reshaping biotech's landscape. It is helping accelerate changes already underway ranging from globalization to healthcare reform to climate change.

Venture investors, often the umbilical cord of food/fuel for this industry, long provided the lifeblood for young biotechnology companies, but the model is more challenged today. Venture investors don't have the ability to provide round after round of financing and wait 10 to 15 years for a return. Fortunately, with the evolution of a more "virtual" business model where companies can piece together the resources they need as they go, venture investors are now looking for alternative structures with which to build value (virtual companies, proof-of-concept financing of projects, building companies post POC, etc.). Today, there will be a movement to finance individual projects rather than companies. Teams will come together around an idea of a potential product, raise money to advance it, and then sell it to an established company.

We are already seeing significant evidence that companies are adapting – particularly in the pharmaceutical industry. There is recognition that their blockbuster business model is now a thing of the past and we are seeing a paradigm shift in their strategies to evolve towards targeted medicines in line with the era of personalized medicine. Pharmas are also adapting because they are facing massive erosion to their bottom lines as blockbuster drugs fall out of patent to be replaced by generics that have been eagerly waiting in the wings.

For their part, biotech companies are adapting to accommodate the fact that their essential partners are on this track. With capital in short supply, many biotech companies have been showcasing their best assets and shelving others. We have seen a huge surge in partnering deals as a result and expect this to continue. Add to the mix healthcare reform, globalization, biogenerics and an investment community that, by and large, has removed the word "risk" from its vocabulary and you have a world like no other we have experienced.

Innovative products will not only be judged by how well they perform, but whether they represent true improvements over existing products and do so on a cost-effective basis. It will not be enough just to do something better. Pharma, biotech and medical device companies will also need to provide true economic savings and better patient outcomes. And, as regulators put increasing attention on safety concerns, this will also raise the bar for new technologies. This could make it harder for companies pursuing unproven technologies, particularly new classes of drugs, to raise money. It also threatens to increase the cost of development to meet more rigorous clinical trial standards.

The dizzying pace of innovation promises to redefine healthcare as we know it. The walls that have traditionally separated scientific disciplines are falling and the combination of biotechnology with nanotechnology and information and communication technologies (visualization, improved power supplies, wireless communication and access globally) will transform the way doctors monitor, diagnose, and treat disease. The question is to what extent new technologies will be embraced for their ability to cut healthcare costs by shifting the paradigm of medicine away from treating disease toward keeping people well.

There is no doubt that the business of biotechnology is conducted on a global scale. Until comparatively recently the prominence of China, India, Latin America have come as the result of biotech/pharma industries in the US, Europe and Japan taking advantage of access to comparatively cheap labor, production and untapped populations to conduct clinical trials.

Global arbitrage will be extremely important as a survival strategy. Companies will need to look further afield for financing and potential partners such as the rapidly developing BRIC countries – Brazil, Russia, India and China.

To effectively create company value and achieve global competitiveness requires experience and knowledge. I believe that more resources are needed to help motivate our future biotech entrepreneurs.

I am delighted with this book's goals as I am sure it will engage students, scientists and managers interested in putting biotechnology to work to create valuable companies and manage risk. My own company's analysis reveals those companies that are the most responsive to the changes around them will thrive and be successful. It will be important to understand this new environment in which biotech companies now operate.

The business of biotechnology is as exciting today as when I started out over 40 years ago and I cannot wait for it to unfold into new and opportunistic areas of endeavor.

Preface

Sir William Osler, one of the founding fathers of the North American medical education system at the turn of the 20th century, famously quipped: "To study the phenomenon of disease without books is to sail an uncharted sea, while to study books without patients is not to go to sea at all." The same holds true for students of business, particularly in the emerging and complex biomedical industry. Simply studying the anatomy and physiology of companies, without actually working in them, might give students the knowledge they need, but not the skills and attitudes derived from actually participating in a startup, spin-out or emerging biopharmaceutical, diagnostic or device company. Biotechnology-based businesses are truly unique in that companies of all sizes are creating and exploring fundamentally new scientific insights while simultaneously developing novel commercial products.

Case studies capture the complexity and context of the biotech industry, and effectively create a bridge between studying books and working in a business. We created this book to fill that gap in biotechnology entrepreneurship and management education. The case method of teaching is a proven and engaging pedagogical approach that is part and parcel of medical, business and legal education. By simulating and analyzing problems, students have the opportunity to practice decision-making in uncertainty with little business risk (although possibly substantial academic risk), create alternative solutions, and gain the benefit of seeing how their recommendations compare to the actual outcomes.

We all, particularly scientific entrepreneurs and managers, learn more from doing. Simulating successes and failures, as well as how to cope and learn from those experiences, is a critical survival skill that can be honed by case study analysis. We hope this volume helps to fill the void in life science entrepreneurship and management case books and provides faculty and students with not only the charts, but the simulated experience of sailing the turbulent and exciting oceans of the biomedical industry towards creating significant value for patients and society.

ABOUT THE EDITORS

Mark J. Ahn is associate professor, global management at the Atkinson Graduate School of Management, Willamette University; and principal at Pukana Partners, Ltd. that provides strategic consulting to life science companies. Dr. Ahn's teaching and research interests include global strategy and innovation management; biotechnology and life science industries; leadership and transformational change; and social entrepreneurship. Dr. Ahn was founder, president, and chief executive officer of Hana Biosciences, and also held senior management positions at Genentech, Amgen and Bristol-Myers Squibb Company. Dr. Ahn also serves on the board of directors of public and private biotechnology companies.

Michael A. Alvarez is founding director of the Stanford School of Medicine Career Center. His professional experience spans 17 years working at the cross-section of business and academia, including coaching and teaching at Boston College, business consulting as part of Accenture's NYC practice, founding UCSF's Office for Career & Professional Development, and consulting to life science companies in the areas of strategy and business development. He is widely sought for input and expert opinion and has published numerous articles on medical research and training, entrepreneurship, technology commercialization, and labor market economics. Michael holds bachelors and master's degrees from Boston College, a certificate in IP management from the U of Washington, and studied business at IESE in Barcelona, Spain. He is a member and vice chairman of the board for Bay Bio and the Bay Bio Institute respectively.

Arlen D. Meyers is professor of otolaryngology (Ear Nose and Throat and Facial Plastic Surgery), engineering and dentistry at the University of Colorado Denver. He is an award winning clinician, researcher, educator and bioscience entrepreneur. Dr Meyers is the founder of three bioscience and healthcare companies, consults to industry and leads several global bioentrepreneurship education initiatives. He has published over 300 books and journal articles. Dr. Meyers received his BS from Dickinson College, an MD from Jefferson Medical College, completed his residency at the University of Pennsylvania, and his MBA from the University of Colorado. He is a former Harvard-Macy fellow, a National Library of Medicine Fellow and a Fulbright Scholar (bioentrepreneurship).

Anne S. York is associate professor of strategy and entrepreneurship at Creighton University. She is director of Creighton's NSF-funded bioscience entrepreneurship program and the professional science masters degree in bioscience management. Her articles have appeared in a range of journals, including the *Strategic Management Journal, Academy of Management Journal, Organization Science, the Journal of Commercial Biotechnology, Journal of International Business Studies, Journal of Accounting and Public Policy, and Long Range Planning*. She is past chair and program chair of the U.S. Association of Small Business and Entrepreneurship's Technology and Life Science Entrepreneurship division.

Introduction

YALI FRIEDMAN
thinkBiotech LLC

Educating biotechnology founders and managers presents a novel challenge. These individuals generally come from scientific or management educational programs. These educational paths have traditionally been incomplete or inefficient; scientific doctoral programs prepared students for leadership positions in scientific research and discovery, but left a void in preparations for the difficult decisions in product development and general management which must be made using different mindsets and without sufficient information. Management education programs taught students how to deal with decision-making and general management, but often failed to adequately prepare students for the challenges of managing R&D. A case-based teaching approach, such as the one in this book, is intended to help address the gaps in both educational systems.

This book is written as a complement to *Building Biotechnology*, which is the product of more than a decade spent in biotechnology education. More than ten years ago, while managing the first blog on the business of biotechnology I realized that there was a significant knowledge gap between individuals trained in science and those trained in business in trying to apply their knowledge and skills to the biotechnology industry. Seeking to fill this gap I embarked on a project to write a primer on the biotechnology industry. *Building Biotechnology* was the result of that effort, and was rapidly adopted by management and science programs for classes on biotechnology management. As I interacted with educators using the text, it developed rapidly and grew both in the depth and breadth of coverage. In continuing to develop the text and in producing several related products I became aware of a new gap: there was a strong need for a targeted set of cases addressing biotechnology management. Accordingly, I was delighted when the editors of this book came to me with the idea of just such a compilation.

The basic structure of this book follows the scheme of *Building Biotechnology*. Cases are presented in the categories of Science; Laws, Regulations and Politics; and the Business of Biotechnology. This arrangement covers the pillars of biotechnology. As I describe in *Building Biotechnology*, impediments in all these categories must be satisfied for successful biotechnology commercialization. These elements are seldom satisfied at the outset of a project—if they were, there would be little work to be done—so a biotechnology company's mission may be to engage in research and development to address scientific shortcomings, to obtain domestic regulatory approval to market products approved elsewhere, to find a business model enabling commercialization of established science, etc.

Beyond the literal presentation of events in each case, there is another important lesson in this book. The biotechnology industry is dynamic and, unlike other dynamic industries, there are no formalized continuing education requirements. The challenges faced by companies will change over time, as will the solutions. As you read the cases, look not only for the tactics employed in different circumstances, but also the strategies. While the ideal tactics will change with context, the strategies will have greater general utility. The cases in this book therefore should lay a foundation for readers to examine industry developments and to understand the drivers of change, the levers with which to respond, and the potential outcomes, enabling you to build resilient companies and maintain leadership positions.

How to Teach Bioscience Using the Case Discussion Method

TRENT WACHNER AND ANNE S. YORK
College of Business, Creighton University

The important thing in science is not so much to obtain new facts as to discover new ways of thinking about them.
—Sir William Henry Bragg

WHAT IS A CASE AND THE CASE TEACHING METHOD?

Cases are stories with a message. While cases may be entertaining, their primary purpose is to educate. Consequently, the use of cases gives a teacher an immediate advantage; he or she has the attention of the audience. Cases have been used extensively across many fields of study including law, medicine, business and more recently, sciences, to practice analyzing real world problems using actual, and many times historical data. Typically, cases have evolved as a supplement to lecture in teaching, as adult education literature increasingly suggests alternative, active learning methodologies for adult learners. As such, it applies especially to graduate professional education.

The case teaching method adds a critical element to these educational stories. Also referred as the discussion method, the case teaching method introduces stories with an educational message that *refuse to explain themselves* (Ellet, 2007). These particular cases add ambiguity and uncertainty to stories that are at the heart of this learning methodology. Cases used with the case teaching method require a significant issue, enough information to base deductions in uncertainty, and no stated conclusions. Many times students must sort through "noise" or irrelevancies with limited information in the case, often times not obviously linked, to find subtle and often unstated meanings.

The case teaching method differs in two ways from those used in most other academic programs: (1) it requires students to work almost exclusively with primary source material; and (2) a typical class is built around an active dialogue between teacher and student about what lessons can be learned from interpreting the case, as opposed to absorbing facts through a straightforward class lecture. The initial use of case teaching was in the military setting (e.g., classics such as Sun Tzu's *Art of War*, Miyamoto Musashi's *The Book of Five Rings*, Carl von Clausewitz's *On War*, Erwin Rommel's *Infantry Attacks*). That is, strategy is situation specific and rich lessons are to be learned from case studies. Thus, strategy may be characterized as a practice which, when executed at its highest level, is an art form. West Point historian Nye pointed out that the greatest lesson for future commanders is to "prepare for the unknown by studying how others in the past have coped with the unforeseeable and the unpredictable" (1993:156). Today, aspiring officers in military war colleges routinely study lessons learned in historic battles and how they could be applied to modern leadership demands of strategic warfare.

An underlying but important requirement for successful case teaching is that students read and prepare an analysis of the material ahead of the class discussion, which is not necessarily a requirement for a lecture. In a case discussion class, learning occurs in the class and is assessed by whether a student actively participates in class and, in the end, whether that student can independently perform a case analysis and deliver a coherent recommendation. Thus, assessment also differs from a lecture-based course, in which performance is often assessed by a test on content.

A COMPARISON OF TEACHING METHODS

Broadly speaking, all teaching methods fall into one of two categories: 1) the teacher-centered approach or 2) the active learning approach (Christensen, Garvin & Sweet 1991). The teacher centered approach, or lecture approach, in science focuses on facts, principles, phenomena and substantive content. This approach makes perfect sense since being successful in the field of science is dependent on a thorough understanding of a codified base of knowledge. So it is not surprising that studies have found up to 75% of classroom time is spent in this fashion (Murray & Lang, 1997). The teacher-centered approach is dependent on the transfer of information from an expert (the teacher) to the novice (a student). The expert typically defines and controls the process by means of the syllabus, presentation structure, defined outcome, objectives, etc. Consequently, the bulk of the exchange between the teacher and student is confined to question and answer sessions, and clarifying points of the information presented. Thus, the teaching method approach provides an efficient transfer of large amounts of facts and figures from the teacher to the student.

Since a one-way transfer of information characterizes the lecture method it is, by definition, a passive process. Knowing that passive learning is a poor conductor of retention by students (although still more efficient in transmitting codified knowledge than most other methods), many teachers are under the false assumption that simply adding educational stories (our first definition of cases) changes lectures from a passive to an active learning environment. Adding cases alone may make the class more interesting and engaging for student, but if this case goal is to arrive at a specified conclusion, the process is still passive. This distinction, while subtle, is critical to understanding the case or discussion method of teaching and how it differs from traditional teaching.[1]

Lectures, while great at disseminating knowledge, do not encourage students to think about the content and apply it to real life situations. Herreid (2006) describes the three dimensions of scientific literacy to include 1) the substantive facts; 2) the nature of scientific inquiry; and 3) how science interacts with society. The first point, Herreid argues, is covered by the teaching method, while the second and third points are pretty much left alone under traditional approaches. Therefore, understanding the nature of scientific inquiry and how science interacts with society cannot be taught by a textbook style organization of knowledge. Rather, it requires either 1) real world experience or 2) an alternative teaching model. Fortunately, this alternative teaching model is well accepted under the heading active learning and is the premise of the case teaching method.

The case or discussion method of learning requires students to actively participate in the knowledge creation process. No longer do students receive knowledge, but with the guidance of a skilled instructor, they create knowledge. New knowledge is created by means of dialogue between other students or with the instructor, with the instructor many times acting merely as a facilitator. When students take ownership of the outcome of the dialogue, students have a new sense of learning responsibility.

---

1   *In defense of great teachers, using a passive method of learning does not mean lectures or teachers cannot be engaging and interesting. Quite the contrary, many instructors are quite engaging through being dynamic, credible and authoritative; as well as using multiple presentation modes including relevant examples, video, guest speakers, labs and other resources (this could be called the "carrot approach"). Additionally, instructors engage students using exams, quizzes, random call-outs in class and such (thus the "stick" approach). Nothing seems to engage a student more than when told "this information will be on the exam". But the end goal of the lecture format is the same: transferring knowledge from the transmitter (e.g., instructor) to the receiver (e.g., student) as intact as possible whether or not stories are used to illustrate points.*

The case method has the unique ability to provide context and application to learned knowledge. In a world where traditional lectures are logical, linear and reasoned, the case method forces students to apply that knowledge to situations that are ambiguous, uncertain and in a sense, messy—much like the real world where we are sending them! Cases allow students to "practice" real world decision making in a much more controlled and safe environment of the classroom.

Much of the case method of discussion is centered on the Socratic method where the instructor uses leading questions, advocates opposing viewpoints, and pushes students to defend their positions. Assessment here is based on the process of how the students defend their position, not necessarily the ultimate conclusion. The Socratic method is designed to teach critical thinking and rational decision-making skills, not to transfer specific knowledge. From this perspective, cases depend less on the content expertise of the instructor, but more on instructor's ability to create meaningful dialogue with the students.

A common misconception is that cases and lectures are mutually exclusive (i.e., you choose one alternative or the other). Cases, however, are not designed as an isolated teaching instrument in the classroom. For a case to be effective, an acquired body of knowledge, or at least background knowledge, must be present. That is why cases are used in many capstone courses, where students are expected to arrive knowledgeable of the subject matter. Lectures are a much better tool for dissemination of knowledge, while cases are a great tool to provide context, application, and critical thinking skills to be used to apply that knowledge. One path is to use a lecture to first transfer knowledge, followed by a case method of discussion to apply that knowledge.

A CASE FOR THE CASE METHOD

Why is the context and application of learned knowledge versus knowledge alone so important? Further, why is it important to understand the nature of scientific inquiry and how science interacts with society? After all, isn't just knowing the substantive facts about science enough?

Thomas Friedman (2005) points out that global competition and the "flattening" of the world has dramatically increased competition in the sciences due to a perfect storm of technology and political factors. For the first time in history, talent is becoming more important than geography in determining a person's opportunity in life. In the context of this case textbook, large biopharmaceutical firms were found to be agnostic in terms of global location in terms of establishing alliances and acquiring new biotechnology (Ahn, Davenport, Meeks & Bednarek, 2009). Thus, all students are now not only competing with the most brilliant minds in the world for positions, but resources as well. This should serve as a call to action for scientists, where many fields that have a fixed base of knowledge (e.g., digitization of biology via the human genome) can now be replicated anywhere in the world.

Friedman argues that job security and competitiveness no longer depend on having narrow, vertical specialties, but instead depend on multiple, often unrelated, specialties *and* the ability to synthesize, leverage and explain your knowledge. In other words, knowing scientific facts is no longer enough. Steve Jobs, in a commencement address at Stanford University, attributes his success with being able to "connect the dots" of two disparate fields—computer science and calligraphy—to create the Macintosh computer, and ultimately personal computing as we know it today.

Keeping this in mind, we have an obligation as educators to prepare our students for this flattened world, where knowledge is necessary, but no longer sufficient. By comparing the pedagogical properties of the lecture format and the case discussion method of teaching, we emphasize lecture's efficiency at transmitting knowledge, with the case discussion's effectiveness at providing context and critical-thinking opportunities to that learned knowledge. In making a case for cases, we argue the best way for educators to help students "connect the dots" is through a combination of lecture and case method discussion which results in what IDEO CEO Dan Brown calls T-Shaped people who are observant, empathetic, and creative—traits which the competitive marketplace demands (Figure 1). As Brown

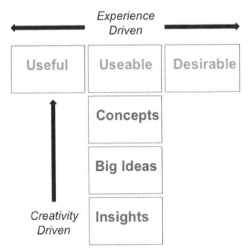

Figure 1: Work force of the future: T-shaped people
Source: Adapted from Brown (2005)

noted: "You just can't stand in your own shoes; you've got to be able to stand in the shoes of others. Empathy allows you to have original insights about the world. It also enables you to build better teams" (2005:2).

A SHORT HISTORY OF CASES

Cases are the primary method of teaching law in the United States. Their use was pioneered at Harvard Law School by Christopher Langdell in 1870 and is based on the principle that rather than studying legal rules, the best way to learn law is to read actual judicial opinions. New decisions, cases, and laws are built upon old decisions or jurisprudence. Students learning the profession must study the cases of the past and use them as examples of judicial reasoning. Typically, professors require students to prepare the cases and then ask questions to determine whether they identified and understood the correct rule of law from the case. Initially described as "an abomination," the case teaching method has been adopted and expanded by other disciplines over time, especially by professional science programs and graduate business schools.

The use of cases in medicine is similar to law. In practice, physicians' work consists of cases in which general physiological systems have broken down in some way. In the late 1940's, Arthur Conant, a chemistry professor at Harvard, returned from World War II convinced that the U.S. educational system in the sciences was flawed. Determined to correct this academic oversight, Conant began what he called "case study teaching" using the lecture method. Conant would take an important historical event such as the discovery of oxygen or the overthrow of the phlogiston theory, and painstakingly describe the steps and misadventures of the protagonists in the setting of the time.

Modern medical education in the United States, taking its lead from Conant, prepares students for their future responsibilities by having them spend two years taking basic courses in anatomy, physiology, embryology, and biochemistry before graduating them into the clinical setting where they are allowed contact with patients: that is, real cases. Recently, an increasing number of medical schools have revolutionized their curricula and set up physician education completely around the study of cases.

Thus, in both medicine and law, cases are real stories dealing with people in trouble. Students attempt to figure out what went wrong and how to fix it. The cases are chosen because they serve to illustrate general principles and good practices; correct answers and facts have a high priority. Once again, cases are real stories—examples to study and appreciate, if not emulate.

In business the case teaching method, or discussion method, evolved from Harvard professors inviting business professionals into the classroom to tell students about actual problems. The students held discussions and offered solutions, and the process gradually evolved into a model that is emulated around the world with thousands of individual cases, journals and books now available. A typical business case may devote fifteen pages plus appendices documenting a business dilemma (e.g., a decision whether to diversify into a new product line or whether founders should exit through an IPO or acquisition). The student is expected to prepare for the class discussion by closely analyzing the background data leading to the decision. The class then discusses the case in an organized way with the instructor moving through various critical topics while outlining the problem and developing options with help from the students.

Business cases employed in today's classrooms are typically real and told in narrative form. Instructors give the cases to the students in an incomplete state and have the class analyze and discuss them to determine what action should be taken. The purpose of the case teaching method in business is to produce managers who can make decisions based on frameworks (e.g., business theories and models), facts (e.g., cash flows, net present value), and less tangible factors (e.g., ethics, culture). The late Harvard business school professor C. Roland Christensen states: "When successful, the case method of instruction produces a manager grounded in theory and abstract knowledge, and more important, one who is able to apply those elements." Cases allow the student to deal with situation-specific dilemmas. Consequently, a business school instructor never expects a discussion to go the same way twice; there are always novelties and unexpected turns in the conversation. However, always, as in law and medicine, the discussion is based only on the facts presented in the case at the time of the case.

Two characteristics distinguish the use of cases in law (and medicine) from business (Merseth, 1991). The first is that law and medical practice depend upon a well-defined knowledge base as a starting point, whereas in business much of the knowledge is in flux. Conditions are always changing in the business environment. Second, business education stresses the human condition and the subjective view. According to Christensen, "any problem may well be understood differently by individuals and groups and that perception changes." In law and medicine, deductive logic and lines of precedent make the individual or specific context of a case less compelling. Merseth concludes, "to a remarkable extent, the purposes and uses of the case method turn on the nature of the body of knowledge that exists in the professional field" (1991:9).

The most recent disciplinary entrant to the case teaching fold is science, a relative newcomer, having embraced the method in the classroom only since the 1990's as an antidote to what were perceived as shortcomings of the traditional teaching method of lecturing. Concepts of case teaching were borrowed from law, business and medicine to create a unique method applicable to the sciences.

Textbook science is a body of facts, concepts, principles, and paradigms that forms the core of scientific knowledge. It is important for science students to know a substantial amount of information, such as how the heart works, laws of thermodynamics, the chemical elements, and so on. Instructors have traditionally taught biology, chemistry and physics primarily through lecture and, thus, may need substantial work and creativity to come up with cases to enrich these scientific topics.

On the surface, the situation appears to demand close-ended cases with correct answers. This is not necessarily so. Instead, many fact-driven cases are open-ended and have multiple solutions because the data are inadequate, personalities and emotions are involved, or ethical or political decisions are at stake. Cases involving the participation in experimental programs to cure genetic disorders, along with cases involving government decisions on global warming, pollution control, human cloning, or NASA space probe funding to find life on Mars, all can be loaded with facts. However, the decisions are not always clear due to the multiple stakeholders involved. Moreover, in cutting edge science, often called "frontier science," different scientists looking at the same empirical evidence can reach different conclusions. So, most proponents of case teaching in science believe that science professors should have access to a wide array of case types to illustrate how science-based decisions are made in practice.

Thus, in summary, we can say the following about the case teaching method as it applies to law, medicine, business and science. First, it is evident that in all instances cases are stories, usually real stories. Second, there is no one right way to teach cases. Teaching methods depend on course objectives and on the style best suited to the teaching. Conant, for example, used cases integrated into his lectures. Law professors use Socratic questioning to elicit key issues in a case. Business school instructors use discussion leadership techniques and reciprocal dialogue with and among students. Science professors may combine elements of all three methods, depending on the context of the problem, and as such, they are probably more similar to business professors than to medicine or law. Third, the subject matter definitely determines the nature of the cases and their expected conclusions. Some cases (and perhaps the method of teaching) are fact driven and deductive (i.e., there is a correct answer). Other cases are context driven, in the sense that multiple solutions are reasonable, and the best answer depends upon the situation at the moment.

The following sections describe a general structure for leading a case discussion and then address ways to motivate student preparation, differences in teaching styles and learning objectives, and, finally, the assessment process.

STRUCTURING A CASE DISCUSSION AND ANALYSIS

Before we give specific recommendations on structuring a case discussion, it is important to remember that leading a discussion is very different than giving a lecture. Leading a discussion in many ways is more like an art or craft than a defined process. The skills of being a good lecturer do not always transfer over to being a good discussant. Structuring a case discussion requires an instructor to facilitate and guide learning and not to disseminate knowledge. For many of us, this is a sharp change in mindset and results in an uncomfortable loss of control in the classroom. As such, we hope to provide a few suggestions to assist in the transition.

The following section melds suggestions from primarily the science and business literature regarding case teaching. As mentioned above, a primary goal of the case teaching method is to help students learn structured ways to organize and analyze complex problems, ultimately arriving at a decision that is logically consistent with both the facts and the broader context of the case. The following seven steps are typically followed with the goal of the faculty modeling the analytical process so that gradually students are able to take over the faculty role themselves.

1. **Structure Baseline of Knowledge:** At the beginning of the case discussion make sure that all students in the class understand the basics of the science underlying the case. You can do this in several ways. First, you might call on someone in the class with industry or science background in the same or a related area. A second, and sometimes complementary way is to ask someone who does not have a relevant background to explain their understanding of the science and make sure that others in the class agree. If there is an inadequate explanation of the background science, ask the students to visit Wikipedia and other web sites to develop an understanding of the relevant science as part of their preparation to discuss the case. While it is always, in every discipline, important for students to understand the facts behind the problem or context they are analyzing, it is even more critical in the sciences, in which the problems tend to be more technical and complex. Failure to do the homework to truly understand the science behind the issue can result in embarrassing and sometimes catastrophically poor decisions and recommendations.

2. **Identify Key Issues:** Ask one or more students to identify what they believe to be the main problem or set of problems facing the organization. Usually this will not be a technical, science-based issue, largely because the timing of discoveries and such

are less predictable than business, law, political, and regulation-based decisions. List these on the board. Let the class engage in a conversation with each other about which are the most critical issues and why. In this process, the professor should serve in the role of scribe and facilitator, not the commander. Note that it is vitally important that students used facts and quotes from the case to defend their arguments. This sounds easy but is the most important part of the class. Often, because of their training, science students will focus more on the facts than the politics, ethics, personalities, or culture. One way to show them the fallacy is to begin class with a classic Harvard Business School case: Honda A and Honda B. In one case, students are presented with the facts from archival sources after the fact. In the other case, students are presented with information taken from interviews with the people involved during the case. In many cases, what you deduce from the retrospective, impersonal "facts" contradict the actual experiences of the decision makers involved. By dividing the class in half and asking half of the students to read and analyze one case and the other half to do the same with the second case, the lesson of paying attention to what you think are the "facts" comes through loud and clear.

3. **Direct Discussion:** Observe/note which students are prepared with facts to back up their arguments. This will form the basis of a strong class discussion. If "experts" in the class fail to participate, which can happen often in classes with science students more used to lecture than to class discussion, using your knowledge of their backgrounds and "cold calling" them is often key to bringing them and their ideas into the class discussion. The types of facts used to back up an argument in science-based cases may vary from more traditional business cases. For example, it may not be possible to identify competitors using traditional market research because the product is in too early a stage to find that sort of information. The closest future competitor data may be found in research papers and scholarly presentations based on university research or in corporate 10-k filings with the Securities and Exchange Commissions (SEC). If early-stage research is sponsored/funded by a publicly-traded, for-profit firm, then that firm must be considered a competitive threat.

4. **Define Analysis, then Identify Possible Solutions:** Once the problem set is identified, ask for a possible set of solutions. Often there is time to fully develop just one line of logic during the class, depending on time. It is important that students, from the beginning, understand that there are different solutions that can be equally well defended. This keeps them from being nervous about offering their solutions in the future. However, it is important also for the faculty to make clear that even though there is no one right answer, some solutions are better than others, and yet others are clearly bad solutions. It is the facts and context of the case that determine which is which, so again, the ability of students to support their arguments with facts and logic is primary here. One thing to keep in mind is that if the organization does not possess the skills or resources needed to execute the solution successfully, then it cannot be a workable solution, no matter how technically brilliant. Likewise, if the best solution is one that the founders and/or management of the organization are not willing to carry out, due to clashes with personal values or corporate culture, then that, too, is not a workable solution.

5. **Reflect and Challenge Assumptions:** It is important to have the class correct their own errors in the discussion, if possible, and also to challenge the assumptions and logic of their peers if they don't agree. This will only happen if you, as case discussion

leader, are respectful of students' comments and limitations. While it is not necessary to tolerate clear lack of preparation, it is important to hear students out in terms of their ideas, thought processes and suggestions. It is often helpful to arrange the class seating so that students can talk to each other rather than the professor. Sometimes, their thinking will completely change your mind about a case that you believed you understood well. That's when you as a professor will learn! Be open to that.

6. **Implement Solutions:** If there is time, it is helpful to have a discussion about how the firm might start to implement a recommended solution. This serves as a way to drill down into more details, but also to reinforce the facts and quotes initially used to come to the solution.

7. **The Rest of the Story:** Finally, it is always good to provide the students with "the rest of the story." All cases are by nature history, some more recent than others. While it isn't necessary to have all cases be within the last few years, be aware that if you use older cases, you will need to preface the discussion with the reasons why they are still relevant. When discussing a case, you ask the students not to read anything about what happened after the time of the case because that is not really the point. It is not whether a particular solution is "right" but rather about the process taken to develop a logical solution from the facts at hand. As such, case analysis is a way to simulate managerial decision making in a somewhat simplified classroom setting. It is important that the students understand that the "real world" is even more complicated because you have, as a manager, to develop your own information set and will always be making decisions with incomplete information.

One final note. While you may walk into class with a clear plan for how you want the class to go, rarely will that plan fit what actually happens in the actual classroom. As Napoleon said, when asked his definition of strategy, "engage … and see what happens." Some remarkable learning can take place if you are not too determined to command the class and rather are willing to sit back and let it take on a natural flow. Of course, this doesn't mean allowing the class to venture off on tangents. But one of the interesting aspects of the case method is that rarely will two discussions be exactly the same, largely because of the diversity of backgrounds, interests and perceptions of the students in a particular class. It's important to embrace that diversity and learn to use it to enhance class creativity.

HOW TO MOTIVATE STUDENTS TO PREPARE FOR A CASE DISCUSSION

As mentioned above, the most critical element of having a strong case discussion is the level of preparation of the students. Professors who use the case teaching method do this in different ways. Some ask for students to complete a generic outline of the issues, solutions and supporting facts and quotes prior to class and to turn these in, typed, at class end. These analysis sheets serve as the basis for part of the participation grade.

A second method is to engage in cold calling rather than voluntary comment. Cold calling is obviously not a favorite technique of students, especially those not used to speaking in class. However, it helps those who are not comfortable (e.g., cultural background) to gain class participation points, as long as the questions to be answered are provided in the syllabus ahead, so that those who are prepared can demonstrate their preparation. It's important to keep up with whom you've called on and provide all students with relatively equal opportunities to participate in this way. Also, it is helpful to provide the students with a free pass or two; all of them will have stressful times when they can't prepare for class due to a key test or paper that is due in another class.

A final way to make sure the class is prepared is to either ask students to write up a brief for each case prior to class to be turned in by the end of class; or to ask them to be prepared and then randomly choose a set of students to write up the brief during the class. Each of these methods has obvious advantages and disadvantages. Choose what best fits the time constraints that you have for grading and that the students have for preparation. Or develop your own technique! In science-based cases, where knowledge of facts, terms, and concepts are key, a short quiz at the beginning may be a good way to ensure that students are prepared.

DIFFERENCES IN TEACHING STYLES

Clearly, the most entertaining professors are the best lecturers, but the best case discussion teachers often are more like coaches. They may call a "time out" if the group is getting off track, but they focus on preparing the class ahead and then letting them play out the discussion, with occasional comments from the sidelines. Doing a good job as a case discussion leader requires not only knowing the cases in great detail and preparing your analysis ahead of time, but also knowing the backgrounds and perspectives of students in the class prior to leading the discussion. The most difficult case discussion to lead is a one-time class with a set of students you don't know and may never see again. However, it can be done. This is pretty much what every case discussion professor faces every semester when a new course begins with a new set of students. Do your homework and find out as much as you can at the beginning of class. Study the information and take notes after class about what you have learned about each student's participation style, background and personality. Clearly, this is easier to do in a small seminar setting than in a large cohort of 50 or more, but the more you know about the students, the better the class discussion. Most professors who are converting from lecture methodology to case teaching will need to relinquish control gradually and, as such, will have more structured discussions (i.e., combinations of lecture and cases) until they become more comfortable with the methodology. Each professor must find his or her own comfort zone. The concentration required to focus on each student throughout a class and to weave their interactions into a coherent discussion is demanding but also exhilarating.

DIFFERENCES IN LEARNING OBJECTIVES

Because cases are stories and may used to illustrate a variety of points, especially those that are rich in facts (which are often found in exhibits, which is why it is critical to assure that students not only read the body of cases but also the exhibits carefully), it is important to determine ahead of time what specific issues fit into your learning objectives for the case. For example, if the goal is to introduce students to political and ethical issues, facts may be less important to the analysis and political and ethical decision models more relevant. If so, students will need to be exposed to those models in order to prepare the case, and all should be able to participate in the discussion. Prior preparation may be directed through readings assigned along with cases or through a lecture covering those models in the class preceding the case discussion. If the goal is to understand, on the other hand, how a firm should be valued, then careful examination and analysis of the numbers, valuation models, and valuations of comparable firms may be more of the focus. In these situations, sometimes assigning different students or student teams to tackle various options and then present to the class can be helpful. Also, students appreciate your including possible discussion questions in the syllabus, along with specific assignments. This helps them prepare for cold calls. Discussion questions along these lines are included in the teaching note for each case in this book.

ASSESSMENT TECHNIQUES

Assessment of learning in case teaching is much more complex than assessing content mastered in lecture courses, largely because there is no one "right" answer; rather there are tightly argued essays backed up by consistent facts and opinions as compared to poorly-argued and poorly-supported essays. As such, students often complain (both throughout the course and at the end in their teaching ratings). Sometimes students can love the course and professor, but hate the grading rubric. No matter how many cases you as the professor taught or essays that you have graded over time or how many class discussions you have led, some students will complain that your grading of their essays and participation is subjective. There is no way to completely eliminate these complaints; however, there are ways to minimize them. In the case writing arena, one way that professors address this issue is to share with the students early in the course sample cases of poor and strong papers. Be sure that you collect these at the end of class or have students use them in your office if you are planning to teach the case again any time soon. The danger of this is that students will then complain that their case was "just like" the sample, so it is important if you choose this technique that you provide them with multiple, highly dissimilar analyses.

A second way is to provide a relatively detailed grading rubric ahead of time that is used for all cases analyses (an example is included in the appendix). A third way is to allow students to write a practice case, receive feedback, and then rewrite the case. Some faculty prefer to have classes choose a subset of cases to write, which allows them to pick cases that best suit their interests or workload. In this scenario, if students are not happy with a grade, they can write an additional case. One relatively happy outcome of having case briefs as a primary source of grading is the likelihood that more than one student will write a case nearly exactly the same is very low. Thus, cheating is rare and relatively easy to detect. On the participation front, it is important that the class participation portion of the grade be significant (at least 25%). A grade should be assigned for each student as soon as possible following every class. Often, especially in larger classes, it is helpful in large classes to use a seating chart to remind you who said what. Let students know that these daily grades are available to them allows them to check on how they are doing throughout the course and thus make positive adjustments to their participation style. Jotting down a comment for particularly insightful comments is a good idea, along with "me, too" repetitive comments. Telling students that game-changing comments count more and "me, too" comments detract from grades, hopefully will help you use class time more effectively. One final note is that it is critical to provide feedback on student case essays and participation prior to the final exam. This is the only way that you can accurately assess whether students are learning to analyze cases.

CONCLUSION AND BOTTOM LINE VALUE OF CASE TEACHING

We leave you with a few final thoughts. While melding case teaching of business and science presents challenges, it is possible to do, and, when successful, provides both students and professors with a creative teaching model that provides fresh and complex insights into both standard and cutting edge problems and material. It also helps students develop critical and interdisciplinary thinking skills, and tools that are increasingly high prized and necessary, especially in the business of bioscience. A highly recommended guide to embracing the case method of teaching is the book, *Education for Judgment*, published by Harvard Business School Press (Christensen *et al.*, 1992). This collection of readings covers every topic mentioned above in greater detail. Especially relevant to the types of cases in this book are "Changing Ground: A Medical School Lecturer Turns to Discussion Teaching" by Goodenough, "Teaching Technical Material" by Greenwald, and "A Delicate Balance: Ethical Dilemmas and the Discussion Process" by Garvin. We hope that these thoughts and observations, collected over several decades of engaging in case teaching and of apprenticing with outstanding case teachers, will help the teaching process go smoothly and will give you confidence in the classroom.

APPENDICES

APPENDIX A

CRITERIA FOR BIOSCIENCE CASE CLASS PREPARATION AND CASE BRIEF GRADING RUBRIC

Name: _____Case: _____Total Points: _____

1. Identified what you believe to be the key strategic issue facing the company in the first paragraph or two, using various strategy and technology models presented in class. Stated clearly your recommendation to the firm. (10 points)

2. Demonstrated a clear knowledge of the technology central to the case, using information from the case and other background information (such as Wikipedia, etc.) as needed. Can state the concept clearly in a paragraph or two (10 points).

3. Provided a strong analytical argument to justify strategic recommendation (should be most of the paper – 30 total points, broken down as follows)

    a. Considered external environment (including industry trends, competitors and technological trajectories) and firm resources as appropriate (10 points)

    b. Included supporting facts from case to back up recommendation (including numerical and financial analysis – 10 points)

    c. Made a logical argument, using the facts, in favor of your recommendation (10 points)

4. Provided an implementation plan and addressed potential threats to strategy (last few paragraphs – 10 points)

5. Effectively produced and discussed high-value-added exhibits useful in supporting the case analysis (up to one page – 10 points)

6. Organized and edited case as directed (include only your student ID number, kept to page length limit, included supportive exhibits, good grammar and no misspelled words, paper was readable, etc. – 10 points)

7. Other Comments/Questions

REFERENCES

Ahn, M., Davenport, S., Meeks, M. & Bednarek, R. 2009. Exploring Technology Agglomeration Patterns for Multinational Pharmaceutical and Biotechnology Firms, *Journal of Commercial Biotechnology*, 16: 17-32.

Brown, T. 2005. Strategy by Design. *Fast Company*. Accessed 11 Feb 2010, http://www.fastcompany.com/magazine/95/design-strategy.html.

Christensen, C.R., Garvin, D.A., Sweet, A. 1991. *Education for judgment: The artistry of discussion leadership.* Harvard Business School Press: Boston, MA.

Ellet, W. 2007. *The Case Study Handbook.* Harvard Business School Press: Boston, MA.

Friedman, T. 2005. *The World is Flat: A Brief History of the 21st Century*, New York: Farrar, Strauss, and Giroux.

Herreid, C.F. 2006. *Start with a story: The case study method of teaching college science*, Natl Science Teachers Association.

Merseth, K.K. 1991. *The Case for Cases in Teacher Education*, AACTE Publications.

Murray, H. and Lang, M. 1997. Does classroom participation improve student learning? *Teaching and Learning in Higher Education*, 20: 2-9.

Nye, R. 1993. *The Patton Mind*. Avery: Garden City, New York.

Cases
Part I: Science

Medarex: Realizing its Potential?

MARK J. AHN[1], ALAN LEONG[2], WEI WU[1], AND MASUM RAHMAN[1]

[1]*Atkinson Graduate School of Management, Willamette University;* [2]*University of Washington, Bothell*

- **Summary and key issue/decision:** Medarex, Inc., a Princeton, New Jersey-based development-stage biotechnology company was founded in 1991. Medarex had partnered with Bristol-Myers Squibb for its Phase III product MDX-010 (ipilimumab), and had licensed its technology to other leading industry partners including Pfizer, Johnson & Johnson, Novartis and Amgen for various antibody-based licensing products. This decision-based case assesses the strategic options facing Howard H. Pien, chairman and chief executive officer of Medarex, on how to respond to a proposed acquisition offer from Bristol-Myers Squibb in 2009.

- **Companies/institutions**: Medarex, Bristol-Myers Squibb

- **Technology:** UltiMAb Human Antibody Development System for fully human antibody-based products in multiple therapeutic areas (e.g., cancer, autoimmune diseases) that has generated over 30 drug candidates to date.

- **Stage of development at time of issue/decision:** Medarex's lead drug, MDX-010 (ipilimumab), an immunotherapy for advanced melanoma, was in the final stage of Phase III trials. The company also had one of the deepest pipelines of human monoclonal antibodies in the biotechnology industry: seven of thirty-four were in Phase III clinical trials, eight were in Phase I or I/II, with the remainder in Phase I or pre-clinical stages of development.

- **Indication/therapeutic area:** The primary developmental focus of ipilimumab was for the treatment of metastatic melanoma. MDX-010 was also being studied as a monotherapy and combination treatment in a number of indications including prostate, breast, renal, and ovarian cancers.

- **Geography:** USA

- **Keywords:** monoclonal antibody, acquisition, immunology, oncology

---

* *This case was prepared as a basis for class discussion rather than to illustrate either effective or ineffective handling of an administrative situation.*

INTRODUCTION

After only two years of becoming president, CEO and chairman of Medarex, Inc., Howard H. Pien was deliberating on how to proceed with merger and acquisition negotiations with Bristol-Myers Squibb Company (BMS). Medarex's UltiMAb human antibody and antibody-drug conjugate technology platforms, as well as its promising monoclonal antibody product pipeline, were generating multiple products and alliance deals. Numerous large multinational pharmaceutical companies were interested in collaborating with Medarex to strengthen their pipelines to accelerate growth and mitigate patent expiry risk. Licensing discussions between Medarex and BMS, initiated in December 2008, evolved into acquisition negotiations by March 2009 and Wall Street was closely following the deliberations. As negotiations began heating up, Mr. Pien reflected, "Our core technology has been validated, we have a diverse set of partnerships with global biopharmaceutical companies, and we're well financed." He glanced at his computer screen and noted, "Merger discussions have already increased our share price, but are we selling out too early?"

MEDAREX COMPANY HISTORY AND FINANCIALS

Medarex, Inc., a portmanteau of <u>Me</u>dical College of <u>Dar</u>tmouth/Ess<u>ex</u> Chemical, is a Princeton, New Jersey-based biopharmaceutical company that utilized monoclonal antibody-based therapeutics to develop treatments for cancer, immunology and infectious diseases. Medarex was founded in 1987 by Drs. Michael W. Fanger, Paul M. Guyre, and Edward D. Ball of Dartmouth College through an equity joint venture between Dartmouth College and Essex Vencap, the venture capital arm of Essex Chemical Company. Charles Schaller and Donald L. Drakeman represented Essex Vencap in the deal. Drakeman, a Dartmouth graduate, brought the parties together and served as the company's first chief executive officer.

While at Dartmouth, the founding scientists developed a new approach to monoclonal antibody technology in the early 1980s, whereby engineered antibody cells would attach themselves to both tumor cells and human macrophages.[1] This process would initiate the release of so-called killer cells that destroyed disease cells. Contrary to other researchers who focused on a single application, their approach was aimed at developing "magic bullet" antibodies that could attack a multitude of diseases (see Appendix B for a summary of monoclonal antibody technology and products).

Initially, Medarex sold its reagents to universities, hospitals, and research institutes for use in laboratory research. However, the laboratory business generated minimal income and Medarex needed capital to pursue drug development, so the firm completed an initial public offering (IPO) in June 1991. In 1992, after Medarex's first product candidate for leukaemia initiated clinical trials, Medarex launched a $7 million secondary public offering, but the process had to be halted after allegations of fraud and misconduct—allegations that were later proved to be unfounded. Another round of stock offering in July 1994 netted the company about $3.2 million.

Although none of its early drug candidates received Food and Drug Administration (FDA) authorization, Medarex successfully generated funds for research by issuing stock and entering collaborative research agreements with multiple partners. For example, E. Merck bought $3.2 million in stock after reaching a partnership agreement in 1994. In 1995, Novartis purchased $4 million of Medarex stock at a $1.1 million premium as part of a collaboration agreement. In November 1995, institutional investors <u>paid another $9.7</u> million for a private placement of stock (see Figure 2 for Medarex's financing history).

---

1   *Monoclonal antibodies (MAb) are highly selective fused protein molecules that recognize and bind to specific targets (antigens) which have been developed as research tools, diagnostics, and therapeutics. MAbs are a highly versatile biotechnology therapeutic platform which can be used alone or to deliver drugs, toxins, or radioactive materials. Their ability to directly target specific diseased cell populations make MAbs an ideal technology platform while avoiding healthy normal cells (Brekke, Sandlie, 2003).*

Drakeman's strategy for generating cash flow was a flexible approach to ensure the firm's survival and sustainability, thereby preventing the need to sell Medarex to a larger firm from a position of weakness.

GROWTH THROUGH ACQUISITION

Drakeman also used Medarex's stock to pay for the acquisition of other firms. In 1997, Medarex embarked on a series of acquisitions by first acquiring Houston Biotechnology Inc., a Texas-based biotechnology company focused on monoclonal antibodies and products for treatment of cataracts. Houston Biotechnology research was important as the acquisition provided Medarex with its first product in Phase III clinical trials. Unfortunately, success was short-lived: in November 1998, the clinical trials were terminated when patients experienced severe reactions to the drug.

In 1997, Medarex acquired GenPharm International, Inc., based out of San Jose, California, through a $62.7 million stock purchase deal. GenPharm's bioengineered mice provided a major technology capability to Medarex, which provided new streams of revenue through the sale of mice and the licensing of the technology behind them. With the premise that animal milk would contain human proteins, and therefore, could be used as ingredients for drug development, Nils Lonberg, a Harvard educated molecular biologist, invented the bioengineered mice by developing a way to introduce human genes into animal DNA.

Lonberg began his research career at GenPharm, which, like other monoclonal antibody research firms, used mice to create antibodies but encountered the same obstacle of mouse proteins causing adverse side effects to humans. To circumvent the problem, Lonberg created two different sets of mice: the first set was incapable of producing its own antibodies while the second set had genes for producing human antibodies. By breeding together these mice, Lonberg eliminated the fractious mouse proteins that plagued monoclonal antibodies because the mouse offspring were now more "human" in genetic terms than regular mice. While other researchers endeavored to create mice antibodies with limited success, Lonberg had focused on simply engineering the mice themselves, which proved successful. After the acquisition, Lonberg sat on Medarex's board as science director and his transgenic mice became Medarex's engine for identifying and generating promising molecules for drug development.

As researchers experimented with new ways to use the mouse proteins in new drug development, this refined approach to monoclonal antibody technology development helped create a strong resurgence of scientific interest in the field. Within a short period, six monoclonal antibodies received FDA approval and a pipeline for hundreds of other drugs using this technology was established. Advances from Medarex's acquisition of GenPharm spurred renewed multinational pharmaceutical interest and in 1998 alone, the company established partnerships with seven major firms including Bristol-Myers Squibb, Novartis, and Schering. During this period, Medarex sold more than $31 million of its transgenic mouse technology to its research partners. "Our basic strategy," Drakeman explained, "is breed them, feed them, and mail them." Also, Bay City Capital, a San Francisco-based investment bank, took a 12 percent stake in the company for $27 million. 1998 was also a great year for Medarex's stockholders as the company's shares gained over 1,000 percent during that year alone.

BUILDING A ROBUST PIPELINE

Medarex's core antibody technology platform, known as UltiMAb, was considered the most state-of-the-art technology solution in the marketplace for developing fully human monoclonal antibodies using transgenic mice. These genetically engineered mice possess a unique immune system, which is capable of harnessing the natural *in vivo* immune response and affinity maturation processes of the mice to rapidly produce high-affinity, fully human antibodies. This state is achieved by deactivating the mouse antibody genes and essentially replacing them with human antibody genes, while leaving intact the other components of the mouse immune system. Since the transgenic mice produce human genes that are stable, these genes are passed on to the future generations of mice. Thus, the mice can be bred indefinitely at relatively low cost without additional engineering.

UltiMAb enabled Medarex to develop antibodies that were fully human, were of considerable affinity, and could be manufactured relatively quickly and efficiently. Since its inception, this single UltiMAb technology had enabled Medarex to generate over forty antibody product candidates that were either undergoing human clinical trials, or had regulatory applications submitted for such trials. This technology platform created relatively strong bargaining power with pharmaceutical partners as it promised a vast product portfolio and huge potential for revenue generation. As a result of the broad demand and potential utility of UltiMAb, Medarex established collaboration agreements with more than thirty-five pharmaceutical and biotechnology companies to either jointly develop and commercialize products or to license other companies to use its proprietary UltiMAb Human Antibody Development System in their development of new therapeutic products. Indeed, Medarex possessed one of the broadest and deepest pipelines of human monoclonal antibodies in the industry (see Figure 1).

Of note, Medarex had partnered with Bristol-Myers Squibb on multiple programs, mostly focused on oncology. Under a co-promotion and profit sharing agreement, Medarex and BMS were developing MDX-010 (ipilimumab), a fully human anti-CTLA-4 antibody with a novel mechanism of action that down-regulated T cell activation, accelerating cell-mediated response and thus providing a boost to the immune system response against patients' tumors. MDX-010 had promising early results in a number of diseases including prostate, breast, renal, ovarian and other cancers. The joint collaboration's main developmental focus of ipilimumab had been metastatic melanoma, which was being studied in Phase III trials as both second line monotherapy and in front line combination therapy with dacarbazine (DTIC), a chemotherapy agent.

RAISING CAPITAL TO FUEL GROWTH: EARLY 2000S

In 2000, Medarex expanded its research activity by establishing research and development facilities in California and New Jersey. The firm funded these projects by raising $388 million in a stock offering. In addition, Genmab, a joint venture with a Danish firm that Medarex had formed in February 1999 to

Product Candidate	Indication	Clinical Status	Partner/Licensee
STELARA™ (ustekinumab) anti-IL12/IL23	Psioriasis	Approved in Canada and Europe	Centocor Ortho Biotech Milestones/Royalties
SIMPONI™ (golimumab) anti-TNFα	Rheumatoid Arthritis, Psoriatic Arthritis, Ankylosing Spondylitis	Approved in Canada and U.S.	Centocor Ortho Biotech Milestones/Royalties
Ilaris (canakinumab) anti-IL1β	Cryopirin-associated Periodic Syndromes	Approved in U.S	Novartis Pharma Milestones/Royalties
Ofatumumab (HuMax-CD20) anti-CD20	Chronic Lymphocytic Leukemia, Rheumatoid Arthritis	BLA/MAA Filed	Genmab/GlaxoSmithKline Equity Interest
Zalutumumab (HuMax-EGFR) anti-EGFR	Head and Neck Cancer	Phase 3	Genmab Equity Interest
MDX -066 (CDA-1) + MDX-1388 (CDA-2) anti-C, difficile Toxin A and Toxin B	C. difficile Infection	Phase 2	Merck & Co. Co-development

Figure 1: Medarex selected partnerships
Source: Company reports

further develop some of its technology to create human antibody therapeutics products and to serve as Medarex's European distribution arm, also went public, generating another $250 million.

The early 2000s marked a spur in research activity: during this period, more than 150 products derived from the company's technology, either by Medarex itself or by its corporate partners, were beginning preclinical studies and a number of products had already started human clinical trials. Medarex continued to harness their momentum through new collaboration agreements, over a dozen of which were forged in 2003 alone. In 2004, Medarex successfully collaborated with Pfizer, Inc., by signing a ten-year, $500 million research agreement. A couple weeks after the Pfizer deal, Medarex reached a partnership agreement on its experimental treatment for metastatic melanoma with Bristol-Myers Squibb Company, which included a $25 million equity stake in Medarex and up to $530 million in cash, royalties, and milestone payments. The metastatic melanoma treatment began Phase III trials in 2004.

Investors in the financial markets were also treating Medarex favorably with the price of the company's stock experiencing 60 percent gains in 2003 and another 70 percent in 2004. Another batch of Medarex-based products rolled into human trials in 2005, bringing the total to 31 potential new drugs. Medarex's steady progress in research and development, as well as collaboration with global pharmaceuticals, allowed the company to complete ten financings totalling $1.3 billion over the 1994-2008 period (see Figure 2).

GOVERNANCE CRISIS AND LEADERSHIP CHANGE

In 2006, the Securities and Exchange Commission (SEC) and the US Attorney's Office for the District of New Jersey started investigating the company's stock option grants and practices and related possible accounting fraud during 1996 to 2006. As a result, Medarex was forced to restate its financial statements for 2003 to 2006 (Datamonitor, 2009b). The investigations had management fallout—in November 2006, Donald Drakeman ended his 19-year tenure as president and CEO of Medarex Inc. amidst the investigation into backdated stock options. Nevertheless, investors were not spooked by the

Financing Type	Date Completed	Amount Raised	Notes
Secondary	1-Feb-08	$151,800,000	Medarex sold 2.5 million shares of Genmab A/S (CSE:GEN, Copenhangen, Denmark) for net proceeds of $151.8 million. As a result, Medarex's stake in Genmab was reduced to about 5.1%.
Secondary	20-Feb-07	$152,000,000	MEDX sold 2.5 million shares of Genmab A/S (CSE:GEN, Copenhangen, Denmark) for $152 million. As a result, MEDX's stake in GEN was reduced to about 11%.
Follow-on	7-Apr-06	$135,125,000	MEDX proposed the offering on April 4, when its share price was $13.22
Sr convert notes	28-Apr-04	$150,000,000	The notes mature in 2011, bear 2.25% interest and convert into MEDX stock at $13.72, a 30% premium over MEDX's April 27 close of $10.55. On April 26, MEDX proposed to raise $125 million in the note deal.
Convert notes	18-Jul-03	$125,000,000	The notes bear 4.25% interest, mature in 2010 and convert into common stock at $6.72, which represents a 27.5% premium over MEDX's July 17 close at $5.27. The proceeds include the sale of the entire overallotment on July 22.
Convert notes	22-Jun-01	$175,000,000	The notes bear 4.5% interest, mature in July 2006 and convert into MEDX stock at $28.84
Follow-on	3-Mar-00	$412,663,088	On Jan. 28, MEDX filed to sell 1.75 million shares when its price was $65.188
Direct public offering	8-Nov-95	$11,000,000	
Common	10-Oct-94	$2,300,000	
Follow-on	7-Jul-94	$3,750,000	

Figure 2: Medarex public financing history (1994-2008)
Source: BioCentury (2009)

news (indeed, the stock added $0.80 to $12.99), for Drakeman had grown the company from a start-up to a solid $1.6 billion market capitalization with 33 antibodies in the clinic, either alone or with partners. Irwin Lerner, one of the board's members, became interim CEO and president until Howard H. Pien was appointed to the position of CEO and president on June 14, 2007 (BioCentury, 2007).

Howard H. Pien became the chairman of the board and chief executive officer of Medarex, Inc. in 2007. Prior to that, Mr. Pien was a private consultant from 2006 to 2007. Prior to that, he served as president and chief executive officer of Chiron Corporation, a biopharmaceutical company, from 2003 to its acquisition by Novartis AG in 2006. Prior to that, he served in various executive capacities at GlaxoSmithKline plc (GSK), a pharmaceutical company, and its predecessor companies, culminating in his tenure as president of GSK's International Pharmaceuticals business from 2000 to 2003. Prior to joining SmithKline Beecham plc (a predecessor of GSK), Mr. Pien worked for six years at Abbott Laboratories and for five years at Merck & Co., Inc. in positions in sales, market research, licensing and product management. Mr. Pien was also a director of Vanda Pharmaceuticals, Inc. and ViroPharma Incorporated. Mr. Pien's experience in small and large biopharmaceuticals, as well as mergers and acquisitions, proved to be particularly useful in negotiations with Bristol Myers Squibb.

BRISTOL-MYERS SQUIBB COMPANY: A GLOBAL BIOPHARMACEUTICAL COMPANY WITH SIGNIFICANT PATENT EXPIRY EXPOSURE

Bristol-Myers Squibb Company (BMS), founded in 1887 and based in New York City, was a global biopharmaceutical company with revenue of $20.6 billion and net income of $5.2 billion in 2008. In 2008, the company's leading products were Plavix® (clopidogrel) with sales of $5.6 billion, Abilify® (aripiprazole) $2.2 billion, Avapro/Avalide® (irbesartan) $1.3 billion, Reyataz® (atazanavir) $1.3 billion, and Sustiva $1.1 billion (see Appendix D for BMS financials).

In addition to being a leading biopharmaceutical company, BMS had recently sold or spun off its medical imaging unit, ConvaTec wound healing division and Clairol consumer products division. BMS also spun off 83% of its Mead Johnson nutritionals business into a separate company listed on the New York Stock Exchange—resulting in an increasingly tight focus on its core business of prescription medicines.

BMS faced pressure to seek new sources of growth and consolidation. Mergers and acquisitions (M&A) had become a major force in shaping the biopharmaceutical industry in recent years due to such factors as reduction in research and development productivity (e.g., drug approvals had fallen by half during 1999-2006 while research and development costs had doubled), patent expiry (e.g., over 50% of revenues would be subject to patent expiry by 2012) (Singer, 2009), and government cost containment (e.g., Medicare Prescription Drug Modernization Act of 2003 and Deficit Reduction Act of 2005).

With a renewed focus on its core business and the need to counter revenue losses from its top performing products due to loss of exclusivity, BMS began strategically aligning with small and mid-sized drug developers and biotech companies through its so-called *String of Pearls* strategy, which was formalized in 2007. The strategy envisioned threading together a library of compounds and portfolio of technologies for accelerating the discovery, clinical development and commercialization of new therapies across a broad range of therapeutic areas. According to BMS:

> *"String-of-Pearls [seeks] to accelerate the discovery and development of new therapies, the company is complementing and enhancing its internal capabilities with a suite of innovative alliances, partnerships and acquisitions with small and large companies (Bristol-Myers Squibb, 2009)."*

BMS PIPELINE DRYING UP

In spite of having several blockbuster drugs, BMS faced significant patent expiry exposure. First, lead product Plavix with $5.6 billion in revenues would lose patent protection in November 2011. BMS had already twice felt the financial implications of a weakened pipeline when blockbusters Glucophage® (metformin) lost patent protection in 2002 and Pravachol's US patent expired in 2006. The subsequent generic drugs penetration significantly weakened the company's financial position. Though the company rebounded from the downturn, it remained hostage to the future of its flagship blockbuster drugs. More than 40% of BMS's pharmaceutical revenues were sourced from patented bestsellers Plavix and Abilify.

As a result of this patent expiry exposure, BMS was actively seeking external sources of growth in already served markets such as oncology, cardiology and immunology. For example two biotherapeutics, Erbitux® (cetuximab) and Orencia® (abatacept), were licensed and securely positioned for market leadership. BMS expected to launch two new biologic drugs in the next few years, a therapeutic protein for organ rejection, belatacept; and a human monoclonal antibody for melanoma, ipilimumab. Ipilimumab was being developed in partnership with Medarex (Datamonitor, 2009a). Indeed, successfully achieving its pledge to its customers, employees, partners and shareholders meant seeking new sources of growth more than ever before (see Figure 3).

Our company's mission is to extend and enhance human life by providing the highest-quality biopharmaceutical products.

We pledge — to our patients and customers, to our employees and partners, to our shareholders and neighbors, and to the world we serve -- to act on our belief that the priceless ingredient of every product is the honor and integrity of its maker.

To Our Customers: We pledge excellence in everything we make and market, providing the safest, most effective and highest-quality medicines. We promise to continually improve our products through innovation, diligent research and development, and an unyielding commitment to be the very best.

To Our Colleagues: We pledge personal respect, fair compensation and honest and equitable treatment. To all who qualify for advancement, we will make every effort to provide opportunity. We affirm our commitment to foster a globally diverse workforce and a companywide culture that encourages excellence, leadership, innovation and a balance between our personal and professional lives. We acknowledge our obligation to provide able and humane leadership and a clean and safe work environment.

To our suppliers and partners: We pledge courteous, efficient and ethical behavior and practices; respect for your interests; and an open door. We pledge to build and uphold the trust and goodwill that are the foundation of successful business relationships.

To our shareholders: We pledge our dedication to responsibly increasing the shareholder value of your company based upon continued growth, strong finances, productive collaborations and innovation in research and development.

To the communities where we live and work, the countries where we do business and the world we serve: We pledge conscientious citizenship, a helping hand for worthwhile causes and constructive action that supports a clean and healthy environment. We pledge Bristol-Myers Squibb to the highest standard of moral and ethical behavior and to policies and practices that fully embody the responsibility, integrity and decency required of free enterprise if it is to merit and maintain the confidence of our society.

Figure 3: The Bristol-Myers Squibb pledge
Source: www.bms.com

MEDAREX: A POTENTIAL SOLUTION FOR RESEARCH PIPELINE

When Medarex entered new collaboration discussions with BMS in early December 2008, CEO Howard Pien had no idea that by the middle of March 2009, BMS would initiate merger and acquisition talks. An SEC filing on July 28, 2009 stated that an initial offer of $9-$10 on April 9 would have valued Medarex at $1.2-$1.3 billion on a non-diluted basis, a 67-86% premium to its $5.39 share price on April 8. Howard Pien rejected the offer on April 24, claiming that Medarex, "could not proceed with discussions regarding a potential strategic transaction at the proposed offer price" (BioCentury, 2009b).

As BMS management studied their smaller partner's portfolio, they became increasingly interested in Medarex. From R&D to manufacturing and commercialization, the addition of Medarex to Bristol-Myers could provide a boost in the biologics space for BMS. Jeremy Levin, SVP of strategic transactions at BMS aptly put it in a nutshell, "There are different kinds of acquisitions. This one is not just about assets, but about capabilities. That includes people and technologies. What you are accessing is an intellectual reservoir of deep understanding of biologics. I call these 'enriching' assets because they help enrich your own thinking" (BioCentury, 2009a). He noted that a potential Medarex acquisition would supplement BMS research and development capacity, accelerate product development, and provide a robust addition to their pipeline of potential blockbuster drug candidates. Specifically, Medarex was an attractive investment for the following reasons:

- **Two independent technology platforms for biologics discovery.** The crown jewel in Medarex's technology arsenal was the UltiMAb human antibody development system, which had already created three marketed products and at least sixteen more in clinical development. Medarex's newer technology platform was a next-generation antibody-drug conjugate (ADC) technology.
- **Steady stream of royalties from approved products.** Medarex had a steady income from royalties on three approved products: Simponi golimumab from Johnson & Johnson and Schering-Plough Corp. to treat rheumatoid and psoriatic arthritis and ankylosing spondylitis; Stelara ustekinumab (CNTO 1275) from Johnson & Johnson to treat moderate to severe plaque psoriasis; and Ilaris canakinumab from Novartis AG to treat cryopyrin-associated periodic syndrome (CAPS).
- **Equity stake in Genmab A/S.** Medarex had a 5.1% equity stake in Genmab, although Medarex was not entitled to license fees, milestone payments or royalties from Genmab.
- **Multiple wholly owned mAbs in the clinic.** Including ipilimumab, a human mAb against CTLA-4 receptor in Phase III testing to treat melanoma and hormone-refractory prostate cancer (HRPC), Medarex was developing five more in the clinic with collaborative partners. In addition, at least four more out-licensed mAbs were already in the clinic.
- **Multiples mAbs in pre-clinical development.** At least sixteen more mAbs in pre-clinical development were expected to generate one or two investigational new drugs (INDs) each year.
- **Pipeline focused on cancer and autoimmune diseases.** Medarex's pipeline had potential to provide a huge thrust to BMS' early clinical pipeline in those areas.
- **Boost for BMS's biologics pipeline.** Medarex would also more than double the proportion of BMS' pipeline that is biologics, from 17% to 35%.
- **Increased manufacturing capabilities.** Medarex's biologic manufacturing facility and personnel would complement BMS' new $750 million facility in Devens, Mass., as well as the BMS's biologics R&D headquarters in Syracuse, New York (BioCentury, 2009a).

BMS RESPONDS TO THEIR REJECTED OFFER

BMS responded that it could not proceed with a revised offer unless it had more detailed data relating to the companies' 2005 partnership covering ipilimumab. Since, under the partnership contract, BMS could not consider information for purposes outside the collaboration without an additional agreement to use the data for due diligence, both companies signed an agreement at the end of April. On May 13, BMS increased its offer to $12 per share, an 88% premium to Medarex's share price of $6.40 on May 12.

Still, Pien reiterated that the Medarex board would view the price as "inadequate." BMS then returned with a higher offer to $15 on May 20 comprised of $11 plus a $4 per share contingent payment tied to the "post-closing success of ipilimumab." Pien, after talking to his board, responded on May 22 with the suggestion of a $13 per share all cash offer; BMS countered that it would need yet more due diligence material.

The two companies met on June 3-4 and Medarex gave "detailed presentations regarding its pipeline assets, technology platform and capabilities with respect to discovery, research and development and manufacturing." On June 15, BMS offered $13 per share plus $1 per share tied to milestones for the development and commercialization of ipilimumab. In the interim, the Medarex's share price rose 7%, from $6.65 to $7.13, and Medarex said the offer now "was unattractive." BMS came back with a $15 bid at the end of June (BioCentury, 2009b).

ARE WE SELLING TOO EARLY?

As Howard Pien looked out of his office window, he pondered about how to respond to BMS. So far, negotiations had gone in his company's favor: the share price had experienced positive impact when BMS first approached him and already he had forced BMS to raise the offer price to $15 per share. With dexterous ability to provide information in exchange for higher offers, Howard Pien had helped Medarex climb the wall of desirability successfully; indeed Pien was 'selling the sizzle, not the steak'. BMS was already salivating for more information and was apparently very eager to buy Medarex.

Pien contemplated his options. Firstly, he could further boost the deal price to $16 per share, valuing it at $2.1 billion ($2.4 billion on a fully diluted basis). That would represent a 90% premium to the $8.40 share price that Medarex was currently trading at and would be a handsome 75% increase in valuation compared to the lower end of BMS' first offer. It would also represent a corresponding stock price premium 23 percent higher than the April starting point. Secondly, he could refuse to sell Medarex, noting the acquisition offer was too low. However, that might prove to be a risky move for Medarex because it could cause a potential plummeting of share price and investor confidence. As he thought about his decisions, Pien knew he had to come up with the right answer to the fundamental question: "We have so much potential and we're on the verge of completing several late stage clinical trials which could accelerate our value. Are we selling the company too early?"

APPENDICES

APPENDIX A: MEDAREX FINANCIAL STATEMENTS

MEDAREX, INC. AND SUBSIDIARIES
CONSOLIDATED STATEMENTS OF OPERATIONS
(in Thousands, except share data)

	For the Year Ended December 31			
	2008	**2007**	**2006**	**2005**
Contract and license revenues	$ 34,509	$ 33,823	$ 26,736	$ 30,226
Contract and license revenues from Genmab	1,765	2,083	1,553	4,067
Reimbursement of development costs	16,018	20,352	20,357	17,162
Total revenues	52,292	56,258	48,646	51,455
Cost and expenses:				
Research and development	194,861	198,317	194,512	136,940
General and administrative	44,386	46,925	51,928	28,969
Acquisition of in-process technology	-	6,900	-	8,447
Total costs and expenses	239,247	252,142	246,440	174,356
Operating loss	(186,955)	(195,884)	(197,794)	(122,901)
Equitiy in net loss of affiliate	(10,092)	-	(1,037)	(6,323)
Interest, dividend income and realized gains	17,971	20,290	17,352	14,740
Gain on sale of Genmab stock	151,834	152,143	-	-
Impairment loss on investments in partners	(5,298)	(2,141)	(5,170)	(33,347)
Interest expense	(6,183)	(6,162)	(4,709)	(4,233)
Minority interest - Celldex	-	4,699	6,891	4,410
Non-cash gain on loss of signinficant influence in Genmab	-	-	3,202	-
Loss before provision (benefit) for income taxes	(38,723)	(27,055)	(181,265)	(147,654)
Provision (benefit) for income taxes	(258)	12	436	358
Net loss	$ (38,465)	$ (27,067)	$ (181,701)	$ (148,012)
Basic and diluted net loss per share	$ (0.30)	$ (0.21)	$ (1.50)	$ (1.34)
Weighted average number of common shares outstanding - basic and diluted	128,152	126,665	121,126	110,309

Source: Company reports

APPENDIX A, CONTINUED

MEDAREX, INC. AND SUBSIDIARIES
CONSOLIDATED BALANCE SHEETS
(in Thousands, except share data)

	For the Year Ended December 31			
	2008	2007	2006	2005
ASSETS				
Current assets:				
Cash and cash equivalents	$ 72,482	$ 37,335	$ 34,511	$ 90,602
Marketable securities	281,186	311,437	304,983	260,705
Marketable securities—Genmab	-	152,000	150,000	-
Prepaid expenses and other current assets	21,793	29,013	22,271	31,608
Total current assets	375,461	529,785	511,765	382,915
Property, buildings and equipment:				
Land	6,780	6,780	6,780	6,795
Buildings and leasehold improvements	86,901	87,217	85,123	82,338
Machinery and equipment	70,314	68,729	61,076	54,130
Furniture and fixtures	4,932	5,122	5,025	4,553
	168,927	167,848	158,004	147,816
Less accumulated depreciation and amortization	(101,773)	(87,923)	(73,663)	(61,832)
	67,154	79,925	84,341	85,984
Marketable securities—Genmab	87,428	139,165	344,382	-
Investment inGenmab	3,047	-	-	3,255
Investments in, and advances to, other partners	790	6,040	8,141	6,400
Segregated securities	1,300	1,530	1,477	2,033
Other assets	1,675	3,415	4,587	6,289
Total assets	$ 536,855	$ 759,860	$ 954,693	$ 486,876
LIABILITIES AND SHAREHOLDERS' EQUITY				
Current liabilities:				
Trade accounts payable	$ 5,721	$ 7,579	$ 7,154	$ 4,939
Accrued liabilities	30,516	47,194	42,250	29,371
Deferred contract revenue—current	28,062	26,872	21,032	20,872
Total current liabilities	64,299	81,645	70,436	55,182
Deferred contract revenue—long-term	73,577	85,103	94,115	106,827
Other long-term liabilities	4,670	4,351	3,689	4,032
2.25% Convertible senior notes due May 15, 2011	145,430	143,505	141,581	150,000
Minority interest	-	-	4,699	11,590
Commitments and contingencies	-	-	-	-
Total Liabilities	$ 287,976	$ 314,604	$ 314,520	$ 327,631
Shareholders' equity:				
Preferred stock				
Common stock, $.01 par value	1,285	1,275	1,243	1,118
Capital in excess of par value	1,192,709	1,145,453	1,107,487	943,245
Treasury stock	(85)	(85)	(111)	(215)
Deferred compensation	-	-	-	(599)
Accumulated other comprehensive income	84,156	289,334	495,208	(2,351)
Accumulated deficit	(1,029,186)	(990,721)	(963,654)	(781,953)
Total shareholders' equity	248,879	445,256	640,173	159,245
Total liabilities and shareholders' equity	$ 536,855	$ 759,860	$ 954,693	$ 486,876

Source: Company reports

APPENDIX B: EMERGENCE OF MONOCLONAL ANTIBODIES (MAB) AS A MAJOR BIOTECHNOLOGY PLATFORM

While the first generation of biotechnology products used recombinant DNA methodologies to manufacture recombinant protein therapeutics to replace naturally occurring factors in which patients were deficient (e.g., insulin for diabetes, EPO for red blood cells for anaemia, G-CSF for white blood cells to fight infections, human growth hormone for dwarfism), monoclonal (mAb) antibodies have emerged over a lengthy and costly development to become a dominant second generation biotechnology platform in the late 1990s-2000s. However, interest in using antibodies (nature's own defense system against foreign or harmful germs) or the patient's own immune system for the treatment of cancer dates back more than 100 years, when immunologist Paul Ehrlich (1900) first proposed the use of targeted antibodies as "magic bullets" against malignant tumors. However, it was not until Kohler and Milstein (1975) developed the breakthrough technology for making monoclonal antibodies, for which they never pursued filing a patent but were awarded the Nobel Prize in 1984 (Alkan, 2009), that mAbs began their evolution towards broad experimentation and commercialization.

Early clinical development with therapeutic monoclonal antibodies faced significant hurdles, due to limited efficacy (e.g., short half-life or circulating time, inadequate penetration into diseased tissue) and side effects (e.g., allergic reactions, adverse reactions which limited repeat dosing) (Oldham, 1983; Oldham and Dillman, 2008; Waldman, 1991). In the 1980s a *Journal of Clinical Oncology* review article bleakly concluded "...there has been little evidence to support the use of monoclonal antibody alone with clinically apparent tumor burdens" (Oldham, 1983, p. 583). Indeed, despite a substantial increase in the number of products and over 40 clinical trials in the 1980s, only one product based on murine antibodies—Orthoclone OKT3® (muromonab-CD3) for transplant rejection—was approved in 1980s (Waldman, 1991; Reichert, 2001). Consequently, mAbs were predominantly viewed as a high risk, esoteric approach to drug development and as a result, they fell in and out of favor with venture capital firms; and large multinational pharmaceutical companies largely ignored them. However, from these difficult early beginnings, in the last decade mAb technology has progressed to overcome early technical difficulties to become the dominant platform in the biotechnology industry.

As an example of the dramatically expanded efficacy of mAbs consider Rituxan® (rituximab), based on the clinical work Miller, Maloney, Warnke and Levy (1982). Rituxan is a breakthrough innovation for treating NHL, a disease that went over 25 years without a change in standard of care and survival (Coffier, 2002). Due to the innovative nature of Rituxan, the FDA (Food and Drug Administration) approved the treatment in less than 7 years after it was discovered versus the industry average of 12 years (Grillo-López, 2000). Five years after Rituxan was launched a clinical review of NHL therapies concluded: "because of its activity, coupled with a favorable toxicity profile, rituximab has become almost ubiquitous in the treatment of most B-cell malignancies (Cheson, 2003)." Rituxan's commercial success reflects its clinical efficacy in improving survival — sales went from $5 million in 1997 to $2.9 billion in 2008, which represented 30% of Genentech's product sales in that year.

Venture capital investors and large multinational pharmaceutical firms have, in turn, shown significant interest in this technology segment through investment and strategic alliances. Today, mAbs represent the most dominant platform in terms of commercial interest and success in the biopharmaceutical industry [e.g., 21 US FDA approved mAbs by 2008 including eight multibillion dollar blockbusters such as Avastin, Rituxan, Enbrel, Erbitux (see Figure 4)].

Further, there are more than 200 mAbs in clinical trials (63% cancer, 16% immunology, 11% anti-infective, 5% cardiovascular and 5% other), and, as an indicator of increased legitimacy, are attracting increasing high valuations even at the earlier stages of development (Nelson and Reichert, 2009). Consider, for example, multinational pharmaceutical Merck's license of a Phase II (development stage) monoclonal antibody product candidate from Medarex, Inc. and the Massachusetts Biologic Laboratories (MBL) of the University of Massachusetts Medical School. Merck paid $60 million up

Antibody	Brand name	Approval date	Type	Approved treatment(s)
Muromonab-CD3	Orthoclone OKT3	1986	murine	Transplant rejection
Abciximab	ReoPro	1994	chimeric	Cardiovascular disease
Daclizumab	Zenapax	1997	humanized	Transplant rejection
Rituximab	Rituxan, Mabthera	1997	chimeric	Non-Hodgkin lymphoma
Basiliximab	Simulect	1998	chimeric	Transplant rejection
Infliximab	Remicade	1998	chimeric	Several autoimmune disorders
Palivizumab	Synagis	1998	humanized	Respiratory Syncytial Virus
Trastuzumab	Herceptin	1998	humanized	Breast cancer
Gemtuzumab	Mylotarg	2000	humanized	Acute myelogenous leukemia (with calicheamicin)
Alemtuzumab	Campath	2001	humanized	Chronic lymphocytic leukemia
Adalimumab	Humira	2002	human	Several auto-immune disorders
Efalizumab	Raptiva	2002	humanized	Psoriasis
Ibritumomab tiuxetan	Zevalin	2002	murine	Non-Hodgkin lymphoma (with yttrium-90 or indium-111)
Tositumomab	Bexxar	2003	murine	Non-Hodgkin lymphoma
Bevacizumab	Avastin	2004	humanized	Colorectal cancer
Cetuximab	Erbitux	2004	chimeric	Colorectal cancer, Head and neck cancer
Omalizumab	Xolair	2004	humanized	mainly allergy-related asthma
Natalizumab	Tysabri	2006	humanized	Multiple sclerosis and Crohn's disease
Panitumumab	Vectibix	2006	human	Colorectal cancer
Ranibizumab	Lucentis	2006	humanized	Macular degeneration
Eculizumab	Soliris	2007	humanized	Paroxysmal nocturnal hemoglobinuria

Figure 4: Monoclonal antibody product approvals by the US Food and Drug Administration, 1986-2009
Source: US FDA

front and will pay up to $165 million in development and regulatory milestones, plus sales milestones and double-digit royalties (BioCentury, 2009c). In addition, monoclonal antibodies have experienced significant pricing power reflecting the breakthrough clinical value. For example, mAbs indicated for cancer treatments have a 94% premium compared to all other cancer treatments during the 1992-2008 period (Figure 5).

Further, the evolution in legitimacy of the monoclonal antibody platform has evolved with the development of its ability to move from a murine (mouse) to a fully human form (see Figure 5). Publication of the seminal technology of how to isolate and immortalize mouse antibody producing cells (mouse hybridomas) in 1975 paved the way for the emergence of therapeutic mAbs (Kohler, Milstein, 1975). During the 1980's, efforts were directed towards the development of murine mAbs for imaging and therapeutic purposes. By 1986, muromonab-CD3 was the first murine mAbs product to receive approval by the U.S. FDA for treating acute rejection of kidney transplants. However, muromonab's success was dependent on the fact that patients receiving a transplant were already immunosuppressed, thereby increasing its efficacy. Between 1980 and 1992, murine mAbs dominated the clinical-trial landscape, reaching a peak of eight products during 1991 (Reichert, 2001). Regardless, as murine mAbs progressed through clinical trials, their limited therapeutic potential became apparent, due to their short serum half-life, incompatibility and immunogenicity (Khazaeli *et al.*, 1994). To overcome the limitations associated with murine mAbs, new strategies were employed to make mAbs more "humanlike", by utilizing recombinant DNA technology to generate chimeric variants, that is, antibodies possessing human protein sequences (Morrison *et al.*, 1984; Boulianne *et al.*, 1984). Although chimeric mAbs have shown to be less immunogenic than their murine counterparts are, significant immunogenicity nonetheless has been observed, resulting in their rapid clearance from circulation and limited efficacy

(Bell and Kamm, 2000). As the legitimacy of antibody-based therapeutics evolved, subsequent efforts focused on an alternative to mouse hybridomas by producing humanized or fully human antibodies through generating transgenic mice carrying human immunoglobulin genes, or screening combinatorial phase-display libraries containing a large collection of variant antibody-like molecules (Jones *et al.*, 1986; Skerra and Pluckthun, 1988; McCafferty *et al.*, 1990). The production of fully human mAbs has been possible since the early 1990's, but patent disputes for the technologies hindered their commercial development, until being resolved in 1997. Compared to their murine counterparts, fewer humanized mAbs entered clinical-trials between the late 1980s and early 1990s, however this class of mAbs has been the main source of antibody-based products since 1997 (Reichert, 2001). Currently, 21 mAb products are approved for marketing in the US (three murine, five chimeric, eleven humanized and 2 fully human), with more than 200 potential products in clinical development (Figure 6).

Mouse monoclonal antibodies are generated using mouse hybridoma technology. Recombinant DNA technology is used to advance the development of chimeric, humanized and fully human antibodies. Chimeric antibodies are generated by cloning mouse variable genes into human constant region genes. The introduction of mouse complementarity-determining regions (CDR) onto human constant and variable domains produces humanized antibodies. *In vitro* screening of phage-display libraries and human hybridoma technology is employed for the development of fully human antibodies.

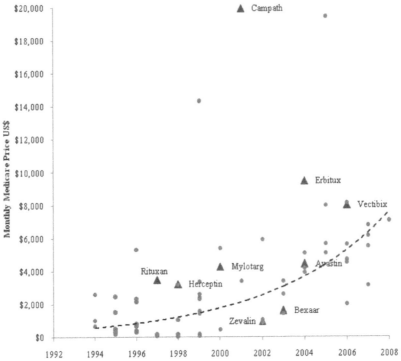

Figure 5: Monoclonal antibody therapies pricing—approved oncology drugs and pricing (▲ = Mabs) 1992-2008
Source: Adapted from NEJM

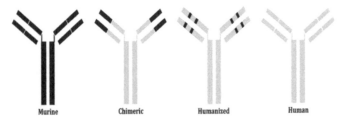

Figure 6: Evolution of monoclonal antibody platform technology

APPENDIX C: SHARE PRICE CHARTS FOR MEDAREX AND BRISTOL-MYERS SQUIBB

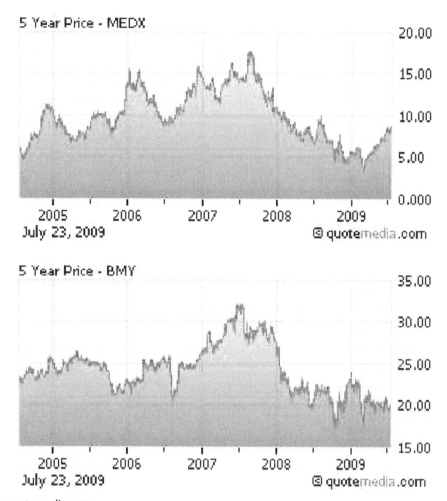

Source: www.quotemedia.com

APPENDIX D: BRISTOL-MYERS SQUIBB COMPANY FINANCIAL STATEMENTS

Bristol-Myers Squibb
CONSOLIDATED STATEMENTS OF EARNINGS
Dollars and Shares in Millions, Except Per Share Data

	Year Ended December 31			
	2008	2007	2006	2005
EARNINGS				
Net Sales	$ 20,597	$ 18,193	$ 16,208	$ 18,605
Costs of products sold	6,396	5,868	5,420	5,737
Marketing, selling and administrative	4,792	4,516	4,469	4,989
Advertising and product promotion	1,550	1,415	1,304	1,464
Research and development	3,585	3,227	2,951	2,678
Acquired in-process research and development	32	230	-	-
Provision for restructuring, net	218	183	59	32
Litigation expense, net	33	14	302	269
Gain on sale of product lines and businesses	(159)	(273)	(200)	(569)
Equity in net income of affiliates	(617)	(524)	(474)	(334)
Gain on sale of ImClone shares	(895)	-	-	-
Other expense, net	191	351	292	35
Total Expenses, net	15,126	15,007	14,123	14,301
Earnings from Continuing Operations Before				
Income Taxes and Minority Interest	5,471	3,186	2,085	4,304
Provision for income taxes	1,320	682	431	870
Minority interest, net of taxes	996	763	440	592
Net Earnings from Continuing Operations	$ 3,155	$ 1,741	$ 1,214	$ 2,842
Discontinued Operations:				
Earnings, net of taxes	113	424	371	145
Gain on Disposal, net of taxes	1979	-	-	13
	2092	424	371	158
Net Earnings	$ 5,247	$ 2,165	$ 1,585	$ 3,000

Source: Company reports

Bristol-Myers Squibb
CONSOLIDATED BALANCE SHEETS
Dollars and Shares in Millions, Except Per Share Data

	December 31,			
	2008	**2007**	**2006**	**2005**
ASSETS				
Current Assets:				
Cash and cash equivalents	$ 7,976	$ 1,801	$ 2,018	$ 3,050
Marketable securities	289	424	1,995	2,749
Receivables, net of allowances of $128 and $180	3,710	3,994	3,247	3,378
Inventories, net	1,765	2,162	2,079	2,060
Deferred income taxes, net of valuation allowances	703	851	649	776
Prepaid expenses	320	310	314	1,594
Assets held for sale	-	560	-	-
Total Current Assets	14,763	10,102	10,302	13,607
Property, plant and equipment, net	5,405	5,650	5,673	5,693
Goodwill	4,827	4,998	4,829	4,823
Other intangible assets, net	1,151	1,330	1,852	1,921
Deferred income taxes, net of valuation allowances	2,137	2,716	2,577	1,808
Other assets	1,269	1,130	342	286
Total Assets	$ 29,552	$ 25,926	$ 25,575	$ 28,138
LIABILITIES				
Current Liabilities:				
Short-term borrowings	$ 154	$ 1,891	$ 187	$ 231
Accounts payable	1,535	1,442	1,239	1,579
Accrued expenses	2,936	2,951	2,332	2,321
Deferred income	277	201	823	1,056
Accrued rebates and returns	806	763	411	125
U.S. and foreign income taxes payable	347	296	444	538
Dividends payable	617	614	552	547
Accrued litigation liabilities	38	205	508	493
Liabilities related to assets held for sale	-	35	-	-
Total Current Liabilities	6,710	8,398	6,496	6,890
Pension, postretirement, and postemployment liabilities	2,285	782	942	804
Deferred income	791	714	354	241
U.S. and foreign income taxes payable	466	537	-	-
Other liabilities	474	552	544	631
Long-term debt	6,585	4,381	7,248	8,364
Total Liabilities	17,311	15,364	15,584	16,930
STOCKHOLDERS' EQUITY				
Preferred stock	-	-	-	-
Common stock, par value of $.10 per share	220	220	220	220
Capital in excess of par value of stock	2,828	2,722	2,498	2,457
Restricted stock	(71)	(97)	-	-
Accumulated other comprehensive loss	(2,719)	(1,461)	(1,645)	(765)
Retained earnings	22,549	19,762	19,845	20,464
	22,807	21,146	20,918	22,376
Less cost of treasury stock	(10,566)	(10,584)	(10,927)	(11,168)
Total Stockholders' Equity	12,241	10,562	9,991	11,208
Total Liabilities and Stockholders' Equity	$ 29,552	$ 25,926	$ 25,575	$ 28,138

Source: Company reports

REFERENCES

Advent Ventures. 2009. *Don Drakeman Venture Partner.* http://www.adventventures.com/team/full. cfm?ref=34, Accessed December 2009.

AllBusiness. 2009. *Medarex Partners With Pfizer In Potential $510M Antibody Deal.* November 9, 2009. http://www.allbusiness.com/company-activities-management/company-structures-ownership/13401142-1.html, Accessed December 2009.

BioCentury. 2009a. Medarex Pipeline. *The Bernstein Report on BioBusiness.* July 27, 2009.

BioCentury. 2009b. Bidding on Medarex. *The Bernstein Report on BioBusiness.* August 3, 2009.

BioCentury. 2009c. *BioCentury – Company Overview.* http://www.biocentury.com/BCApp/BioCenturyCommon/BCAbout.aspx?ss=1, Accessed May 13, 2009.

BioCentury Extra. 2007. *Company News: Medarex Names Pien CEO.* May 16, 2007.

Bio-Medicine. 2009. *Jeremy Levin Joins Bristol-Myers Squibb as Senior Vice President of External Science, Technology and Licensing.* September 27, 2009. http://www.bio-medicine.org/medicine-news-1/Jeremy-Levin-Joins-Bristol-Myers-Squibb-as-Senior-Vice-President-of-External-Science--Technology-and-Licensing-2233-1/, Accessed December 2009.

Boulianne, G.L., Hozumi, N., Shulman, M.J. 1984. Production of Functional Chimeric Mouse/Human Antibody. *Nature,* 312: 643-646.

Brekke, O. H., Sandlie, I. 2003. Therapeutic Antibodies for Human Diseases at the Dawn of the 21st Century. *Nature Group,* 11.

Bristol-Myers Squibb. 2009. *String of Pearls.* http://www.bms.com/news/features/2009/Pages/string_of_pearls.aspx, Accessed December 2009

Cheson, B. 2003. Radioimmunotherapy of Non-Hodgkins Lymphomas. *Blood,* 391-398, Vol. 101, No. 2.

Coffier, B. 2002. CHOP Chemotherapy plus RituximAb Compared with CHOP Alone in Elderly Patients with Diffuse Large B-cell lymphoma. *New England Journal of Medicine,* Vol. 346, No. 4, 235-242.

Datamonitor. 2009a. *Bristol-Myers Squibb Company Profile.* April 2009. 10: 8

Datamonitor. 2009b. *Company Spotlight – Medarex, Inc.* July, 2009. 23–28.

Frost & Sullivan. 2001. *Monoclonal Antibody Cancer Frost 2001.* San Jose: Frost & Sullivan.

Frost & Sullivan. 2005. *U.S. Therapeutic Monoclonal Antibodies Markets.* Frost 2005. Palo Alto: Frost & Sullivan.

Frost & Sullivan 2008. *US Monoclonal Antibodies Frost 2008.* Palo Alto: Frost & Sullivan.

Grillo-Lopez, A.J., Hedrick, E., Rashford, M., Benyunes, M. RituximAb: Ongoing and Future Clinical Development. *Semin Oncol 2002,* Supplement 2 :105-112.

Hazlett, R. 2009. *Bristol-Myers Squibb Adds Another Biotech.* BMO Capital Markets.

International Directory of Company Histories. 2007. *Medarex, Inc.* Vol. 85. St. James Press. Reproduced in Business and Company Resource Center. 2009. Farmington Hills, Mich. Gale Group. http://galenet.gale-group.com/servlet/BCRC Accessed November 9, 2009, 3.15 PM.

Jeffries & Company. 2009. *Narrower 2Q Loss; BMY Tender Offer Expires on August 24.*

Jones, N.H., Clabby, M. L., Dialynas D. P., Huang, H. S., Herzenberg, L.A., Strominger, J. L. 1986. Isolation of complementary DNA clones encoding the human lymphocyte glycoprotein T1/Leu-1. *Nature,* 323: 346 - 349

Journal of Oncology. 2009. *Clinical Cancer Advances 2009: Major Research Advances.* November 6, 2009. http://jco.ascopubs.org/cgi/reprint/JCO.2009.26.6171v1.pdf, Accessed December 12, 2009.

Khazaeli, M.B., Conry, R.M., LoBuglio, A.F. 1994. Human Immune Response to Monoclonal Antibodies. *Journal of Immunotherapy,* 15: 42–52

Kohler, G., Milstein, C. 1975. Continuous Cultures of Fused Cells Secreting Antibody of Predefined Specificity. *Nature,* 256: 495-497

Kuilman, J., Li, J.T. 2009. Grades of Membership and Legitimacy Spillovers: Foreign banks in Shanghai, 1847-1935. *Academy of Management Journal,* 52: 229-245.

McCafferty *et al..* Phage Antibodies: Filamentous Phage Displaying Antibody Variable Domains. *Nature,* 348:552-554 (1990)

McKendrick, D. G.; Carroll, G. R.. 2001. On the Genesis of Organizational Forms: Evidence from the Market for Disk Drive Arrays. *Organizational Science,* 661-683.

Medarex, Inc. 2009. http://www.medarex.com/Development/Pipeline.htm Accessed November 8, 2009.

Miller, R.A., Maloney, D.G., Warnke, R., Levy, R. 1982. Treatment of B-cell Lymphoma with Monoclonal Anti-Idiotype Antibody. *New England Journal of Medicine*, Vol. 306, No. 9: 517-822.

Morningstar. 2009. *Bristol-Myers, Armed With More Cash, Ready To Make Deals.* 12 November, 2009. http://news.morningstar.com/newsnet/ViewNews.aspx?article=/DJ/200911121209DOWJONESDJONLINE000649_univ.xml, Accessed December 2009.

Morrison, H., Bernasconi, C., Pandey, G. 1984. A Wavelength Effect on Urocanic Acid E/Z photoisomerization. *Photochemistry and Photobiology.* 40:549-50.

Nelson, A.L.; Reichert, J.M. 2009. Development Trends for Therapeutic Antibody Fragments. *Nature Biotechnology*, 27, 4: 331-337.

Oldham, R.K. 1983. Monoclonal Antibodies in Cancer Therapy. *Journal of Clinical Oncology.* Vol. 1, No. 9: 582-590.

Oldham, R., Dillman, R.O. 2008. Monoclonal Antibodies in Cancer Therapy: 25 Years of Progress. *Journal of Clinical Oncology.* Vol. 26, No. 11: 1774-1777.

Pena, S. D. 2007. *Of Mice and MAbs: Medarex.* OBR Corporate Profile.

Pipeline 2007. 2007. http://media.haymarketmedia.com/documents/1/pipeline07_245.pdf, Accessed December 2009.

Reichert, J.M. 2001. Monoclonal Antibodies in the Clinic. *Nature Biotechnology*, 19: 819–822.

Sethi, C. 2009 Bristol-Myers Squibb Buys Medarex. July 23, 2009. *BioPharm International.* http://biopharminternational.findpharma.com/biopharm/News/Bristol-Myers-Squibb-Buys-Medarex-for-24-billion/ArticleStandard/Article/detail/613502, Accessed December 2009.

Singer, Natasha. 2009. Bristol-Myers's Reliance on Three Drugs Casts Doubt on Strategy. *The New York Times.* January 27, 2009.

Skerra, A., Plückthun, A. 1988. Assembly of a functional immunoglobulin Fv fragment in Escherichia coli. *Science*, 4855:1038-41.

Tenthoff, E. A., Messer, C. J. 2009. Derivative Calls on Medarex Acquisition. *PiperJaffray.*

Tirrell, M., Pettypiece, S. 2009. Bristol-Myes Squibb Slits Off Rest of Mead Johnson. November 16, 2009. *Bloomberg.* http://www.bloomberg.com/apps/news?pid=20601202&sid=ai5f1jXINRUY, Accessed December 2009.

Waldman, T.A. 1991. Monoclonal Antibodies in Diagnosis and Therapy. *Science.* 252: 1657-1662.

FoxHollow Technologies: The SilverHawk® Cuts Open a New Market

ERIK MILLER, DINA FINAN, AND MICHAEL ALVAREZ
Stanford University

- **Summary and key issue/decision**: FoxHollow Technologies (FHT) was formed with a mission to remove plaque from coronary arteries. Just as FHT was completing the design of their SilverHawk® device and beginning early clinical trials, initial results from coronary drug-eluting stents were being reported and the outcomes were extremely positive. FHT had to face the fact that with limited capital resources, intense competition from the largest players in the industry, and a competitive product the likes of which had never been seen in the medical device field, the future of the company was clearly in doubt. Robert Thomas, the CEO of FHT, had to make a quick decision to change strategies while still trying to satisfy investors and the company's founders and employees - many of whom believed that the new drug-eluting stent data were flawed. Thomas' grandfather had recently undergone two amputations for vascular disease in his legs, and this inspired an alternative application for the SilverHawk. Breaking into the peripheral arterial disease (PAD) market seemed to be the answer to satisfy all interested parties, and most importantly, to address a huge unmet clinical need.

- **Company**: FoxHollow Technologies, Inc.

- **Technology**: SilverHawk®, a device for performing atherectomies to remove plaque from the inside of arteries to restore normal blood flow.

- **Stage of development at time of issue/decision**: The initial product was developed as a device to treat the coronary arteries. There was approval in Europe only, and some trials had been performed with mixed results. The SilverHawk® device to treat the vasculature in the legs had not yet been developed, nor did it have any associated approvals.

- **Indication/therapeutic area**: The SilverHawk device to treat the vasculature in the legs ultimately gained U.S. FDA approval for treatment of all arteries in the legs, starting with the Superficial Femoral Artery in the thigh area all the way down to the tibial arteries in the lower extremities.

- **Geography**: Redwood City, California

- **Keywords**: Silverhawk, PAD, Superficial Femoral Arteries, atherectomy, drug-eluting stents

---

* *This case was prepared as a basis for class discussion rather than to illustrate either effective or ineffective handling of an administrative situation.*

INTRODUCTION

Cardiologist and entrepreneur Dr. John Simpson had been thinking about one idea for a long time: instead of crushing arterial plaque to the side with a balloon or wedging arteries open with a wire mesh cage, why not remove the plaque completely to restore healthy blood flow? That was the goal he had in mind when he founded FoxHollow Technologies, Inc. (FHT) in 1996: to develop a catheter-based system with a small, rotating blade that would treat atherosclerosis by cutting plaque away from the walls of coronary arteries.

Although the company was making progress in the development of its device, new competitors were arriving on the scene. The first generation of stents, the wire mesh devices used to prop open arteries after angioplasty, had problems that left much room for improvement. Arteries would often scar and reclose at the point of stent insertion, a process called restenosis, which made further treatment even more difficult. FHT was attempting to develop a new technology to compete with stents in the hopes of developing a better treatment. In 2003, Johnson and Johnson received US FDA (Food and Drug Administration) approval for the Cypher® drug-eluting stent, and Boston Scientific introduced the Taxus® stent the following year. These devices slowly released drugs that prevented the development of scar tissue at the treatment site, significantly reducing the rate of restenosis. FHT faced a choice: continue to compete with major, established companies in the crowded coronary atherosclerosis market or develop an alternative application for their product.

Robert Thomas, the CEO of FHT, was inspired by his grandfather to develop an alternative application for the company's technology. Peripheral arterial disease (PAD), loss of circulation in the legs due to plaque formation in the femoral arteries, forced doctors to amputate both of his grandfather's legs. His grandfather died shortly thereafter, experiencing a great deal of pain and a vastly reduced quality of life. The SilverHawk® device could be introduced into the large, straight peripheral arteries more easily than the coronary arteries, and stent treatments had been much less successful in the treatment of PAD. While the market for PAD was thought to be smaller than that for coronary atherosclerosis, the initial results for the SilverHawk were very promising, and the competition was less intense.

Bart Beasley, Marketing Director of FHT at the time, reflected back on that decision, "There was an idea that there was this big [coronary] market, and we're going after that, but it's higher fruit. But then the decision was, no, let's go for the lower hanging fruit; maybe it's a smaller market but it's a great place to start testing it out, start getting exposure in the United States… and then we can grow that market. We can pursue that bigger, higher hanging fruit later, as soon as possible, as soon as we have the right clinical, technical solution. That was the assumption. But in retrospect you could make a different argument. There was more opportunity in the peripheral market - it was less crowded and nobody else was pursuing that. It was untapped… and then really, in retrospect, that did prove itself to be true. At least if you look at the three, four years of just explosive sales growth of FoxHollow, there was definitely an unmet need that nobody had tapped into."

COMPANY HISTORY

When FoxHollow Technologies was founded, its mission was to develop a device for the treatment of coronary atherosclerosis. In 1997, over one million angioplasty interventions were performed worldwide (http://www.ptca.org/history_timeline.html). While less invasive than coronary bypass surgery, which required opening the ribcage and transferring the patient to a heart and lung machine, angioplasties suffered from a high rate of restenosis (reclosing of the artery). Balloon angioplasty resulted in restenosis rates of 30-50% of patients, and the addition of a stent cut this restenosis rate in half. As each angioplasty procedure cost about $12,000, any technologies that could reduce the number of additional interventions required would be extremely valuable to health care providers (Watanabe, 1997).

Early efforts at FHT were focused on trying to remove plaque from coronary arteries. These initial efforts failed to produce a device that could be tested in a patient, even though the company had spent over $13 million and grown to 43 employees (Levin, 2003). The product was not working consistently, was difficult to use, and even more difficult to make. FHT underwent a major reorganization in 2001, with Robert Thomas taking over as CEO and the company scaling back to 16 employees, most of them focused on engineering and design. They essentially tore down and rebuilt the company: "We kept the engineers, assemblers, people in the machine shop, and we said O.K., now let's go fix the technology."

The next few months of working around the clock and going through rapid, iterative cycles of manufacturing and testing paid off, and safety studies for SilverHawk use in coronary applications began in Europe in 2002. FHT obtained CE mark approval for the SilverHawk® in October of the same year. At that time, FHT also began a clinical trial in the US to support FDA approval, but the trial was placed on hold in December 2003. "It had, I think it's safe to say, mixed results," said Robert Thomas. "There were some actually very good cases. We treated both native coronary arteries as well as people with in-stent restenosis, and got very good results. We also got some very poor results, and it was clear to us that the product was not ready for full-scale commercialization. It needed some changes before we could commercialize."

The other major application for the SilverHawk® device was in peripheral arterial disease (PAD), the development of atherosclerosis in non-coronary arteries; mainly, the large arteries in the legs and pelvis. In 1997, the peripheral vascular stent market was $137 million (Frost and Sullivan, 1998). While the market for peripheral procedures was significantly smaller than the market for coronary procedures (Figure 1), it had two major advantages. First, the peripheral atherectomy device was classified by the FDA as a 510(k) device, while the coronary device required the more complicated and lengthy premarket approval.[1] Second, there was much less competition in the PAD market, particularly for the legs. Robert Thomas described the state of the market, "The only thing people knew was that nothing worked down there. So the best technologies for the coronary, the renal, for all other areas of the body just didn't work in the legs." While FHT would need to develop technology that could beat the performance of stents in coronary applications, devices for the treatment of PAD would be breaking new ground.

Clinical trials for peripheral use of the SilverHawk® began in January 2002, and the device performed extremely well in this application. Robert Thomas described the initial results. "We went over to Germany. We did our first case in Hamburg, and I remember being there with a doctor. We pulled out hundreds of milligrams of arterial plaque that was obstructing blood flow on a patient that was scheduled for amputation. So we remove all that, we restore the blood flow, and the doctor was in shock. We were all in shock. And the family outside in the waiting room, when told the news that their father, their grandfather would not need an amputation, they started crying, and it was a very celebratory moment." FHT received its CE mark for peripheral treatment in May 2003, and 501(k) clearance

---

1   FDA defines a medical device in the Federal Food, Drug and Cosmetic Act as "…an instrument, apparatus, implement, machine, contrivance, implant, in vitro reagent, or other similar or related article, including a component part, or accessory which is: … Intended for use in the diagnosis of disease or other conditions, or in the cure, mitigation, treatment, or prevention of disease, in man or other animals…" Before a new medical device can be introduced in the market, the manufacturer must generally obtain FDA clearance (510(k) clearance) or pre-market application (PMA) approval. In the United States, medical devices are classified into one of three classes (Class I, II or III) on the basis of the controls deemed necessary by the FDA to reasonably ensure their safety and effectiveness. Class I devices are subject to general controls (for example, labeling, pre-market notification and adherence to the Quality System Regulation). Class II devices are subject to general and special controls (for example, performance standards, post-market surveillance, patient registries and FDA guidelines). Generally, Class III devices are high-risk devices that receive significantly greater FDA scrutiny to ensure their safety and effectiveness (for example, life-sustaining, life-supporting and implantable devices, or new devices which have been found not to be substantially equivalent to legally marketed Class I or Class II devices) (FDA, 2010).

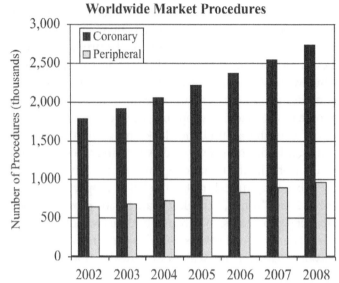

Figure 1: Coronary and peripheral projected market sizes
Source: Levin, 2008

for peripheral use in the US followed in June.

While things were going well with the peripheral applications, still more competition was arriving on the coronary front. Johnson and Johnson, Inc. was developing a new stent that was embedded with the immunosuppressive drug Rapamune® (sirolimus) to inhibit restenosis after implantation. In an initial trial, none of the 48 patients implanted with the drug-eluting stent developed restenosis, compared to 27% of the patients receiving standard stents (Morice *et al.*, 2002). These initial results were met with enthusiasm in the cardiology community, but were forcing many medical device firms to change their strategy or leave the field of restenosis entirely (Stuart, 2001). Johnson and Johnson received approval for the Cypher® drug-eluting stent in 2003, and Boston Scientific began selling the Taxus stent the following year. By dramatically reducing the rates of restenosis, these devices set the bar much higher for competing technologies in coronary applications. FHT needed to change its focus, take the plunge, and direct all of their efforts towards the peripheral market.

"When we were still trying to figure out if we wanted to go into peripheral, that market was known to be a huge one in terms of the number of people who had the disease, but no one quite clearly understood how big it was," said Robert Thomas. Because even the best of technologies did not work in the legs, the disease was often not even diagnosed until it had progressed to the point where amputation was the only remaining option. Indeed, there were 200,000 amputations in the United States and Europe every year due to advanced PAD (Stuart, 2004).

One problem contributing to the under-diagnosis and lack of awareness was that many patients remained asymptomatic until the disease reached advanced stages. Only then do the most recognizable symptoms begin to manifest themselves in the legs: pain, numbness, weakness, sores that do not heal, and changes in skin color and temperature. Many of these symptoms can be attributed to other conditions, and thus PAD was often not diagnosed in a routine doctor's visit. Still, the fact remained that this was a very serious disease, affecting 8-12 million people in the United States, and associated with high morbidity and mortality.

Luckily, a very simple test already existed that could quickly and accurately diagnose PAD. The ankle-brachial index (ABI) is measured by taking the ratio of the blood pressure in the ankle to the blood pressure in the arm. In healthy people, the ratio should be between 0.9 and 1, while a lower number indicates PAD. Studies have shown that the ABI is a very good predictor for the severity and outcome of the disease (Hirsch *et al*, 2001 and Collins *et al*, 2007). The test can be performed by a primary care physician or a nurse in 5 minutes during a routine office visit, and is reimbursed. The challenge, then, was to increase awareness such that this would become a routine procedure and PAD patients could immediately be referred to an interventional cardiologist for treatment.

With a small salesforce (about 20 people initially), FHT began to sell the SilverHawk® to doctors, even without clinical data to support their claims. "We said, it takes someone 60 years to get the artery in their legs filled with plaque… and we're able to clean it out in about 15 minutes… it's probably a pretty good bet," Robert Thomas explained. Immediately, sales took off. Despite the fact that the technology was still being perfected and the devices would occasionally fail to turn on, doctors were buying them faster than FHT could make them. "The doctors were so hungry to get the catheters—they would actually pay for them, remove them, put another one in that worked, get through the case, and were deliriously happy with the result that was never before seen in the peripheral setting."

At the beginning, FHT was limited by its manufacturing capacity. "Our first year of sales, we were expecting about $3-4 million [in revenue]. We actually came in about $31 million. So things just took off", recalled Robert Thomas. It took almost a year to get out of backorder and keep up with demand, even before the company had a real marketing plan. Once the manufacturing capacity developed though, FHT needed to find a way to connect with more doctors and patients to let them know that a treatment existed for this devastating disease.

"We started to see that despite the phenomenal sales and everything we were seeing, we were just scratching the surface," said Robert Thomas. The problem that remained was how to reach more patients and get them to a doctor who would be able to diagnose the problem and perform the procedure. Once the operating efficiency and manufacturing had caught up, the marketing team realized that reaching out to referral sources would help them identify patients long before they ever ended up at a cardiologist's clinic. "One thing we noticed very early on, is that a lot of people would go to the podiatrist… their toenails would start decaying and fall off… people would see their feet turning abnormal colors, and their first stop would be their podiatrist because they thought they had a foot problem. Clearly they had a blood supply problem, but they didn't know that." Other patients would go to wound care centers for treatment of wounds on their legs or feet that wouldn't heal.

FHT salespeople started visiting podiatrists and educating them about PAD. As the sales team got busier and busier, a specifically focused team was created to call on podiatrists, and follow up with them about their patients. Bart Beasley explained, "Podiatry is a practice that is largely overlooked. Podiatrists never get salespeople from industry coming to see them, bringing them doughnuts in the morning, letting them know what their patients can do, following up. Of the ten salespeople who were going to see a cardiologist every week, zero of them would visit a podiatrist. So right away we were welcomed by podiatrists, who wanted to understand the problem better, who wanted to know what to do about it, and wanted to follow up on their patients, and just needed the conduit to make that happen. We provided that conduit."

The effect of reaching out to podiatrists on sales growth was huge. Patients flooded in to see cardiologists, radiologists, and surgeons for atherectomy procedures without the hospitals having to do any of the work to find them. Most of these patients, in addition to PAD, also had coronary disease or carotid disease, and other medical issues that needed attention. Soon hospitals were seeing measurable increases in the number of patients waiting to see specific doctors who performed these procedures. The SilverHawk® device had an initial selling price of about $600, but as its benefits became clear to hospitals, FHT was able to sell the newer devices for $2500 and then $3000. Every time improvements in R&D yielded a new version of the device (which was happening about once a quarter), they increased the price. This, along with the sheer number of units they were selling contributed to the explosive growth in sales and revenue (Figure 2).

There were still a few hurdles to overcome. Many cardiologists considered balloons and stents to be the standard of care that was about as good as it was going to get in treating PAD. This meant that the marketing campaign needed to appear disruptive and shake things up in order to be successful and

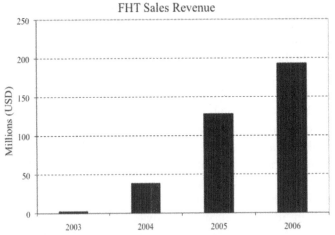

Figure 2: Sales revenue for the SilverHawk® device from 2003 to 2006

get people to take notice. The marketing team, led in part by Bart Beasley, did just that. The branding and coloring they used were aggressive—a black and orange color scheme and a brash logo—which was atypical for medical devices. Even the name of the device, the SilverHawk, was nonstandard in the industry. Salespeople distributed orange silicone bracelets to doctors with the number of amputations per year in the United States engraved on them, to serve as a constant reminder for how devastating the effects of the disease could be; once a patient's limb is amputated, the increase in morbidity is swift.

The marketing plan also involved making users of the SilverHawk® seem like members of an exclusive club. Bart Beasley explained some of the techniques they used: "Get on board, become part of the club, and then you're cool - you use the SilverHawk." Most notable was a groundswell-type marketing campaign to "bring in some key aggressive young interventionalists who were willing to really buy into the concept, and then share with each other. We had some marketing programs to support that… physician peer-to-peer programs, we called them. We would bring physicians together - groups of ten or fifteen at a time - we'd sponsor the event and we would have them watch live surgeries by the most experienced in the group. They were all experienced, but they would watch [the most experienced guy's] surgeries, and they would ask all kinds of questions of each other and of him, and they would go back to their respective cities all across the United States, but they would feel like they were part of this elite club. We found ways to keep them connected and keep them sharing cases with each other and talking to each other. So without some huge footprint or marketing campaign, you have these guys spreading the word."

FHT's entry into PAD treatment was phenomenally successful. The new technology they developed provided an effective treatment for an underserved disease, and their marketing campaign produced explosive growth in sales revenue. Ultimately, ev3 Inc. acquired FHT for $780 million in 2007 and incorporated the SilverHawk® device into its portfolio of vascular intervention devices.

FOXHOLLOW TECHNOLOGY, PRODUCTS, AND PIPELINE

Atherosclerosis is the narrowing and hardening of arteries caused by the buildup of plaque containing fat, cholesterol, calcium, and other materials found in the blood. Plaques can form in many different arteries. The coronary arteries supply blood to the heart muscle tissue, and blockages in these arteries are the cause of heart attacks. The carotid arteries in the neck supply blood to the brain; blockages here result in stroke. Atherosclerosis can affect peripheral arteries as well, restricting blood flow to the limbs or the kidneys.

Two major interventions were available for severe cases of atherosclerosis at the time FHT was founded. Bypass surgery is a procedure where a healthy blood vessel from another part of the body is grafted onto the diseased artery to allow blood to flow around the site of plaque buildup. The other, less invasive option, was angioplasty. In this procedure, a catheter delivers a small balloon to the site of the plaque. The balloon is then inflated to push the plaque aside, increasing the diameter of the artery and restoring blood flow. A small metal mesh cage called a stent is often deployed at the site of angioplasty to prop open the artery and maintain blood flow after the procedure.

FHT was focused on a very different type of intervention, the atherectomy, which is noninvasive and physically removes the plaque from the body to restore blood flow. The SilverHawk® plaque excision system consisted of a catheter, a hand-held control unit, and a carbide cutting blade at the tip of the catheter (Figure 3). A guide wire is inserted into the patient's artery and the catheter travels along the wire to be positioned at the site of the blockage. Upon activation of the device, the blade rotates at approximately 8,000 rpm, and the angle of tip deflection controls the depth of the cut. The hollow nose cone in front of the cutter housing collects the excised plaque for removal. The SilverHawk® was manufactured with a variety of diameters and catheter lengths optimized for different procedures. For example, the coronary arteries are much narrower than the femoral arteries and require the use of smaller devices.

The femoral arteries are long and straight, resulting in some fundamentally different requirements for the treatment of atherosclerosis. Plaques in the femoral artery tend to be longer and more diffuse, which makes them harder to treat with angioplasty and stents. Also, stents placed in the leg are prone to break under the stress from the surrounding muscles and larger range of motion in the leg (Stuart 2004). The same conditions that make angioplasty and stenting difficult, however, are ideal for atherectomy. Robert Thomas compared the SilverHawk® applications, "Our device really did have good applications in other areas of the body. It was not that the technology was horrible; it was actually quite good, but in a different place. And being that the arteries in our legs are straight, as opposed to the manipulations that you have to go through and the curves and turns you need to go through with a coronary device, the issues that we were having with coronary went away in a heartbeat when you started treating the legs."

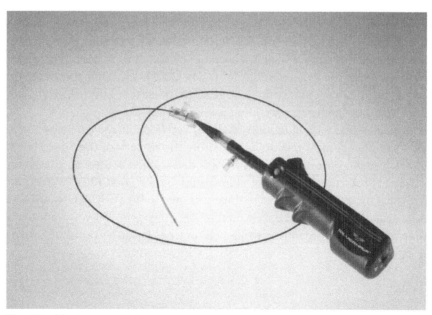

Figure 3: The SilverHawk® plaque excision system
Image courtesy of ev3 website

CONCLUSIONS

Dr. John Simpson founded FoxHollow Technologies with one mission: to develop an atherectomy device that would replace the coronary stent. He believed that placing a chunk of metal in a patient's coronary artery was an inferior method of treatment when it was possible to remove the arterial plaque that was causing the disease. Restenosis was a major problem after the placement of a stent. Then, not only was there a blocked coronary artery to deal with, but there was also a metal cage embedded in the tissue making treatment much more difficult. In addition to removing the arterial plaque, atherectomy procedures had the advantage of leaving all options for future treatment on the table.

The remarkable success of the drug-eluting stents released by Johnson and Johnson in 2003 and by Boston Scientific in 2004 made this mission significantly more difficult. Interventional cardiologists had grown up performing angioplasties and stent placements, and drug-eluting stents nearly eliminated restenosis, the major complication of the standard procedure. Convincing physicians that an alternative procedure would be safer and more effective would be a tough business strategy.

FHT's technology had another medical application, the treatment of PAD. This market was smaller and less developed at the time of the company's founding, but Robert Thomas' personal experiences had made him mindful of the need for some form of treatment of this debilitating disease. FHT focused on this market with their device and had spectacular results. Patients recovered circulation in their legs and sales and revenue grew explosively. From initial investments totaling $51 million, FHT grew into a publicly traded company that was eventually acquired by ev3, Inc. for $780 million (see Fig 4 for share prices). Financially, FHT was a tremendous success. According to Robert Thomas, "It should be noted that the companies who did put money in at that point [during the 2002 coronary trials] got an incredible return, the likes of which, I've been told, have not been repeated in medical devices either before or after, so they were able to get very cheap equity which turned into a phenomenal return for them. So they took a huge risk and they got a huge reward."

The financial success made it difficult for FHT to remain focused on its goal of eliminating the devastating effects of advanced PAD. As CEO, Robert Thomas could see the change in the company. "I think one of the downsides to being so successful is that, in the early days I would walk down the hallway and I would see engineers on their screens with medical devices and trying to come up with new things, and people talking about how do we take better care of patients, and towards the end of my time there I would walk down the hall and I would see people's stock screens with our stock there and I could see them doing the mental calculations to what their net worth was. We had lost our focus, and I have to take responsibility for that." The marketing efforts were successful, but PAD was still taking a heavy toll and interventional treatment had not become the standard of care. "I'm angered about what happened to my grandfather. I'm angered that to this day [2009], 150,000 patients get their legs cut off because doctors hide behind, well, there's not a large swath of double-blind clinical studies as they cut the patients legs off and they ignore that, and that is an outrage."

Could clinical studies make the adoption of the SilverHawk® more extensive? While there were discussions of performing a clinical trial to demonstrate the efficacy of the SilverHawk® device, FHT never sponsored a full study. Part of the difficulty was designing a control as a basis for comparison. Robert Thomas didn't think running a double-blind study was a good idea. "What you're asking the patient to do then is use unapproved devices that were known to break, which would result in amputation. That's the one arm of this study, and we would be the other arm. Do you want to sign up for that, or do you want one of your family members to sign up for that? Because frankly, I have a moral, ethical issue with doing that."

APPENDICES

APPENDIX A: KEY PEOPLE

- **Dr. John Simpson, Ph.D., M.D.** founded FoxHollow Technologies in September 1996 and served as a member of the board of directors since September 1996. From October 1996 to July 1997, Dr. Simpson served as president. Before founding FHT, Dr. Simpson founded several other interventional cardiology companies, including Perclose, a manufacturer of femoral artery access site closure devices, Devices for Vascular Intervention, a manufacturer of atherectomy devices, and Advanced Cardiovascular Systems, a manufacturer of balloon angioplasty devices.

- **Robert Thomas** joined FoxHollow Technologies in 1998 as vice president of operations. In 2000 he became president and chief operating officer, and was also a member of the board of directors. Before coming to FHT, from March 1997 to May 1998, Mr. Thomas served as vice president of operations at Conceptus, a manufacturer of contraceptive medical devices.

- **Bart Beasley** started working at FoxHollow Technologies in 2002 and spent about a year in the Milan, Italy office, working with doctors there on the initial coronary applications for the SilverHawk. He then moved to Redwood City, CA and worked as director of marketing of FHT. Previously he worked for Perclose, a company that also developed and manufactured cardiovascular devices, and has worked at several other medical device companies.

APPENDIX B: TIMELINE OF MAJOR MILESTONES AND EVENTS

September 1996	The company is incorporated in Delaware as ArterRx, Inc. Initial capital of $3 million is invested by founder John Simpson and angel investors.
October 1996	The name is changed to FoxHollow Technology.
June 1998	Robert Thomas joins the company as vice president of operations.
March 2000	Robert Thomas becomes president and CEO.
2001	The company undergoes major reorganization, and downsizes from 43 employees to 16. $13 million dollars in capital is raised. The process of iteration to improve the SilverHawk® technology gets underway.
October 2002	CE mark approval is received for coronary applications and sales to European distributors begin.
June - July 2003	FDA approves 510(k) clearance for the SilverHawk® for peripheral applications, and sales to several large medical centers are made.
December 2003	Clinical trials for coronary applications are voluntarily placed on hold.
January 2004	A full commercial launch of the SilverHawk® in the US is started.
October 2004	The company becomes publicly traded, at $14 a share.
2005	A partnership with Merck begins, attempting to use excised plaque to develop diagnostic tests and biomarkers.
December 2005	Robert Thomas leaves the company.
July 2007	A merger with ev3 is announced, and FxH becomes an official subsidiary of ev3, Inc. with an acquisition price of $780 million.

APPENDIX C: FOXHOLLOW SHARE PRICE

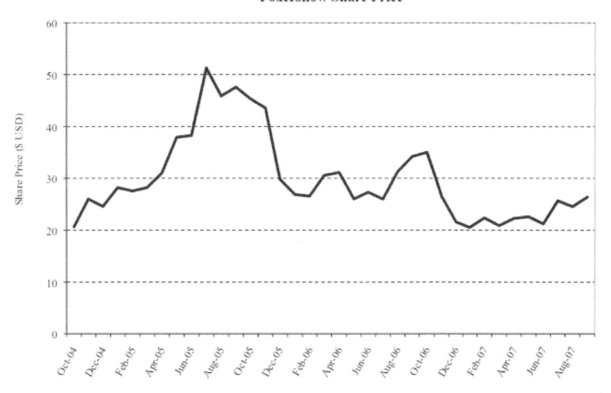

Figure 4: The price of FHT shares, from going public in October 2004 to the ev3 acquisition
Sources: CRSP, S&P Compustat, WRDS

APPENDIX D: GLOSSARY OF TERMS

Angioplasty: The mechanical widening of a blood vessel through the inflation of a balloon inserted along a catheter in the obstructed vessel.

Atherectomy: The removal of material from within an artery using a scraping blade deployed through a catheter.

Atherosclerosis: Obstruction of the arteries due to the buildup of cholesterol and fatty materials and calcification of the deposit.

Coronary arteries: The arteries that supply blood flow to the muscle tissue of the heart.

Drug-eluting stent: A stent carrying cytostatic drugs to prevent the local tissue from growing back and causing restenosis.

Peripheral arterial disease: Loss of circulation in the legs due to atherosclerosis in the femoral or downstream arteries.

Restenosis: The reoccurrence of stenosis (the narrowing of a blood vessel) after treatment.

Stent: A self-expanding tube inserted into a body vessel to counteract local restrictions of fluid flow.

REFERENCES

Collins, T. C., Beyth, R. J., Nelson, D. B., Petersen, N. J., Suarez-Almazor, M. E., Bush, R. L., Hirsch, A. T., and Ashton, C. M. 2007. Process of care and outcomes in patients with peripheral arterial disease. *Journal of General Internal Medicine*, 22: 942-948.

Frost and Sullivan. 1998. *U.S. peripheral vascular stent and stent-graft market.* September 18.

Hirsch, A. T., Criqui, M. H., Treat-Jacobson, D., Regensteiner, J. G., Creager, M. A., Olin, J. W., Krook, S. H., Hunninghake, D. B., Comerota, A. J., Walsh, M. E., McDermott, M. M., & Hiatt, W. R. 2001. Peripheral arterial disease detection, awareness, and treatment in primary care. *Journal of the American Medical Association*, 286:1317-1324.

Levin, S. 2003. FoxHollow: Reviving atherectomy. *IN VIVO*, April.

Morice, M. C., Serruys, P. W., Sousa, J. E., Fajadet, J., Hayashi, E. B., Perin, M., Colombo, A., Schuler, G., Barragan, P., Guagliumi, G., Molnar, F., & Falotico, R. 2002. A randomized comparison of a sirolimus-eluting stent with a standard stent for coronary revascularization. *The New England Journal of Medicine*, 348: 1773-1780.

Stuart, M. 2001. Hijacking the restenosis market. *Start-Up*, November.

Stuart, M. 2004. Atherectomy devices: At the cutting edge of peripheral vascular disease. *Start-Up*, December.

US FDA (Food & Drug Administration). 2010. *Device Advice: Device Regulation and Guidance.* Accessed 14 February 2010, www.fda.gov.

Watanabe, M. 1997. Restenosis: Multi-billion-dollar market tempts many. *Start-Up*, September.

OraPharma: Reformulation of an Existing Product

PAYAM BENYAMINI[1], MARK J. AHN[2], AND DANIELLE HATHAWAY[2]

[1]*University of California, Los Angeles;* [2]*Atkinson Graduate School of Management, Willamette University*

- **Summary and key issue/decision:** In 2002 OraPharma Inc. was a six-year old Warminster, Pennsylvania-based company commercializing products that maintained and restored oral health, received an acquisition offer from Johnson & Johnson for $85 million. OraPharma's initial product Arestin®, a therapeutic for the adjunct treatment of periodontal disease, was approved for marketing by the FDA in February 2001 and launched on April 2, 2001. OraPharma's new product programs also included OC-1012, a compound for the prevention of mucositis, as well as OC-1016, a bone regeneration compound, both of which Orapharma began Phase 1 trials in 2001. The decision-based case assesses the strategic options of Michael D. Kishbauch, founder, president and chief executive officer to either stay independent or sell OraPharma to Johnson & Johnson.

- **Companies/institutions:** OraPharma, Johnson & Johnson

- **Technology:** Locally Applied Antibiotics (LAA). Arestin® is a microsphere encapsulated antibiotic (minocycline hydrochloride, 1 mg) in the tetracycline family, administered locally into the infected areas around teeth, that comes in powder form. This powder is bio-adhesive when placed inside infected periodontal pockets and releases the antibiotic for up to 14 days. A dentist or dental hygienist can place Arestin˙ after scaling and root planning, a common procedure used to treat beginning and moderate periodontal disease.

- **Stage of development at time of issue/decision:** Johnson & Johnson acquires OraPharma one year after its lead product, Arestin®, was approved for marketing by the FDA. During that same year, OraPharma was also developing compounds for bone regeneration and prevention of mucositis, for which the Company began Phase 1 trials.

- **Indication/therapeutic area:** Periodontal disease is a complex infection that involves plaque-associated bacteria, host genetic factors, and environmental features. Symptoms can be observed in an estimated 47% of Americans over the age of 35 (10% of adults worldwide) and range from simple inflammation (gingivitis) to extensive destruction of the tissue and bone that surround and support the teeth (chronic periodontitis). While symptoms are primarily confined to the oral cavity, recent studies suggest that the effects of periodontal disease are more widespread, possibly contributing to heart disease, stroke, systemic infections and premature births.

- **Geography:** US

- **Keywords:** Arestin®, Locally Applied Antibiotics, periodontal disease

---

* *This case was prepared as a basis for class discussion rather than to illustrate either effective or ineffective handling of an administrative situation.*

INTRODUCTION

After two years of reviewing strategic options with a consultant to accelerate value and growth, Michael Kishbauch, founder, president and chief executive officer of OraPharma, Inc. entered into co-promotion[1] discussions with multinational pharmaceutical giant Johnson & Johnson (J&J). After several months of negotiating co-promotion scenarios, Kishbauch and Mark Carbeau, vice president of corporate development, had a dinner meeting with Michael Sneed and Ronald Stricklin, the president and the vice president, worldwide business development, respectively, of Personal Products Company, a division of McNeil-PPC, Inc., a subsidiary of J&J on October 10, 2002. During the course of this meeting, the parties discussed a variety of potential business arrangements, including a possible acquisition of OraPharma by J&J.

After months of due diligence, Sneed called Kishbauch on October 10, 2002 to propose terms of an all cash merger with J&J for $7.41 per share or $85 million, which represented a 63% premium to their share price of $4.54, subject to due legal and scientific diligence. At that time, OraPharma was only a six-year old Warminster, Pennsylvania-based company commercializing products that maintain and restore oral health, with sales of their lead product Arestin® of $7.8 million in 2001 and $14.8 million for the first nine months of 2002.

On one hand, the offer from an iconic multinational company such as J&J was a fulfilling validation of the hard work that propelled OraPharma from start up in 1996 to acquisition candidate in 2002. More importantly, the research and development, sales and marketing, and global distribution capabilities of J&J would contribute significantly more resources to scale OraPharma's technologies to their maximum potential.

On the other hand, while the offer represented a premium to the current share price, the J&J offer was significantly below its IPO (initial public offering) price of $18.00 or its high of $32.125 per share. Kishbauch asked himself, "J&J can help grow OraPharma, but are we selling too early or at the wrong time? I have a responsibility to shareholders and employees to position our company for success. However, we've only been a public company for 17 months and I also have a responsibility to protect the interests of investors who supported us in the formative stages of our growth. Are we getting full value for our products and pipeline?" Kishbach's mind wandered as he looked out the window at the rapid change in scenery brought on by the fall season and wondered what changes would be ahead for OraPharma.

ORAPHARMA BACKGROUND

Founded in 1996 by president and CEO Michael D. Kishbauch, OraPharma, Inc. is a specialty pharmaceutical company engaged in discovering, developing, and commercializing pharmaceutical products, with an initial focus on oral health care. Prior to OraPharma, Mr. Kishbauch held senior management positions with MedImmune, Inc. Mr. Kishbauch holds an MBA from the Wharton School of the University of Pennsylvania and a BA in biology from Wesleyan University.

The business model of Orapharma was created to discover, develop and commercialize therapeutics for oral health care, oncology and orthopedics, with their premier product to be released in the emerging field of oral health care. Initially a venture capital backed start-up, OraPharma quickly progressed to having an approved product and IPO in only five years from company inception. OraPharma's first product Arestin® is a locally administered antibiotic indicated for the adjunctive treatment of adult periodontitis. Arestin® was approved for marketing by the FDA (Food and Drug Administration) in February 2001 and launched on April 2, 2001. OraPharma's most recent product programs also included OC-1012, a compound for the prevention of mucositis, as well as OC-1016, a bone regenera-

---

1 *Co-promotion is a marketing practice where a company uses another company's sales force, in addition to its own, to promote the same brand or range of brands.*

Financing type	Date completed	Amount raised	Investors	Underwriters
IPO	9-Mar-00	$72,000,000		Robertson Stephens, Piper Jaffray, Gerard Klauer Mattison
Venture capital	14-Dec-98	$16,000,000	Oak Investment Partners, Canaan Partners, Senmed Medical Ventures, Biotechnology Investments Ltd., HealthCap, Domain Associates, Frazier & Co., TL Ventures	

Figure 1: OraPharma financing history
Source: Biocentury (2009)

tion compound, for both of which the Company began Phase 1 trials. Additionally, in connection with a distribution agreement with Kensey Nash Corporation, OraPharma distributed the Epi-Guide® Periodontal Barrier Matrix and Drilac® Surgical Dressing.

OraPharma efficiently accessed capital markets to build their pipeline and operations. First, they completed a venture capital-backed financing of $16 million in 1998. In 1999, the company filed for an IPO of up to $56.4 million. Due to strong demand, OraPharma was able to also raise an overallotment resulting in $72 million priced at $18 per share. On March 13, 2000, the day after the IPO, OraPharma's closing share price rose to $32.125, which raised the market capitalization of the company to $404.8 million.

With additional capital, OraPharma scaled its operations and rapidly grew to 163 employees in 2002. For the nine months ending September 30, 2002, OraPharma reported revenues of $14.8 million, and a net loss of $16.7 million or $1.23 per share. For the full year 2002, OraPharma announced a revenue goal of $21 million and a net loss of $21 million or $1.55 per share (see Appendix A for financial statements).

ORAPHARMA'S TECHNOLOGY, PRODUCTS, AND PIPELINE

Periodontal disease is characterized as a chronic multi-bacterial infection that has progressed to involve the oral tissues that retain and support the teeth. Clinical signs of the disease range from superficial inflammation (gingivitis affecting 80% of US adults) to extensive destruction of the underlying connective tissues and bone (periodontitis affecting 30% of adults). While the inflammatory effects of the disease are primarily confined to the oral tissues, recent studies suggest that the effects of periodontal disease are more prevalent, possibly contributing to heart disease, stroke, diabetic complications and premature births. Conventional treatment of periodontitis involves meticulous mechanical removal of biofilm via scaling and root planing (SRP), in conjunction with pharmacological approaches including Locally Applied Antibiotic's (LAA's). Arestin®, a microsphere encapsulated tetracycline antibiotic that comes in a powder form and is administered locally into the infected areas around teeth, is the current market leader for periodontal indications.

DISEASE COURSE AND PROGNOSIS

Periodontal disease is a complex infection that involves plaque or biofilm-associated bacteria, host genetic factors, and environmental factors (e.g., tobacco use) (American Academy of Periodontology, 2005). Dental biofilms undergo a succession of microbial colonization changes in from the transition from healthy to diseased tissue. The early plaque-biofilm consists predominantly of gram-positive organisms, and if left undisturbed undergoes a process of maturation resulting in a more complex and predominantly gram-negative flora (Haffajee & Socransky, 2002, 2005; Socransky & Haffajee, 2005). In contrast to most infections, periodontal disease is a multi-microbial infection normally caused by anaerobic bacterial pathogens capable of colonizing the gingival sulcus. Periodontitis is the destructive form of the disease and is characterized by (a) formation and maturation of subgingival plaque, (b) induction of local inflammatory processes, (c) apical detachment of the junctional epithelium and formation of periodontal pockets, and lastly, (d) destruction of soft tissues and alveolar bone that surround and support the teeth. Recent evidence from epidemiological studies suggest that there are associations between periodontitis and certain systemic disorders, notably cardiovascular disease (Hujoel, Drangsholt, Spiekerman & DeRouen, 2000; Genco, Offenbacher & Beck, 2002; Geerts, Legrand, Charpentier, Albert & Rompen 2004), neurological conditions (Kamer, Craig, Dasanayake, Brys, Glodzik-Sobanska & de Leon, 2008; Watts, Crimmins & Gatz, 2008), respiratory disease (Scannapieco & Ho, 2001; Scannapieco, Wang & Shiau, 2001), diabetes (Saremi *et al.*, 2005; Mealey & Oats, 2006), osteoporosis (Krall, 2001), pre-term birth and low birth infants (López, Da Silva, Ipinza & Gutiérrez, 2005).

EPIDEMIOLOGY

Periodontal disease is prevalent in developed and developing countries. Periodontal diseases are estimated to be among the most common bacterial infections in the United States today. Current surveys estimate that approximately 30% of the adult population is affected with destructive periodontitis (10% with advanced disease (AAP Position Paper, 2005)). Moreover, according to the World Health Organization (WHO), periodontal disease is a significant health burden in both developed and developing countries, with advanced disease affecting approximately 15% of adults worldwide (Petersen & Ogawa, 2005). It has long been assumed that all periodontal infections are similar no matter where one lives in the world. In a study of more than 300 patients with chronic periodontitis from Kenya, Sweden, the United States, Brazil and Chile, investigators found clear geographical differences in the bacterial content of the oral biofilms obtained from periodontal lesions (Dahlén, Manji, Baelum & Fejerskov, 1989; Cao, Aeppli, Liljemark, Bloomquist, Bandt & Wolff, 1990; Schenkein *et al.*, 1993; Craig *et al.*, 2001; Colombo *et al.*, 2002; Haffajee, Bogren, Hasturk, Feres, Lopez & Socransky, 2004; López, Socransky, Da Silva, Japlit & Haffajee, 2004). Comparing the subgingival plaque obtained from patients with chronic periodontitis, significant variations were found in the microbial profiles from country to country. These findings suggest that because different microbial profiles are found in oral biofilms throughout the world and among individuals, treatment responses with a given broad-spectrum anti-

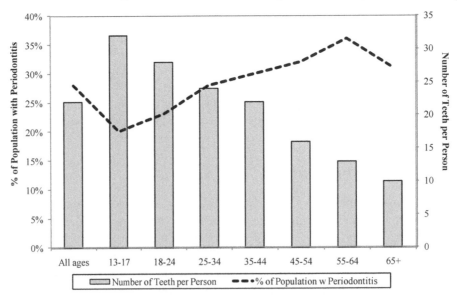

Figure 2: Percent of population with periodontitis and number of teeth per person
Source: Adapted from Brown, Burnelle & Kingman (1996); Marcus, Drury, Brown & Zion (1996)

biotic therapy should be similar in different geographical locations and among individuals.

PERIODONITIS DISEASE MANAGEMENT

The management of periodontal disease involves a different treatment protocol than other oral diseases. While many oral diseases can be treated with proper hygiene, fluoride, pastes, washes and rinses, periodontal disease is often more retractile to treatment. This is due to the differences between the oral and periodontal cavities. The oral cavity is essentially an aerobic environment, constantly perfused by saliva. In contrast, the periodontal cavity is more anaerobic, and is perfused by plasma filtrate, known as "crevicular fluid." As the disease progresses, the periodontal environment becomes more anaerobic, and the flow of crevicular fluid increases (Lawter, Lanzilotti, Bates & Hunter, 2005). This microenvironment provides an ideal niche for growth of hard to treat periodontal pathogens.

Current treatment options for advanced periodontal disease fall into two categories: non-surgical treatment (e.g., scaling and root planing (SRP) with or without antibiotics) and surgical options to repair or regenerate supportive tissues. Although, mechanical therapy is effective for majority of periodontal patients, it rarely results in complete removal of periodontal pathogens. Presently, a broad-spectrum of systemic antibiotics are used to adjunctively manage periodontal infections, including tetracyclines, macrolides, and quinolones (Jorgensen & Slots, 2002). However, these pharmacological efforts to treat periodontal disease have met with limited degrees of success. This is mainly due to inaccessibility of agents to the periodontal cavity; the increased outflow of crevicular fluid that accompanies periodontal disease inhibits therapeutic agents from entering and remaining in pockets (American Academy of Periodontology, 2000). Furthermore, the potential rise in resistant pathogenic strains has limited

---

Periodontal Disease Progression

The figures below illustrate the different stages of periodontal disease. With time, plaque can spread and grow below the gingiva. Toxins produced by the bacteria in plaque irritate the gingival tissue. The toxins stimulate a chronic inflammatory response in which the body in essence turns on itself, and the tissues and bone that support the teeth are broken down and destroyed. Gingiva separate from the teeth, and more gingival tissue and bone are destroyed. Eventually, teeth can become loose and may have to be removed.

Gingivitis is the mildest form of periodontal disease. The gum becomes irritated and swollen (inflamed). The space between the gum and tooth gets deeper, forming a pocket. Gums become red and may bleed. Left untreated, it can progress to periodontitis.

Periodontitis is an infection which induces further inflammation to the bone supporting the teeth and causes gums to recede. Pockets between the teeth become deeper and harder to clean. Redness, swelling, and bleeding develop as infection begins to destroy the bone and loosen teeth.

Chronic Periodontitis represents advance disease where pockets significantly deepen and can fill with pus. Around the roots of the teeth, bone loss continues. The teeth feel sensitive to heat or cold and hurt when brushed. In some cases, teeth may need to be removed to keep the disease from spreading.

The cost of periodontal disease to the healthcare system is significant. In the US alone, periodontal disease accounts for $14 billion of the total $85 billion dental market with approximately 177,000 dentists.

the use of systemic antibiotics in treating periodontal infections due to the significant side effects to the digestive system. Lastly, systemic antibiotic therapy also requires frequent dosing and therefore leads to a decrease in patient compliance (Lawter *et al.*, 2005). To that end, in recent years, efforts have focused on local delivery of pharmacological agents directly to periodontal pockets, in some cases, in controlled-release formulations.

LOCALLY APPLIED ANTIBIOTICS

Administration of agents directly to the pockets permits higher local drug concentrations equivalent to systemic administration while minimizing the associated side effects. A major difficulty with local antibiotics is keeping the active agent in place at the site for an extended period of days necessary to treat the infection. Simply rinsing the pocket with antibiotics does little, considering that the gingival crevicular fluid in a 5mm pocket is replaced about 40 times per hour. Thereby flushing the antibiotic out of a pocket in less than 2 minutes. The need for an effective, non-invasive treatment (American Academy of Periodontology, 2000; Maria, 2001; Lawter *et al.*, 2005) has led to the development of four products that are formulated for delivery into periodontal pockets over a sustained period, with Arestin®, manufactured by Orapharma, being the market leader (Spadaro, 2004).

- **Atridox®: CollaGenex Pharmaceuticals, Inc. Newton, PA:** Atridox® is a biodegradable gel (poly(DLLactide), N-methyl-2-pyrrolidone) containing 10% doxycycline hyclate. The medicament is supplied in two syringes that are mixed together chair-side. The mixed solution is placed into one syringe where it is inserted to the depth of the periodontal pocket. The solution is dispensed until it overfills the pocket and begins to set. Upon contact with the moist environment, the liquid rapidly solidifies. The antibiotic is released into the surrounding tissues over 21 days (Johnson & Stoller, 1999).
- **PerioChip®: Dexcel Pharm, Edison, NJ:** 4 x 5 mm biodegradable chip of hydrolyzed gelatin containing 2.5mg of chlorhexidine gluconate. The chip is placed into periodontal pockets greater than 5 mm and requires no retentive system. The chlorohexidine chip completely dissolves in 8 to 10 days (Ciancio, 1999).
- **Actisite®: Originally manufactured by Procter & Gamble:** Actisite® is a thin thread similar to dental floss which is treated with tetracycline hydrochloride. This thread is placed by the treating dentist into the periodontal pockets around the roots of the teeth after a root planning, or sometimes after other surgical procedures. This thread is left in place for ten days and then removed (Newman, Kornman & Doherty, 1994).

Although the above mentioned products may be able to dispense and appropriate the treatment for a time span of a week or more, they have not been widely used. Common issues include difficult and time-consuming application of the product as well as a potential dislodgment by the patient during tooth brushing, flossing or eating (Eschler, 2009).

The aforementioned disadvantages were overcome by the development of Arestin®. Arestin® is a microencapsulated minocycline hydrochloride in a bioabsorbable polymer, poly (D,L-lactic-*co*-glycolic acid (PLGA), resulting in microspheres that are injected in a dry powdered form into periodontal pockets. This powder hydrates and becomes bio-adhesive when placed inside infected periodontal pockets and releases the antibiotic for up to 14 days. A dentist or dental hygienist can place Arestin® after SRP. Arestin® administration results in local antibiotic concentrations of 340 ug/ml for up to two weeks (Spadaro, 2004).

Minocycline, a member of the tetracycline class of antibiotics, has a broad spectrum of activity. It is bacteriostatic and exerts its antimicrobial activity by inhibiting protein synthesis (Stratton & Lorian,

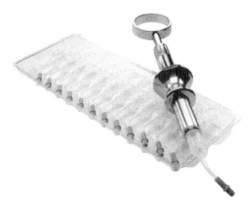

Figure 3: ARESTIN® (minocycline hydrochloride) microspheres, 1 mg is a Locally Administered Antibiotic (LAA) that is indicated as an adjunct therapy to scaling and root planing (SRP) procedures for reduction of pocket depth in patients with adult periodontitis, an infection of the gums that can lead to tooth loss. *Source: www.arestin.com*

1996). *In vitro* susceptibility testing has shown that the organisms *Porphyromonas gingivalis*, *Prevotella intermedia*, *Fusobacterium nucleatum*, *Eikenella corrodens*, and *Actinobacillus actinomycetemcomitans*, which are associated with periodontal disease, are susceptible to minocycline at concentrations of ≤8 µg/mL²; qualitative and quantitative changes in plaque microorganisms have not been demonstrated in patients with periodontitis, using this product (Slots & Rams, 1990; O'Connor, Newman & Wilson, 1990). In addition to their antimicrobial activity, tetracyclines also inhibit matrix metalloproteinases (MMP). These enzymes catalyze the breakdown of periodontal tissue and bone. Taken together, the antibacterial and enzyme inhibitory features of tetracyclines make them an ideal antibiotic for the treatment of periodontal disease (Eschler, 2009).

ARESTIN® CLINICAL STUDIES

In two well-controlled, multicenter, investigator-blind, vehicle-controlled, parallel-design studies (three arms), 748 patients (study OPI-103A = 368, study OPI-103B = 380) with generalized moderate to advanced adult periodontitis characterized by a mean probing depth of 5.90 and 5.81 mm, respectively, were enrolled. Subjects received one of three treatments: (1) scaling and root planing, (2) scaling and root planing + vehicle (bioresorbable polymer, PGLA), and (3) scaling and root planing + Arestin®. To qualify for the study, patients were required to have 4 teeth with periodontal pockets of 6 to 9 mm that bled on probing. In addition, treatment was administered to all sites with mean probing depths of 5 mm or greater. Patients enrolled in the studies were in good general health; those with poor glycemic control or active infectious diseases were excluded. Follow up treatments occurred at 3 and 6 months after initial treatment, and any new site with pocket depth ≥5 mm also received treatment. Patients treated with Arestin® were found to have reduced probing pocket depth. According to published results, these studies revealed patients receiving Arestin® in conjunction with SRP had statistically significant mean probing depth reductions compared to patients treated with SRP alone or SRP + vehicle at 9 months after initial treatment (Williams *et al.*, 2001; OraPharma Inc., 2008).

In these two studies, an average of 29.5 (5-114), 31.7 (4-137), and 31 (5-108) sites were treated at baseline in the SRP alone, SRP + vehicle, and SRP + Arestin® groups, respectively. When these studies are combined, the mean pocket depth change at 9 months was -1.18 mm, -1.10 mm, and -1.42 mm for SRP alone, SRP + vehicle, and SRP + Arestin®, respectively. Taken together, SRP + Arestin® resulted in a greater percentage of pockets showing a change of PD ≥2 mm and ≥3 mm compared to SRP alone at 9 months.

To that end, the application of Arestin® in conjunction with SRP resulted in a statistically significantly 25% reduction in pocket depth compared with those given SRP alone. Additionally, the mean percentages of sites with greater than or equal to 2 mm of probing depth reduction was 40.52% for the SRP + Arestin® group vs. 32.87% for the SRP alone group. Due to the limits of clinical accuracy with a periodontal probe, it is important to note that the "2 mm threshold" is the gold standard that clinicians use to monitor disease progression. Clinical trials demonstrated that Arestin® is much easier to administer compared to its competitors, and is efficacious with minimal adverse effects.

ORAPHARMA PIPELINE

In addition to Arestin®, OraPharma was developing OC-1012 and OC-1016, as well as distributing complimentary oral health products to served markets (see Figure 4). In 1999, OraPharma licensed OC-1012, a metalloproteinase inhibitor in Phase I testing for oral mucositis related to cancer radiation and chemotherapy. The collaboration included an agreement with Biomodels subsidiary Mucosal Therapeutics LLC under which OraPharma would develop a formulation to deliver therapy prior to the onset of mucositis. In exchange, OraPharma would receive exclusive worldwide rights to mucosal reaction intervention technology using the delivery formulation that would be developed by Biomodels at Brigham and Women's Hospital.

In 1999, OraPharma licensed OC-1016 from the Children's Medical Center Corp. OC-1016 (osteointegrin) is a synthetic peptide based on osteopontin protein studied to regenerate bone and tissue for dental implants and orthopedics. Preclinical studies demonstrated that in animals given dental implants treated with OC-1016 there was a significantly greater bone growth after four weeks compared to untreated implants. This was measured by bone-to-implant contact area and bone density. Hence, OC-1016 was being studied as an osteointegrative coating for titanium oral cavity implants as a treatment for bone augmentation and regeneration and for allowing implants to be placed and stabilized during tooth extraction.

In 2002, OraPharma further expanded their product line by entering into an alliance with Kensey Nash Corp. to exclusively distribute in the U.S. and Canada KNSY's Epi-Guide periodontal barrier matrix and Drilac surgical dressing oral care products.

JOHNSON & JOHNSON EXPLORES ALLIANCE WITH ORAPHARMA

Johnson & Johnson (J&J) is a multinational healthcare supply company that serves medical professionals, their patients and consumers and is based in New Brunswick, New Jersey. J&J began as a small medical products company in 1886 by brothers Robert Wood Johnson, James Wood Johnson and Edward Mead Johnson to manufacture and deliver the first commercial sterile surgical dressings to the medical field in the United States. J&J was the first to introduce household products such as band-aids, dental floss, sterile sutures, sanitary napkins and Baby Powder between 1886 and 1926.

Product	Therapeutic indication	Development status	Licensors/research institutions
Arestin	Periodontitis	FDA approved in 2001	American Home Products
OC-1012	Mucositis	Phase 1 clinical trials	Brigham & Women's Hospital/ Mucosal Therapeutics
OC-1016	Bone regeneration for use in dental implants and orthopedics		Children's Medical Center Corporation
Drug delivery systems for small and large molecule drugs	oncology, pain management, and central nervous system conditions		Pre-clinical

Figure 4: OraPharma product pipeline
Source: Company reports

J&J began globalizing operations to Canada in 1919 and England in 1924, and shortly thereafter to Mexico, South Africa, Australia, France, Belgium, Ireland, Switzerland, Argentina and Brazil. After going public in 1944, J&J further expanded both domestically and internationally by forming and acquiring new companies, as well as broadening their product mix to include consumer products such as Tylenol® and Listerine®. Throughout the mid-1960's-1980's, J&J focused on the development of prescription products such as Haldol® (haloperidol for schizophrenia), consumer products such as Monistat® (miconazole for woman's health), medical devices and diagnostics such as the Proximate Linear Stapler® (surgical closure device). In 1985, J&J expanded into China.

Nurturing a values-based corporate culture has been critical to the success and sustainability of J&J. General Robert Wood Johnson, former chairman from 1932 to 1963 and a member of the company's founding family, crafted Our Credo himself in 1943, just before J&J became a publicly traded company. Long before the term "corporate social responsibility" was popularized, Our Credo was created as a one-page document to outline the operating principles that guide the company today (see Figure 5).

From the mid-1980's onwards, J&J has remained an industry leader in the areas of consumer products, medical devices, diagnostics and prescription products. Although J&J has been an innovator with internal development of products, a significant tenet of their corporate strategy is growth through alliances (e.g., licensing, co-promotion, mergers and acquisitions) and decentralized operating units, with each international subsidiary, with some exceptions, managed by citizens of the country in which it is located. The Executive Committee of J&J is the principal management group responsible for the operations and allocation of the resources of the company. This committee oversees and coordinates the activities of domestic and international subsidiaries that span the Consumer, Pharmaceutical and Medical Devices & Diagnostics segments. Today, J&J has more than 250 operating companies in 57 countries under its umbrella and employs more than 117,000 people worldwide.

In 2002, worldwide revenues for J&J were $36.3 billion, an increase of 10% from the prior year; and net income was $6.6 billion, an increase of 16% (Appendix B for financials). Sales by operating segment and geography are shown in Figure 6.

J&J'S PERSONAL PRODUCTS COMPANY

The consumer segment's principal products are personal care and hygienic products, including nonprescription drugs such as Tylenol®, adult skin and hair care products, baby care products, oral care products, first aid products and sanitary protection products. One such operating company is the Personal Products Company, a division of McNeil-PPC, a leader in women's health and sanitary products such as K-Y Brand Liquid®, Stayfree®, o.b.®, Carefree®, and Monistat® consumer products. The Personal Products Company is also responsible for the production of a line of oral health products such as Listerine® mouthwash, Reach® toothbrushes, Efferdent® denture cleansers, and Rembrandt® teeth whitening. Recognizing the oral health market was a fast growing segment due to an aging population and increased awareness of oral health, the Personal Products Company was seeking ways to expand their product portfolio to dentists and periodontists—and identified OraPharma as a potential alliance candidate.

Leading the negotiations between J&J and OraPharma was Michael Sneed, company group chairman, Vision Care, for J&J and a member of the medical devices & diagnostics group operating committee. Mr. Sneed, a native of Chicago who attended Macalester College in Minnesota and has a master's degree in business administration from the Tuck School at Dartmouth College, joined J&J in 1983 as a marketing assistant for Personal Products Company. He held positions of increasing responsibility in marketing through 1990, when he was named marketing manager of the Personal Products Company. Mr. Sneed moved to McNeil Consumer Products as a group product director in 1991 and was promoted to vice president, worldwide OTCs (over the counter) in 1995. In 1998 Mr. Sneed relocated to Europe as man-

Our Credo

We believe our first responsibility is to the doctors, nurses and patients,
to mothers and fathers and all others who use our products and services.
In meeting their needs everything we do must be of high quality.
We must constantly strive to reduce our costs
in order to maintain reasonable prices.
Customers' orders must be serviced promptly and accurately.
Our suppliers and distributors must have an opportunity
to make a fair profit.

We are responsible to our employees,
the men and women who work with us throughout the world.
Everyone must be considered as an individual.
We must respect their dignity and recognize their merit.
They must have a sense of security in their jobs.
Compensation must be fair and adequate,
and working conditions clean, orderly and safe.
We must be mindful of ways to help our employees fulfill
their family responsibilities.
Employees must feel free to make suggestions and complaints.
There must be equal opportunity for employment, development
and advancement for those qualified.
We must provide competent management,
and their actions must be just and ethical.

We are responsible to the communities in which we live and work
and to the world community as well.
We must be good citizens - support good works and charities
and bear our fair share of taxes.
We must encourage civic improvements and better health and education.
We must maintain in good order
the property we are privileged to use,
protecting the environment and natural resources.

Our final responsibility is to our stockholders,
Business must make a sound profit.
We must experiment with new ideas.
Research must be carried on, innovative programs developed
and mistakes paid for.
New equipment must be purchased, new facilities provided
and new products launched.
Reserves must be created to provide for adverse times.
When we operate according to these principles,
the stockholders should realize a fair return.

Figure 5: Johnson & Johnson credo
Source: www.jnj.com

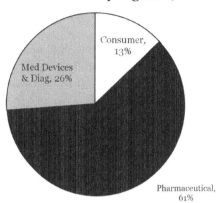

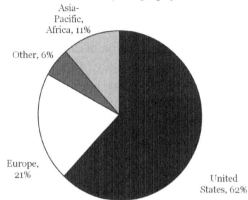

Figure 6: J&J revenues by segment and geography, 2002
Source: Company reports

aging director of McNeil Consumer Nutritionals Europe. He returned to the U.S. as president, McNeil Nutritionals Worldwide, in 2000. In 2002 he was named global president, Personal Products Company.

CONCLUSION

Michael Kishbauch, founder, president and chief executive officer of OraPharma, Inc. reflected on how rapidly discussions with J&J went from co-promotion to outright acquisition. If the merger went through, OraPharma would be come a wholly-owned and operated subsidiary of J&J.

Kishbauch convened a meeting of OraPharma's senior management team and board of directors (Appendix C), as well as Morgan Stanley investment bankers who were hired as advisors, for the proposed merger. There were a number of factors relevant to the merger including the strategic business plan, operations, products, candidates under development, and development programs, financial position, current securities market conditions, and a potential for growth in revenues and earnings.

The Morgan Stanley analyst reported to the board on their valuation analysis that considered a range of scenarios (Appendix D). Morgan Stanley posited that as of November 6, 2002, the implied share price was in the range of $6.06 to $9.05, as compared to the $7.41 per share merger consideration that they considered "fair" consideration.[2]

Kishbauch and the board discussed the pros and cons of the proposed merger. On the positive side the offer price of $7.41 represented a 63% premium to their current share price of $4.54. Based on negotiations to date, the board also felt that J&J had put forth its best offer—and that there was a low likelihood that a third party would pay more for OraPharma. After an extensive review of strategic options, management concluded that OraPharma's share price was unlikely to exceed $7.41 in the near term if the company remained independent. In addition, the J&J offer was cash, as opposed to a transaction in which shareholders received stock. This was a significant consideration because cash provided maximum liquidity to shareholders, which was one of the main goals of OraPharma. Importantly, J&J intended to operate OraPharma as a wholly owned subsidiary, retaining the OraPharma name, organization and employee base.

On the other hand, the board also considered a number of potentially negative factors in its deliberations concerning the merger. Most importantly, OraPharma would no longer exist as an independent company and stockholders will no longer participate in its growth. This was especially significant given that the majority of financing (i.e., $72 million or 82% of all capital raised) into OraPharma to

---

2   A "fairness opinion" is the professional opinion of an investment bank, provided for a fee, regarding the fairness of a price offered in a merger or takeover. In this case, OraPharma paid Morgan Stanley $1.7 million in advisory fees on the proposed merger transaction.

date was via the IPO at $18.00—which meant a 59% decrease at the $7.41 acquisition price. Further, gains from an all-cash transaction would be taxable to the stockholders for U.S. federal income tax purposes.

Kishbauch summed up the discussion, "J&J is a global powerhouse and they can rapidly scale and globalize our products and pipeline. On the other hand, we've only been a public company for 17 months and our share price has fallen significantly below our IPO price, despite our momentum with Arestin® and promising pipeline. We now have to decide whether we recommend to shareholders to vote for or against the J&J offer."

APPENDICES

APPENDIX A: ORAPHARMA FINANCIAL STATEMENTS

INCOME STATEMENT	1996	1997	1998	1999	2000	2001	YTD Sept 2002
Revenues	---	---	---	---	---	$7,748,405	$14,752,454
Operating Expenses							
Cost of Goods	---	---	---	---	---	$1,957,497	$3,661,778
Research & Development	$26,294	$1,706,393	$7,589,000	$9,693,413	$8,205,573	$10,466,790	$6,778,621
Marketing, General & Admin	$408,295	$939,469	$1,604,579	$2,189,577	$8,350,336	$23,488,753	$21,551,984
Operating Loss	($434,589)	($2,645,862)	($9,193,579)	($11,882,990)	($16,555,909)	($28,164,635)	($31,992,383)
Interest Income, net	$641	$504,123	$424,488	$636,957	$4,239,397	$2,478,975	$523,777
Net Loss	($433,948)	($2,141,739)	($8,769,091)	($11,246,033)	($12,316,512)	($25,685,660)	($17,239,929)
Non-cash preferred stock charge	---	---	---	($1,729,651)	---	---	---
Net loss to common stockholders	($433,948)	($2,141,739)	($8,769,091)	($12,975,684)	($12,316,512)	($25,685,660)	($17,239,929)
Basic and diluted net loss per share		($5.05)	($13.72)	($16.74)	($1.13)	($1.91)	($1.23)
Shares Outstanding		424,054	639,339	775,116	10,921,573	13,440,517	13,587,457
BALANCE SHEET							
Cash & Cash Equivalents	$37,704	$10,136,747	$19,236,084	$13,073,803	$75,255,171	$44,121,660	$27,545,131
Total Assets	$61,479	$10,859,584	$20,480,402	$14,771,713	$80,431,518	$57,749,896	$41,004,948
Long-Tem Debt	---	---	$480,978	$288,043	---	---	---
Redeemable convertible preferred stock	$0	$12,822,767	$28,771,713	$32,974,359	$0	$0	$0
Accumulated surplus (deficit)	$0	($435,230)	($2,576,969)	($11,346,060)	($22,592,093)	($60,594,265)	($77,310,417)
Total stockholders' equity (deficit)	($359,071)	($2,446,806)	($11,080,451)	($20,616,829)	$76,472,394	$51,081,201	$34,612,306

Source: Company reports

APPENDIX B: JOHNSON & JOHNSON FINANCIALS

Income Statement (US$ MM)

	1999	2000	2001	2002
Revenue	27,471.0	29,139.0	33,004.0	36,298.0
COGS	8,442.0	8,861.0	9,536.0	10,447.0
Gross Profit	19,029.0	20,278.0	23,468.0	25,851.0
Operating Expenses $Mil				
SG&A	10,503.0	10,875.0	11,992.0	12,216.0
R&D	2,600.0	2,980.0	3,696.0	4,146.0
Other	222.0	34.0	185.0	294.0
Operating Income	5,704.0	6,389.0	7,595.0	9,195.0
Other Income and Expense $Mil				
	1,999.0	2,000.0	2,001.0	2,002.0
Net Int Inc & Other	49.0	233.0	303.0	96.0
Earnings Before Taxes	5,753.0	6,622.0	7,898.0	9,291.0
Income Taxes	1,586.0	1,822.0	2,230.0	2,694.0
Net Income	4,167.0	4,800.0	5,668.0	6,597.0
Diluted EPS$	1.47	1.7	1.84	2.16
Shares	2834	2823	3080	3054

Source: Company reports

Balance Sheet

Assets	1999	2000	2001	2002
Cash and Equiv	2,363.0	3,411.0	3,758.0	2,894.0
Short-Term Investments	1,516.0	2,333.0	4,214.0	4,581.0
Accts Rec	4,233.0	4,464.0	4,630.0	5,399.0
Inventory	3,095.0	2,842.0	2,992.0	3,303.0
Other Current Assets	1,993.0	2,400.0	2,879.0	3,089.0
Total Current Assets	13,200.0	15,450.0	18,473.0	19,266.0
Net PP&E	6,719.0	6,971.0	7,719.0	8,710.0
Intangibles	7,571.0	7,256.0	9,077.0	9,246.0
Other Long-Term Assets	1,673.0	1,644.0	3,219.0	3,334.0
Total Assets	29,163.0	31,321.0	38,488.0	40,556.0

Liabilities and Stockholders' Equity

	1999	2000	2001	2002
Accts Payable	2,003.0	2,083.0	2,838.0	3,621.0
Short-Term Debt	1,806.0	1,479.0	565.0	2,117.0
Taxes Payable	206.0	314.0	537.0	710.0
Accrued Liabilities	3,439.0	3,264.0	4,104.0	5,001.0
Other Short-Term Liabilities	0.0	0.0	0.0	0.0
Total Current Liabilities	7,454.0	7,140.0	8,044.0	11,449.0
Long-Term Debt	2,450.0	2,037.0	2,217.0	2,022.0
Other Long-Term Liabilities	3,046.0	3,336.0	3,994.0	4,388.0
Total Liabilities	12,950.0	12,513.0	14,255.0	17,859.0
Total Equity	16,213.0	18,808.0	24,233.0	22,697.0
Total Liabilities & Equity	29,163.0	31,321.0	38,488.0	40,556.0

Source: Company reports

APPENDIX C: ORAPHARMA SENIOR MANAGEMENT & BOARD OF DIRECTORS

- **Michael D. Kishbauch (50) President, Chief Executive Officer and Director**: Mr. Kishbauch has served as our president and chief executive officer and as a director of OraPharma since September 1996. He served as president and chief operating officer for two business units of Nelson Communications, Inc., an integrated healthcare services firm, from February 1995 to August 1996. He also served as president, chief operating officer and director of MedImmune, Inc., a Maryland-based biotechnology company, from December 1992 to February 1995. From February 1982 to May 1992, Mr. Kishbauch served with the pharmaceuticals division of Ciba-Geigy Corporation in various sales and marketing positions, ending as vice president product planning and promotion. Mr. Kishbauch worked through positions of increasing responsibility in brand management with The Procter and Gamble Company from June 1976 to February 1982. Mr. Kishbauch received a B.A. in biology from Wesleyan University and an M.B.A. from the Wharton School of the University of Pennsylvania.

- **Mark B. Carbeau (39) Vice President, Corporate Development**. Mr. Carbeau has served as our vice president, corporate development since May 1999. From September 1996 to April 1999, he served as general partner in The Lucas Group, a Boston-based strategy consulting and mergers and acquisitions advisory firm, and from January 1995 to September 1996 as a principal in North Atlantic Capital, a private equity firm. Prior to that, he was a consultant and case manager for The Boston Consulting Group, a management consulting firm, from September 1990 through December 1994. Mr. Carbeau held a number of cross-functional positions with Eli Lilly and Company, a pharmaceutical company, from September 1982 to July 1988. He holds a B.S. in industrial engineering from the Pennsylvania State University and an M.B.A. from the Wharton School of the University of Pennsylvania.

- **J. Ronald Lawter, Ph.D (57) Vice President, Chief Scientific and Technical Officer**. Dr. Lawter has served as our vice president, chief scientific and technical officer since he joined us in March 1997. From October 1983 to March 1997, he held scientific and management positions in pharmaceutical product development at American Cyanamid and at American Home Products after its acquisition of American Cyanamid in 1994. While at American Cyanamid, he led the team that developed the drug-delivery technology that is the basis for our lead product, MPTS. From August 1979 through October 1983, he was a senior research scientist in the Advanced drug delivery group at Ciba-Geigy. From 1977 through 1979, he was a research manager in the biomedical division of Abcor, Inc. and from 1972 through 1977, was a consultant with Arthur D. Little, Inc. He received a B.S. in chemistry from the University of South Carolina and a Ph.D. in physical chemistry from the Massachusetts Institute of Technology.

- **Jan N. Lessem, M.D., Ph.D (51) Vice President, Chief Medical Officer**. Dr. Lessem has served as our vice president, chief medical officer since June 1998. From May 1995 to June 1998, he served as medical director and vice president of drug strategy at Takeda America, a pharmaceutical company. Prior to that, he was involved in various clinical research and management roles at several pharmaceutical companies, specifically: SmithKline Beecham from June 1991 to May 1995; Union Chemique Belgique Pharmaceutical in Brussels, Belgium, from May 1990 to May 1991; Syntex Research from January 1986 to May 1990; Bristol Myers from August 1983 to December 1985; and Merck Sharp & Dohme from May 1982 to August 1983. Between 1974 and 1982, he was a fellow, instructor and associate professor in cardiology and geriatrics, at the University of Lund, in Sweden. He is a Fellow of the American College of Cardiology, and a member of the New York Academy of Sciences, as well as The Swedish Medical Association. Dr. Lessem earned an M.D. from the University of Lund in Sweden in 1974, a Ph.D. in clinical cardiology from the same university in 1982, and was board certified in cardiology in Sweden in 1982.

- **James A. Ratigan (51) Vice President, Chief Financial Officer and Secretary.** Mr. Ratigan has served as our vice president, chief financial officer since June 1997 and was named secretary in December 1999. From February 1997 to June 1997, Mr. Ratigan served as the chief financial officer of TL Ventures, one of the initial investors in OraPharma. From September 1996 to February1997, Mr. Ratigan served as the vice president-finance of Robotic Vision Systems, Inc., a publicly-held company widely engaged in machine vision and electronic imaging. From October 1993 to August 1996, Mr. Ratigan served as the executive vice president, chief operating officer and chief financial officer and a director of Perceptron, Inc., a publicly-held company which provides three dimensional machine vision technologies to the automotive, forestry products and aerospace industries. From March 1983 to October 1992, Mr. Ratigan was with the Adler Group, a venture capital fund, where he served in a number of positions including venture manager, chief financial officer, and chief executive officer of a machine vision company controlled by the Adler Group. Earlier, Mr. Ratigan spent eight years with Arthur Andersen LLP, where, as a manager, he focused on entrepreneurial clients. Mr. Ratigan received his B.S. in finance and accounting from LaSalle University, Philadelphia, Pennsylvania and is a CPA.

- **Joseph E. Zack (48) Vice President, Sales and Marketing.** Mr. Zack has served as our vice president, sales and marketing since March 1998. From 1993 to 1998, Mr. Zack held senior management positions of general manager and executive director marketing with Advanced Tissue Sciences, a biotechnology company focused on tissue engineering. Prior to that, he was executive director marketing for Ciba-Geigy from 1987 to 1993, and product director from 1982 to 1987, where he was responsible for a number of successful product launches. From 1973 to 1982, he held positions in sales and new product development with Ciba-Geigy. Mr. Zack obtained a B.A. in biology from Colgate University, and an M.B.A. from St. John's University in New York.

- **James J. Mauzey (51) Director.** Mr. Mauzey has been a director of OraPharma since July 1997. Since March 1999, Mr. Mauzey has been the chief executive officer of Innovex, a division of Quintiles Transnational. From March 1994 through February 1999, Mr. Mauzey was chairman and chief executive officer of Alteon, Inc., a biotechnology company. Prior to that, he spent 22 years in major roles with leading pharmaceutical companies, including as president of the Bristol-Myers Squibb U.S. pharmaceutical division from March 1989 through March 1994 and as the president of the Squibb Corporation U.S. pharmaceutical group and vice president of both U.S. and international operations of Lederle.

- **Christopher Moller, Ph.D. (46) Director.** Dr. Moller has been a director of OraPharma since March 1997. Since 1990, he has served as vice president of TL Ventures, a company which manages a series of private equity funds. Since 1994, Dr. Moller has served as a managing director of the following funds managed by TL Ventures, Radnor Venture Partners, Technology Leaders, Technology Leaders II, TL Ventures III and TL Ventures IV. He is principally responsible for the life science portfolio at TL Ventures, specializing in financing and development of early-stage biotechnology, bioinformatics and e-health companies. Dr. Moller also currently serves as a director on the boards of Adolor Corporation, Assurance Medical, Esperion Therapeutics, Immunicon Corporation, eMerge Interactive, Inc., ChromaVision Systems, Inc. and Genomics Collaborative. Dr. Moller holds a Ph.D. in immunology from the University of Pennsylvania.

- **Eileen M. More (53) Director.** Ms. More has been a director of OraPharma since September 1996. She has been associated with Oak Investment Partners, a venture capital firm, since 1978 and has been a general partner or managing member since 1980. She currently serves as a director of several private companies including Halox Technologies, Psychiatric Solutions and Teloquent Communications Corp. Ms. More was also a founding investor in Genzyme and has also been responsible for early-stage investments in numerous companies including Alkermes, Alexion Pharmaceuticals, Esperion Therapeutics, Inc., KeraVision, Pharmacopeia, Trophix

Pharmaceuticals, Compaq Computer, Network Equipment Technologies, Octel Communications and Stratus Computer.

- **Harry T. Rein (55) Director**. Mr. Rein has been a director of OraPharma since March 1997. He is the principal founder of Canaan Partners and has served as managing general partner since its inception in 1984, with extensive experience working with small and mid-sized companies. Prior to that, he was president and chief executive officer of GE Venture Capital Corporation. Mr. Rein joined General Electric Company in 1979 and directed several of GE's lighting businesses as general manager before joining the venture capital subsidiary. Prior to his GE career, Mr. Rein worked in various capacities with Polaroid Corporation, Transaction Systems, Inc. and Gulf Oil Corporation. In addition to serving on the boards of several private companies, Mr. Rein is also on the board of Anadigics.

- **Seth A. Rudnick, M.D. (51) Director**. Dr. Rudnick has been a director of OraPharma since July 1997. He currently is consulting for several venture capital firms, and serves on the board of NaPro BioTherapeutics, Inc. He was chairman and CEO of Cytotherapeutics, Inc. from 1995 through 1998. Prior to that, Dr. Rudnick served as senior vice president of the R.W. Johnson Pharmaceutical Research Group of Ortho Pharmaceutical Corporation, senior vice president of development with Biogen Research Corporation and director of clinical research with Schering-Plough. Dr. Rudnick has held various faculty appointments with Brown University, the University of North Carolina and Yale University, and received his M.D. from the University of Virginia, with fellowships at Yale in oncology and epidemiology.

- **David I. Scheer (47) Director**. Mr. Scheer has been a director of OraPharma since September 1996. He has been president of Scheer & Company, Inc., a firm with activities in venture capital, corporate strategy, and transactional advisory services focused on the life sciences industry, since 1981. In venture capital, Mr. Scheer has been involved in the founding of our company, as well as ViroPharma, Inc., Esperion Therapeutics, Inc. and Achillon Pharmaceuticals, Inc. and has been a member of the board of directors of Nonlinear Dynamics, Inc. and a series of private and public companies. He has led engagement teams from Scheer providing corporate strategic advisory services to a broad range of companies including Agouron Pharmaceuticals (now a division of Warner-Lambert), American Cyanamid (now a division of AHP), B.F. Goodrich, Pharmacia AB, Pharmacia & Upjohn, Hoffman La-Roche, Eli Lilly, and a range of smaller, publicly- and privately-held companies. Mr. Scheer has also led or played a significant role in a series of transactions involving corporate alliances, licensing arrangements, divestments, acquisitions and mergers in the life sciences. He received his B.A. from Harvard College and his M.S. from Yale University.

- **Jesse I. Treu, Ph.D. (52) Director**. Dr. Treu has been a director of OraPharma since December 1998. He is a managing member of Domain Associates, L.L.C., and has served in this or similar capacities with this firm since 1986. He has served as a director of over 20 early-stage health companies, ten of which have so far become public companies. He is currently a director of Focal, Inc., GelTex Pharmaceuticals, Trimeris Inc. and Simione Central Holdings, Inc. Prior to the formation of Domain, Dr. Treu had 12 years of health care experience at General Electric and Technicon Corporation in a number of research, marketing management and corporate staff positions. Dr. Treu received his B.S. from Rensselaer Polytechnic Institute, and from Princeton University his M.A. and Ph.D. in physics.

APPENDIX D: VALUATION

Morgan Stanley was engaged by OraPharma. Morgan Stanley performed a discounted equity value analysis to determine a range of present values for OraPharma based on financial projections prepared by management. In calculating the discounted equity value per share, Morgan Stanley applied a forward looking multiple range of 10.0x to 20.0x to 2006 projected fully-taxed net income. The equity value for the base-case scenario was discounted to present values using a 25% discount rate. Morgan Stanley calculated equity values per share of between approximately $4.79 and $9.59. Assuming a growth-case scenario, Morgan Stanley observed equity values per share of between approximately $7.50 and $14.99. The equity value for the growth-case scenario was discounted to present values using a 50% discount rate.

Morgan Stanley performed discounted cash flow analysis to determine a range of present values. In the discounted cash flow analysis, Morgan Stanley considered only the value of the Arestin franchise and the value of available cash. In calculating the discounted cash flow equity value per share, Morgan Stanley calculated the unlevered free cash flow estimates for the Arestin franchise for a ten-year projection period from 2003 to 2012 and applied a terminal growth rate of negative 20%-50% to the projected 2012 unlevered free cash flow. The unlevered free cash flows and terminal values were then discounted to present values using discount rates of 12% to 16%. Under this analysis, Morgan Stanley calculated equity values per share of between approximately $6.66 and $8.31.

Finally, the figure below compares the J&J offer to various OraPharma historical share prices since the IPO.

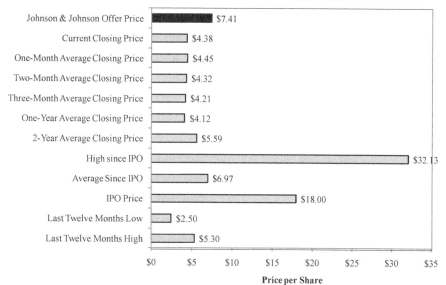

OraPharma: Historical Share Prices v. J&J Offer

Source: Company reports; www.google.com/finance

REFERENCES

American Academy of Periodontology (AAP) Position Paper. 2000. The Role of Controlled Drug Delivery for Periodontitis. *Journal of Periodontology.* 71:125-140.

American Academy of Periodontology (AAP) Position Paper. 2005. Epidemiology of Periodontal Diseases. *Journal of Periodontology.* 76:1406-1419.

Brown LJ, Brunelle JA, Kingman A. 1996. Periodontal status in the United States, 1988-1991: prevalence, extent, and demographic variation. *Journal of Dental Research.* 75:672-683.

Cao CF, Aeppli DM, Liljemark WF, Bloomquist CG, Bandt CL, Wolff LF. 1990. Comparison of plaque microflora between Chinese and Caucasian population groups. *Journal of Clinical Periodontology.* 17:115-118.

Ciancio SG. 1999. Local delivery of chlorhexidine. *Compendium.* 20:427-433

Colombo AP, Teles RP, Torres MC, Souto R, Rosalém WJ, Mendes MC, Uzeda M. 2002. Subgingival microbiota of Brazilian subjects with untreated chronic periodontitis. *Journal of Periodontology.* 73:360-369.

Craig RG, Boylan R, Yip J, Bamgboye P, Koutsoukos J, Mijares D, Ferrer J, Imam M, Socransky SS, Haffajee AD. 2001. Prevalence and risk indicators for destructive periodontal diseases in 3 urban American minority populations. *Journal of Clinical Periodontology.* 28:524-535.

Dahlén G, Manji F, Baelum V, Fejerskov O. 1989. Black-pigmented Bacteroides species and Actinobacillus actinomycetemcomitans in subgingival plaque of adult Kenyans. *Journal of Clinical Periodontology.* 16:305-310.

Eschler B. 2009. Drug Treatment of Periodontal Disease: Locally-Delivered Antimicrobials. *MDA Annual Session.* April:1-3.

Geerts SO, Legrand V, Charpentier J, Albert A, Rompen EH. 2004. Further evidence of the association between periodontal conditions and coronary artery disease. *Journal of Periodontology.* 75:1274-1280.

Genco R, Offenbacher S, Beck J. 2002. Periodontal disease and cardiovascular disease: epidemiology and possible mechanisms. *Journal of the American Dental Association.* 133:14S-22S

Haffajee AD, Bogren A, Hasturk H, Feres M, Lopez NJ, Socransky SS. 2004. Subgingival microbiota of chronic periodontitis subjects from different geographic locations. *Journal of Clinical Periodontology.* 31:996-1002.

Haffajee AD, Socransky SS. 2002. Dental biofilms: difficult therapeutic targets. *Periodontology 2000.* 28:12-55.

Haffajee AD, Socransky SS. 2005. Microbiology of periodontal diseases: introduction. *Periodontology 2000.* 38:9-12.

Hujoel PP, Drangsholt M, Spiekerman C, DeRouen TA. 2000. Periodontal disease and coronary heart disease risk. *Journal of the American Medical Association.* 284:1406-1410.

Johnson LR, Stoller NH. 1999. Rationale for the use of Atridox therapy for managing periodontal patients. *Compendium.* 20:19-25.

Jorgensen MG, Slots J. 2002. The ins and outs of periodontal antimicrobial therapy. *Journal of the California Dental Association.* 30:297-305.

Kamer AR, Craig RG, Dasanayake AP, Brys M, Glodzik-Sobanska L, de Leon MJ. 2008. Inflammation and Alzheimer's disease: possible role of periodontal diseases. *Alzheimer's & Dementia.* 4:242-250.

Krall EA. 2001. The periodontal-systemic connection: implications for treatment of patients with osteoporosis and periodontal disease. *Annals of Periodontology.* 6:209-213.

Lawter JR, Lanzilotti MG, Bates M, Hunter GH. 2004. *Dispensing apparatus and cartridge with deformable tip.* United States Patent Application. Publication No.: US 2004/0152042 A1.

López NJ, Da Silva I, Ipinza J, Gutiérrez J. 2005. Periodontal therapy reduces the rate of preterm low birth weight in women with pregnancy-associated gingivitis. *Journal of Periodontology.* 76:2144-2153.

López NJ, Socransky SS, Da Silva I, Japlit MR, Haffajee AD. 2004. Subgingival microbiota of chilean patients with chronic periodontitis. *Journal of Periodontology.* 75:717-725.

Marcus SE, Drury TF, Brown LJ, Zion GR. 1996. Tooth retention and tooth loss in the permanent dentition of adults: United States, 1988-1991. *Journal of Dental Research.* 75:684-695.

Maria P. 2001. Pharmacotherapy in periodontal therapy. *Access.* Special supplemental issue Sep-Oct:1-10.

Mealey BL, Oates TW. 2006. Diabetes mellitus and periodontal diseases. *Journal of Periodontology.* 77:1289-1303.

Newman MG, Kornman KS, Doherty FM. 1994. A 6-month multi-center evaluation of adjunctive tetracycline fiber therapy used in conjunction with scaling and root planing in maintenance patients: clinical results. *Journal of Periodontology.* 65:685-691.

O'Connor BC, Newman HN, Wilson M. 1990. Susceptibility and resistance of plaque bacteria to minocycline. *Journal of Periodontology*. 61:228-233.

OraPharma Inc. 2008. Arestin product insert. March: R6

Petersen PE, Ogawa H. 2005. Strengthening the prevention of periodontal disease: the WHO approach. *Journal of Periodontology*. 76:2187-293.

Saremi A, Nelson RG, Tulloch-Reid M, Hanson RL, Sievers ML, Taylor GW, Shlossman M, Bennett PH, Genco R, Knowler WC. 2005. Periodontal disease and mortality in type 2 diabetes. *Diabetes Care*. 28:27-32.

Scannapieco FA, Ho AW. 2001. Potential associations between chronic respiratory disease and periodontal disease: analysis of National Health and Nutrition Examination Survey III. *Journal of Periodontology*. 72:50-56.

Scannapieco FA, Wang B, Shiau HJ. 2001. Oral bacteria and respiratory infection: effects on respiratory pathogen adhesion and epithelial cell proinflammatory cytokine production. *Annals of Periodontology*. 6:78-86.

Schenkein HA, Burmeister JA, Koertge TE, Brooks CN, Best AM, Moore LV, Moore WE. 1993. The influence of race and gender on periodontal microflora. *Journal of Periodontology*. 64:292-296.

Slots J, Rams TE. 1990. Antibiotics in periodontal therapy: advantages and disadvantages. *Journal of Clinical Periodontology*. 17:479-493.

Spadaro SE. 2004. Local chemotherapeutics as an adjunct to scaling and root planning. *The Dental Assistant*. November:1088-2886

Socransky SS, Haffajee AD. 2005. Periodontal microbial ecology. *Periodontology 2000*. 38:135-187

Stratton CW, Lorian V. 1996. Mechanisms of action of antimicrobial agents: general principles and mechanisms for selected classes of antibiotics. In *Antibiotics in Laboratory Medicine* (4th ed). Baltimore, MD: Williams and Wilkins.

Watts A, Crimmins EM, Gatz M. 2008. Inflammation as a potential mediator for the association between periodontal disease and Alzheimer's disease. *Neuropsychiatric Disease Treatment*. 4:865-876.

Williams RC, Paquette DW, Offenbacher S, Adams DF, Armitage GC, Bray K, Caton J, Cochran DL, Drisko CH, Fiorellini JP, Giannobile WV, Grossi S, Guerrero DM, Johnson GK, Lamster IB, Magnusson I, Oringer RJ, Persson GR, Van Dyke TE, Wolff LF, Santucci EA, Rodda BE, Lessem J. 2001. Treatment of periodontitis by local administration of minocycline microspheres: a controlled trial. *Journal of Periodontology*. 72:1535-1544.

DesignMedix: Maintain Product Focus or Leverage Technology Platform?

ELIZABETH R. BIVINS-SMITH, BETTINA M. FRANA, AND W. KELLOGG THORSELL
School of Business Administration, Portland State University

- **Key issue(s)/decision(s):** DesignMedix, Inc. is a start-up bioscience company focused on developing treatments for drug-resistant infectious diseases and cancer. David Peyton, DesignMedix's chief scientific officer and professor of chemistry at Portland State University, and his team were in the development stage of a compound used to treat chloroquine-resistant malaria, a disease that claims the lives of over one million people annually. The leadership team of this university-based start-up was faced with the challenge of attracting investors and corporate partners while bringing its lead candidate to market. As typical in a start-up environment, financial resources were limited, and DesignMedix had to effectively utilize its human and financial capital. The company's primary dilemma was whether to focus on therapeutics for malaria, which was a core competency, or extend their technology as a platform to develop treatments for additional drug-resistant diseases. The resolution of this dilemma would influence the company's ability to raise capital, obtain partnerships with leading pharmaceutical companies, and develop its drug portfolio.

- **Companies/institutions**: DesignMedix, Portland State University

- **Technology:** DesignMedix designed a hybrid drug to treat chloroquine-resistant malaria. This compound incorporated chloroquine, a widely used treatment for malaria, linked to a reversal agent, which prevented the malaria parasite from expelling the drug.

- **Stage of development at time of issue/decision:** DesignMedix was in the preclinical stage of development for its lead anti-malarial compound and in the research stage for additional malaria drugs and antibiotics.

- **Indication/therapeutic area:** Malaria affects 350 to 500 million people annually and cause greatest mortality in women and young children. DesignMedix's lead candidate targeted chloroquine-resistant malaria, a growing threat in endemic areas.

- **Geography:** United States, developing countries

- **Keywords:** malaria, chloroquine, hybrid drug, drug resistance, platform technology, pharmaceutical industry, bioentrepreneurship

---

* *This case was prepared as a basis for class discussion rather than to illustrate either effective or ineffective handling of an administrative situation.*

INTRODUCTION

In the spring of 2009, the leadership team of DesignMedix met in a small coffee shop in downtown Portland, Oregon to discuss next steps for their university-based, start-up bioscience company. The cozy cafe was situated in the beautiful Park blocks on the Portland State University (PSU) campus and had come to be known as the company's unofficial headquarters. David Peyton, DesignMedix's chief scientific officer, was a tenured PSU professor of chemistry who developed a promising treatment for drug-resistant malaria. Peyton joined Lynnor Stevenson, chief executive officer, and Sandra Shotwell, chief operations officer, on this day to discuss how to best focus and position the company in order to secure financing and advance DesignMedix's research and development (R&D). As a small start-up, DesignMedix's leadership team was faced with the conundrum of the most effective utilization of their limited financial and human capital. Peyton was focused on bringing the company's potentially life-saving anti-malarial technology to market as quickly as possible. Stevenson and Shotwell concurred and also wanted to create additional value for the company's investors by using Peyton's technology as a platform for other drug-resistant diseases. After taking a long sip of his vanilla soy latte, Peyton asked Stevenson and Shotwell what he should do next. He lamented that new R&D would distract his scientists from their malaria research and that capabilities for new drug targets could not be setup overnight.

The buds on the cherry trees were beginning to bloom, and the discussion at the cafe had come to a crossroads. Should DesignMedix continue to focus on pharmaceuticals for drug-resistant malaria or should it further extend its R&D efforts to identify compounds to treat additional drug-resistant diseases? On the way back to the laboratory, Peyton stopped at the biweekly farmer's market, which was held in the PSU Park blocks, to buy some produce. As he paid a chatty vendor for a carton of fresh eggs, his thoughts turned from his urban surroundings to global health challenges, malaria, and the one million individuals who died annually from the neglected disease[1].

DESIGNMEDIX COMPANY BACKGROUND

DesignMedix was a university-based, biotech start-up focused on overcoming drug resistance to diseases, one of the major challenges impacting human health and drug development since the beginning of the 20th century. The company was founded in 2006 as a Limited Liability Company (LLC). By 2009 it was incorporated, filed for four patents, and received several government grants, including Phase I and II Small Business Technology Transfer (STTR)[2] grants from the National Institute of Health (NIH), which were valued at $1.5 million. In addition to federal grants, the company was successful in raising private funding through the Oregon Angel Fund, the 2009 Angel Oregon competition, Northwest Technology Ventures, and several individual investors. By April of 2009, the company had achieved $1.0 million in a Series A round of financing (Figure 1) (DesignMedix, 2008).

For a second round of financing, DesignMedix sought between $2 and 5 million. In order to provide an attractive value proposition to investors, the leadership team needed to decide on the best strategy for R&D going forward. DesignMedix's core competency was in malaria, which represented a large

---

1   A neglected disease is disease that affects low income populations in developing nations.

2   The STTR (Small Business technology Transfer) program was established by the US Small Business Technology Transfer Act of 1992. STTR is a grant program aimed at stimulating innovation of domestic small business concerns to engage in research and development that has the potential for commercialization. There are three potential phases in the program—Phase I, typically $100K, is to establish technical merit; Phase II, typically $750K, is to establish commercial potential, and Ph III, where appropriate, is for commercialization including outside funding. The unique feature of the STTR program is the requirement for the small business concern applicant organization to formally collaborate with a research institution in Phase I and Phase II. In 2006, STTR grants totalled $68 million (NIH, 2009).

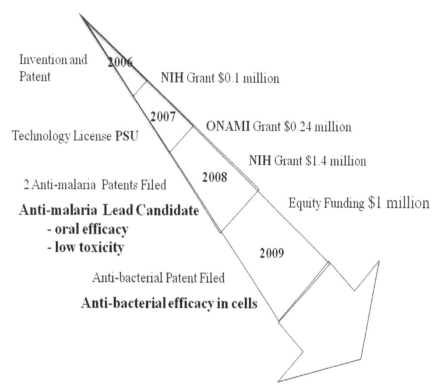

Figure 1: Summary of DesignMedix milestones and funding achievements
Source: Author's elaboration

unmet medical need, but a small commercial market compared to other diseases. Investors wanted DesignMedix to continue to develop their anti-malarial lead candidate, but also explore drug targets for diseases in more lucrative markets. Peyton and his team of scientists were also developing antibiotics to treat drug-resistant bacteria and having success in preliminary tests. The company's future and the commercialization of its anti-malarial technology hinged on its ability to attract additional financing and partnerships with pharmaceutical companies (Stevenson and Shotwell, 2009).

DESIGNMEDIX'S POSITION WITHIN THE BIOPHARMACEUTICAL INDUSTRY

Historically, competition among large pharmaceutical companies was characterized by the race to create and bring blockbuster drugs to market. Firms such as AstraZeneca, Pfizer, and Merck were vertically integrated from R&D to sales (Appendix A). The ability to provide pharmacoeconomic value to payers and providers was only marginally important in the 1990s. New methods of discovery in the past decade (e.g., genomics, high-throughput screening, combinatorial chemistry, and bioinformatics) significantly improved both the speed and reliability of R&D. Despite recent advances in research technology, the pharmaceutical industry was still encumbered by the time and cost intensive process of R&D which continued to be hampered by declining productivity (Ahn & Meeks, 2008).

In an effort to reduce costs and fill product pipelines, pharmaceutical companies adopted a more horizontal organizational approach via mergers and acquisitions (e.g., Pfizer's serial acquisitions of Warner-Lambert, Monsanto, Sugen, Pharmacia, Wyeth) (Appendix A). While sales and marketing were core competencies of large pharmaceutical companies, smaller biotech firms had the ability to concentrate on the more innovative element of drug discovery (Castner *et al.*, 2007). As a result, large multinational pharmaceutical firms shifted away from internal R&D and towards alliances and acquisitions of smaller firms. The R&D necessary to create new products was found in smaller biotech firms, whereas the cash needed for costly clinical trials and mass marketing resided in larger pharmaceutical companies (Palzelt and Brenner, 2008). Small biotech companies generally provided a critical portion of the value chain, and their expertise diminished further up the development and commercialization of the value chain (Appendix A) (Castner *et al.*, 2007).

The process of obtaining a new drug approval is expensive and lengthy. The average development costs per product (including opportunity cost for failed products) are over $1.0 billion and take 12 years from research to approval (Silver, 2009). Only five in 5,000 compounds that enter preclinical testing make it to human testing, with oncology drug candidates twice as unlikely to obtain marketing approval due to the complexity of the disease and setting (Kolchinsky, 2008). New biopharmaceutical products generally progress through the following steps: (1) pre-clinical testing to establish biological activity against the targeted disease; (2) Investigational New Drug Application (IND) filing to allow human clinical trials; (3) Phase I, II and III clinical trials to establish statistically significant safety and efficacy; (4) New Drug Application (NDA) for approval for a specific type and stage of disease (Wierenga & Eaton, 2006; Goldman Sach, 2007). Appendix B provides an outline of the drug development process, success rate of drugs and the length of time each step takes.

MALARIA PREVALENCE, PATHOGENESIS, AND TREATMENT

Malaria is an infectious disease caused by the *Plasmodium* parasite and transmitted by mosquito vector. The life cycle of *Plasmodium* is multi-phasic. Upon injection into the skin by an infected mosquito, *Plasmodium* travels to the liver where it infects hepatocytes. Following parasite replication in the liver, *Plasmodium* travels to the blood where it infects red blood cells. Malaria pathogenesis results primarily from the red blood cell phase, and disease complications include high fever, severe anemia, cerebral malaria, kidney failure, and lung edema. Although the disease has been eliminated in more temperate regions, malaria continues to be a threat in tropical and subtropical areas. In 2006, malaria was endemic in more than one hundred nations, with the majority of cases in Africa and Southeast Asia. Of the approximately 3 billion individuals at risk, 350 to 500 million were infected with malaria. As of 2008, malaria was responsible for up to one million deaths every year worldwide, many of them women and young children (World Health Organization, 2008). 91% of malaria deaths occurred in Africa, and 85% were children under the age of five years. Apart from the human toll, there were significant socio-economic costs associated with the disease. Malaria and poverty were closely linked. Approximately 1.3% of gross domestic product growth was lost annually in countries where malaria was prevalent, which added up to a loss of $12 billion in economic productivity in Africa alone (Chernin, 2009). DesignMedix's leadership team was eager to develop this drug not only for its financial returns, but also for its benefits to those suffering greatest from the ravages of malaria in developing nations.

In the absence of an approved vaccine, malaria prevention included the use of mosquito nets, insecticides, and prophylaxis (World Health Organization, 2008). Prophylactic drugs were required to be taken continuously to reduce the risk of infection and were often accompanied by unwanted side effects. Malaria infections were treated with anti-parasitic medications, such as chloroquine (CQ), sulfadoxine-pyrimethamine, mefloquine, atovaquone-proguanil, quinine, doxycycline, and artemisinin derivatives (Table 1). Chloroquine was one of the safest, most widely distributed, and least expensive anti-malarial drugs and was the treatment of choice in many regions against the *P. vivax*, *P. malariae*, and *P. ovale* (Roll back Malaria Partnership, 2009). Artemisinin-based therapies were generally used to treat *P. falciparum*, because of its pathogenicity and emerging CQ resistance (Opperdoes and Hannaert 2002).

Malarial drug resistance differed widely based on geographical region, but CQ resistance is spreading to almost all malaria-endemic countries (Figure 2). In order to overcome malaria resistance, CQ needs to be administered at high doses, which causes side effects and has proven harmful during pregnancy. CQ-resistant malaria was common in Southeast Asia and Africa, and alternative drugs used

Most Commonly Prescribed Drugs	Market Share (Million Doses)	Company (Potential Partners)	Adults (doses/ days)	Pregnancy	Children	Drug Resistance	Cost
Chloroquine	260	Generic	3/3	Safe	Safe	Strong	Low
Sulfadoxine-Pyrimethamine	140	Roche	3/1	Safe	Safe	Strong	Medium
Artemether-lumefantrine combination	115	Novartis	6/3	No	No	No	High
Artemesinin and derivatives	20	Generic	7/7	Not in 1st trimester	Safe	Developing	High
Total Doses	**535**						
DesignMedix Product Profile	----	----	`	Safe	Safe	No	Low

Table 1: Comparison of DesignMedix technology to current standard of care

Figure 2: Geographical distribution of chloroquine-resistant and sensitive malaria
Source: Canada Communicable Disease Report (2008)

to fight the disease are generally expensive (Appendix B). Resistance to the combination of sulfadox-ine-pyrimethamine, which was present in South America and Southeast Asia, is rapidly emerging in Eastern Africa in 2009. In many parts of the world, few drugs remained effective. Because of increasing drug resistance, the World Health Organization (WHO) advocated treating malaria with combination drug therapies, which contained multiple active agents. In addition, the WHO discouraged market penetration of new drugs and single-mechanism malaria drugs despite the need for novel treatments (World Health Organization, 2008).

DESIGNMEDIX'S TECHNOLOGY

It was long recognized that compounds called reversal agents (RA) could reverse drug resistance by inhibiting drug efflux from the parasite. Peyton discovered that direct chemical bonding of an RA to CQ significantly decreased *Plasmodium* drug resistance in red blood cells (Figure 3) and markedly re-

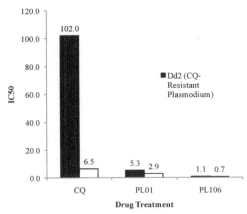

Figure 3: DesignMedix RCQ compounds effective against cq-resistant and cq-sensitive malaria strains. Red blood cells were infected with CQ-resistant *Plasmodium* (Dd2) or CQ-sensitive *Plasmodium* (D6) and treated with CQ, PL01 (DesignMedix's first hybrid drug), or PL106 (DesignMedix's lead candidate following optimization). Inhibitory Concentration 50 (IC_{50}) is defined as the amount of the drug necessary to kill 50% of the parasites.
Source: Company reports. Author's elaboration.

duced the dose of drug necessary, thereby reducing unwanted side effects associated with separate RA therapies. The new library of anti-malarial RA-CQ compounds, termed RCQ compounds, developed by DesignMedix had been tested and demonstrated to cure disease in animals, with no measurable side effects. Mode of action investigations demonstrated heme-binding affinity of RCQ compounds and the interaction of the quinoline portion of the RCQ compounds with heme. Thus, RCQ compounds appeared to share CQ's mode of action. Given a long record of use of CQ and RA compounds in humans, the safety profile for the new therapy was anticipated to be promising (DesignMedix, 2008).

One important feature of DesignMedix's technology was that it provided a hybrid approach for the design of anti-malarial compounds. In addition to a mechanism for overcoming drug resistance to CQ, the drug design readily lent itself as a platform for continuing innovation and the potential for a pipeline of new drugs. As *Plasmodium* developed resistance to a drug based on one RA-CQ combination, another RA could be substituted. In addition, drugs from this new class could be combined with other anti-malarial drugs with different modes of action, so that the development of drug-resistance to an individual compound could be delayed.

DesignMedix licensed exclusive rights to Peyton's anti-malarial technology from PSU, giving the company the right to commercialize the technology in exchange for future royalties on sales (though no royalties would be charged in developing countries given the humanitarian issues associated with treating malaria). The company had three utility patent applications for anti-malarial drugs filed with the U.S. Patent Office. In order to maximize the technology's value, the start-up had to remain mindful of the period of time from lab to market.

THE MALARIA MARKET

Due to limited reporting by nations where malaria was prevalent, determining the size of the malarial market with precision was difficult. The Institute of Medicine of National Academies estimated that between 200 and 400 million courses of anti-malarial drugs were administered annually in Africa; and 100 million treatments in other parts of the world (Arrow, Panosian, Gelband, 2004). As of 2009, the majority of drugs were purchased with public funds, with one-third coming from national governments of endemic countries and one-half from international resources, such as the WHO, World Bank, The United Nations Children's Fund, and the Roll-Back Malaria Partnership (The Roll Back Malaria

Partnership, 2008).

The total market for anti-malarial efforts in 2007 was approximately $1.3 billion (The Roll Back Malaria Partnership, 2008) and segmented into the endemic, traveler, and military markets. An estimated $1 billion was spent annually on malaria treatment in Africa, largely funded by governments and NGOs; and $100 million in Southeast Asia. The traveller and military markets were estimated to be $200 million per year (company estimates); however, the traveler market had significant potential for growth. Over thirty-three million trips to malaria-endemic countries were made by civilians in 2009, yet many of these travelers refused to take prophylactic medications due to their unpleasant side effects. If a medication with fewer or no side effects were to be developed, the size of this market would experience significant growth. The U.S. military deployed 200,000 soldiers to malaria-endemic countries and thus served as an additional potential source of revenue.

COMPETITION IN MALARIA

Vaccine development posed a potential threat to DesignMedix. In 2007, researchers published a Phase I/IIb, double-blind randomized trial of 214 Mozambican infants which demonstrated the safety and immunogenicity of GlaxoSmithKline's S/AS02D malaria vaccine. The vaccine demonstrated an adjusted vaccine efficacy rate of 65.9% (p<0.0001) at 3-month follow-up, and at 6 months there had been 34 (16%) serious adverse events in the intent-to-treat population, 17 in each arm of the trial, and no deaths (Aponte *et al.*, 2007). While these and other efforts were promising, a satisfactory malaria vaccine remains unavailable. However, the creation of one would significantly reduce the need for anti-malarial therapies.

Due to the small commercial market for malaria outside of the developing world, many R&D efforts were pursued through public-private partnerships and non-profit organizations. A number of individual investors and charities funded vaccine research for malaria, and this funding had increased considerably since 2001. For example, the Program for Appropriate Technology in Health (PATH) provided financial support for clinical trials through a grant from the Bill & Melinda Gates Foundation. Medicines for Malaria Venture managed the largest portfolio of potential malaria drugs, with 40 anti-malarial compounds in various stages of development or clinical trials. Included in MMV's portfolio were three joint projects with GlaxoSmithKline (GSK), nine with Novartis, and one with Merck (Medicines for Malaria Venture, 2009).

THE MARKET FOR ANTIBIOTICS

In 2009, antibiotic drug resistance was developing rapidly, and the market for anti-bacterial drugs addressing a wide range of infections was estimated at $30 billion (Kalorama Information, 2003). Growing antibiotic resistance resulted in increased demand for new antibiotics, yet there was limited competition in their development. One report by the Infections Diseases Society of America (IDSA) indicated that new antibiotic drug approvals had decreased by 75% since 1983 (Infectious Diseases Society of America, 2009). According to the IDSA, only ten new antibiotics had been approved by the FDA since 1998, and a recent analysis found that only five of the 506 drugs in the pharmaceutical industry drug pipeline were antibiotics (Hensley and Wysocki, 2005).

In recent years, pharmaceutical firms had elected to focus R&D efforts on drugs for chronic diseases because of their higher margins and greater return on investment. In 2003, Wyeth calculated the net present value of a typical antibacterial drug to be $150 million, compared to $1.15 billion for a drug for a chronic disease like osteoporosis (Hensley and Wysocki, 2005). The same economics that encouraged the development of Lipitor®, a cholesterol drug with approximately $13.6 billion in sales in 2006, drove firms away from the development of antibiotics. Aventis developed Ketek®, which was effective against drug-resistant bacterial strains that caused pneumonia and bronchitis. Ketek® was approved by the

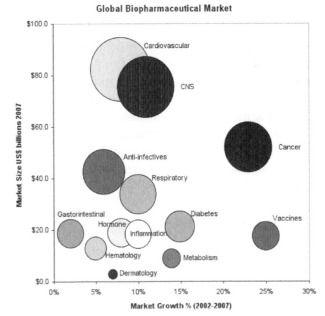

Figure 4: Global biopharmaceutical market
Source: IMS, 2008

United States Federal Drug Administration (FDA) in 2004, after it generated $143 million in revenue in Europe in 2003. Analysts expected the drug would reach over $500 million in annual revenue and predicted that it could eventually exceed $1 billion per year in sales (Hensley, 2004). Although competition in the antibiotic market has diminished in recent the years, the intellectual property surrounding antibiotics, in terms of patents on file with the U.S. Patent office, remained entrenched.

DesignMedix filed one antibiotic patent in 2009 and was working on additional antibiotic drug targets. Preliminary tests demonstrated their drug's effectiveness against a broad spectrum of bacteria including S*taphylococcus aureus.* The market for hospital-acquired bacterial infections, including diseases caused by *S. aureus* (e.g. MRSA), was determined to be $3.3 billion (PharmaInfo, 2010). Once infected, treatment costs for one surgical infection can quickly reach more than $15,000, while MRSA (methicillin-resistant *Staphylococcus aureus*) infections can cost over $30,000 to treat effectively (Frost & Sullivan, 2008).

MARKET FOR CANCER DRUGS

In 2009, cancer drug resistance was the most common reason for cancer treatment failure. It was estimated that up to 50% of cancer cases exhibited multidrug resistance, and new hybrid or combination therapies rather than single-target drugs were needed to combat growing resistance (BioPharm Reports, 2008). The cost of cancer to the healthcare system is significant. The National Institutes of Health (NIH, 2009) estimates that the overall cost of cancer in 2004 was $189.8 billion. This cost includes $69.4 billion in direct medical expenses, $16.9 billion in indirect morbidity costs, and $103.5 billion in indirect mortality costs. According to Reuters (Jackson, 2003), the global cancer market is estimated at $40 billion in 2005. In addition to being a large market, cancer care is also a highly concentrated market, making it ideal for a company to commercialize. Oncologists represent only 1% or 8,400 out of 635,000 total physicians in the US (often further concentrated in major metropolitan areas where specialists practice in teams). Thus, oncologists as a physician group can be promoted to by a specialty sales force (versus primary care therapeutics in areas such as cardiovascular which require thousands of sales representatives).

Antibiotics and cancer represented two new lucrative markets for DesignMedix. Development of an additional drug candidate for an indication other than malaria would make DesignMedix a more attractive partner or acquisition target for a large pharmaceutical company.

MANAGEMENT AND CULTURE: A WINNING COMBINATION

Despite being a small start-up, DesignMedix had a highly experienced management team in infectious disease research, biotechnology entrepreneurship, and technology management. First, David Peyton, Ph.D. was a professor of chemistry at PSU where he had worked since 1987. He earned his Ph.D. from University of California at Santa Barbara. After completing his post-doctoral work at Cornell University Medical College, he came to PSU where he utilized analytical, physical, and organic chemistry to conduct research on drug-resistant malaria.

Lynnor Stevenson, Ph.D. was a successful entrepreneur with over thirty years of experience in start-up bioscience companies. In addition to her scientific training, she was a member of management teams for ten biotech startups, six of which eventually had IPO exits. Stevenson was most recently a managing partner of Alta Biomedical Group, a consulting firm that handled projects in the commercialization of innovative technologies.

Sandra Shotwell, Ph.D. had over twenty years of experience in early stage technology commercialization. She spearheaded biotechnology and pharmaceutical licensing deals for Stanford University, Oregon Health & Science University, and NIH. Most recently, Shotwell served as a managing partner of Alta Biomedical Group with Stevenson (DesignMedix, 2008). Both Stevenson and Shotwell had performed consulting work in malaria prior to DesignMedix and were knowledgeable about the global health concern and need for effective medications.

According to Stevenson, the following three strategic elements had to be in place for any bioscience start-up to succeed: the team, technology, and funding. The risks in the industry were well documented, with only 5% of preclinical drugs making it to market (Napodano, 2009). A team with the experience to evaluate the science, generate the funding, and work through the regulatory and reporting hurdles was essential. Quoting Stevenson:

> "Big pharmaceutical companies like to buy approved drugs, and they bought most of them. They like phase III drugs, and they bought most of them too. So, now they are buying phase I and phase II or even preclinical drugs. What companies are doing is buying perceived future value. What they like to see is that you have a drug in phase I or phase II, that you have technology which enables you to have a series of drugs coming along behind, and that they will likely get some future drugs coming out of that technology" (Stevenson and Shotwell, 2009).

Building perceived value in a biotech start-up was directly related to the prospects of bringing a lead candidate to market. As R&D proceeded through efficacy and safety evaluation to registration and commercialization, the probability of launch and the value of the drug increased significantly. Consequently, in order to create and retain value in the firm, the entity needed to retain as much control as possible, for as long as possible, as the drug progressed to early clinical stages.

Firms also were able to build value through the acquisition of transferable assets related to their business. In 2007, the U.S. congress passed legislation providing incentives for research into neglected diseases, which generally were not well studied due to the existing economics behind pharmaceutical research and development. Signed into law in September of 2007, priority review vouchers would be granted by the FDA to organizations that had a drug approved for a neglected disease. These vouchers entitled the bearer to an expedited regulatory review of a future compound any disease. Additional benefits surrounding the pursuit of neglected diseases were tax credits and FDA assistance in trial design. BioVentures for Global Health quoted estimates of the value of a transferable voucher as great as $300 million, if applied to a blockbuster product (BioVentures for Global Health, 2009). DesignMedix, because of its research on malaria, was one of the organizations eligible to obtain one of these performance vouchers.

The challenge facing all biotech start-ups was to secure the funding necessary to carry out their primary R&D initiatives. DesignMedix had been successful in developing strategic alliances with other organizations and in attracting funds from public health agencies. The company obtained grants from the NIH and the Oregon Nanoscience and Microtechnologies Institute (ONAMI). To validate the efficacy of one of their lead drug candidates, they also collaborated with the highly respected Swiss Institute of Tropical Diseases. DesignMedix had been able to build non-dilutive strategic alliances with public health agencies. By developing their primary scientific initiatives, they put themselves in a position to capture more of the value from an acquisition at some subsequent point in time.

Funding was critical to proceeding through preclinical testing. Standard & Poor's 500 estimated that the average cost to develop a new biologic in 2009 was $1.2 billion (Silver, 2009). DesignMedix was successful in achieving support through government and equity funding and was concentrating next on corporate partnerships. Not only would a corporate partnership provide the financial resources to bring the company's anti-malarial technology to market, it would provide a lucrative exit strategy for DesignMedix's private investors. With its malarial R&D underway, the DesignMedix leadership team next had to determine what would build the greatest perceived value to potential corporate partners.

CONCLUSION: COFFEE SHOP FOLLOW-UP

Two weeks following the C-level meeting, Stevenson and Shotwell met with a consultant, Michael Morgan, to discuss DesignMedix's options further. Peyton was occupied teaching organic chemistry on this morning and trusted Stevenson and Shotwell to draw the appropriate conclusions from their meeting with Morgan.

Morgan began by presenting the company's first possibility, "DesignMedix has built a strong foundation on research in drug-resistant malaria. The timeline for pursuing anti-malarial drugs, compared to a timeline for other drugs, is much more proximate as a lead drug candidate is identified, test results are promising, and studies to enable clinical trials are currently underway. The more developed DesignMedix's malaria platform, the more valuable it will be to a large pharmaceutical firm who can expand its use across their various research interests. In a similar vein, the less life left on the patent, the less valuable DesignMedix will be to a potential acquirer. DesignMedix needs to be careful not to reduce company value by spending too much time exploring new markets rather than focusing on developing its lead candidate. Concentrating on the development of anti-malarial compounds and commercialization of the company's existing technology represents one possible approach for increasing the company's value."

Shotwell remarked, "Yes, and our malaria technology lends itself to continuing innovation. We can substitute reversal agents if the parasite develops drug resistance and are able to streamline the entire discovery and development process."

"Regardless of how clinically successful your malaria platform may become, the commercial value of any anti-malarial drug is modest and highly uncertain." Morgan continued, "DesignMedix is also considering expanding from malaria to additional drug-resistant diseases. Developing a proven platform and expanding its drug portfolio will definitely attract additional funding sources. Expanding research into additional drug-resistant diseases requires increased R&D, personnel, and time—or reducing DesignMedix's existing R&D programs."

"Which option do you recommend for DesignMedix?" Stevenson asked.

"Well, let's have a look at recent acquisitions. Novartis acquired NeuTech Pharma for $575 million for its anti-infective portfolio," Morgan said. "NTP's two key products were Mycograb, an antibody fragment targeting Hsp90, for the treatment of invasive and systemic candidiasis; and Aurograb, a recombinant antibody fragment against ATP binding cassette transporter, for MRSA infection (Webwire, 2007). There was also the Cubist acquisition of Calixa for $92.5 million cash plus another potential $310 million depending on completion of various milestones. Like NeuTech, Calixa had two key products, an intravenous anti-*Pseudomonas* cephalosporin in Phase II clinical trials and the beta-lactamase inhibitor, Tazobactam (Cubist, 2009). I guess the ultimate question is: does DesignMedix's competitive advantage lie in malaria, or can the company manage a balancing act and effectively pursue its lead malaria candidate and develop the technology platform simultaneously?"

APPENDICES

APPENDIX A: RELATIVE INDUSTRY DYNAMICS

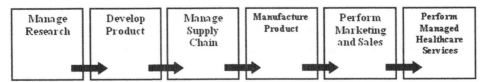

Vertically integrated pharmaceutical company (Castner, Hayes, Shankle, 2007)

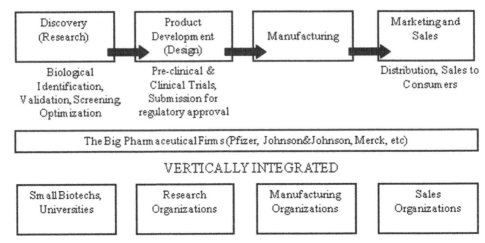

Horizontal organizational structure of pharmaceutical companies (Castner, Hayes, Shankle, 2007)

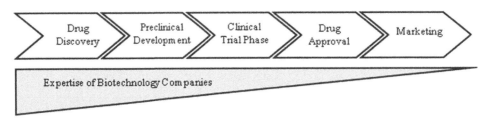

Expertise of biotechnology companies along the value chain (Castner, Hayes, Shankle, 2007)

APPENDIX B: DRUG DEVELOPMENT PROCESS

Clinical Trials									
	Preclinical testing		**Phase I**	**Phase II**	**Phase III**		**FDA**		**Phase IV**
Years	3.5		1	2	3		2.5	12 Total	
Test population	Laboratory and animal studies	File IND at FDA	20 to 80 healthy volunteers	100 to 300 patient volunteers	1000 to 3000 patient volunteers	File NDA at FDA	Review process / Approval		Additional Post marketing testing required by FDA
Purpose	Assess safety and biological activity		Determine safety and dosage	Evaluate effectiveness, look for side effects	Verify effectiveness, monitor adverse reactions from long-term use				
Success rate	5,000 compounds evaluated		5 enter trials				1 approved		

Source: Adapted from Wierenga & Eaton, 2006.

APPENDIX C: MALARIA TREATMENTS APPROVED AND IN DEVELOPMENT

Company Name	Product Names	Description	Latest Stage of Development
GlaxoSmithKline plc	Malarone (Brand), atovaquone/proguanil (Generic), Malarone (Other)	Oral combination of two coumponds that inhibit two different pathways of pyrimidine sythesis	Marketed
Novartis AG	Coartem (Brand), artemether/lumefantrine (Generic)	Fixed-dose artemisinin-based combination	Marketed
sanofi-aventis Group	ASAQ (Compound #), artesunate/amodiaquine (Generic), Coarsucam (Informal), Artesunate-Amodiaquine Winthrop (Other)	Fixed-dose combination of the antimalarial compounds artesunate and amodiaquine	Marketed
Inverness Medical Innovations Inc.	BinaxNOW malaria test (Other)	Rapid test to diagnose malaria	Approved
Sigma-Tau S.p.A.	Eurartesim (Brand), dihydroartemisinin/piperaquine (Chemical), Euratesim (Other)	Antimalarial combination of dihydroartemisinin and piperaquine	Registration
GlaxoSmithKline plc	GSK257049 (Compound #), RTS,S vaccine (Former), Malaria vaccine (Generic), Mosquirix (Informal)	Vaccine utilizing RTS,S peptide and TRAP sporozoite antigen together with QS-21 saponin	Phase III
Pfizer Inc.	Zithromax (Brand), azithromycin (Generic), Zithromax (Other), Zmax (Other)	Macrolide antibiotic	Phase III
Shin Poong Pharmaceutical Co.	Pyronaridine artesunate (Compound #), Pyramax (Informal)	Artemisinin-based combination treatment for malaria	Phase III
Sigma-Tau S.p.A.	Artesunate (Chemical)	Semi-synthetic derivative of artemisinin	Phase III
GlaxoSmithKline plc	SB 252263 (Compound #), tafenoquine (Informal), WR-238605 (Other)	Synthetic analog of primaquine	Phase II
GlaxoSmithKline plc	WR 6026 (Former), sitamaquine (Informal)	8-aminoquinoline	Phase II
Jomaa Pharmaka GmbH	Fosclin (Informal)	Oral fixed dose combination of fosmidomycin and clindamycin.	Phase II
Mymetics Corp.	Malaria Vaccine (Informal), PEV3A (Other)	Malaria vaccine	Phase II
Okairos AG	PlaMavax (Informal)	Genetic vaccine against malaria	Phase II
Ranbaxy Laboratories Ltd.	OZ277/RBx11160 (Compound #), RBx 11160 (Compound #)	Combination of arterolane maleate and piperaquine phosphate	Phase II
sanofi-aventis Group	SSR97193 (Compound #), Ferroquine (Other)	Derivative of chloroquine	Phase II
GenVec Inc.	NMRC-M3V-Ad-PfCA malaria vaccine (Informal), Malaria vaccine (Other)	Two combined recombinant adenovectors expressing CSP and AMA1 antigens	Phase I/II
Crucell N.V.	Malaria vaccine (Informal), rAd35 (Informal)	Recombinant malaria vaccine, vector based on adenovirus serotype 35	Phase I
GlaxoSmithKline plc	GSK932121 (Compound #), 932121 (Other)	Electron transport chain inhibitor	Phase I
Sanaria Inc.	PfSPZ Vaccine (Compound #), Whole-Parasite Malaria Vaccine (Informal)	Sporozoite vaccine containing a weakened form of Plasmodium falciparum	Phase I
Sinobiomed Inc.	Recombinant Malaria Vaccine (Informal), PfCP2.9	Malaria vaccine	Phase I
Affitech A/S	EP1300 (Compound #), Malaria DNA Vaccine (Informal)	A DNA plasmid vaccine encoding 38 CTL and 14 HTL epitopes from CSP, SSP2/TRAP, LSA-1 and EXP-1 proteins	Preclinical
Alnylam Pharmaceuticals Inc.	siRNA against heme oxygenase-1 (HO-1) (Other)	SiRNA against the heme oxygenase-1 (HO-1) gene necessary for malaria infection	Preclinical
Cel-Sci Corp.	CEL-1000 (Compound #), L.E.A.P.S. (Informal)	Small peptide containing a modified CD4 binding domain of the human major histocompatibility complex (MHC) Class II beta chain	Preclinical
Eastland Medical Systems	ArTiMist (Generic)	Sublingual anti-malaria treatment	Preclinical
Genzyme Corp.	Genz-644442 (Compound #)	Chemotype with activity against Plasmodium falciparum and P. berghei	Preclinical
Merck & Co. Inc.	MMV 06/0041 (Compound #), MK 4815 (Former compound #)	Oral compound that targets the mitochondrial electron transport chain of Plasmodium falciparum	Preclinical
Sosei Group Corp.	AD 452 (Compound #)	RS(+) isomer of mefloquine	Discovery

Source: Biocentury (2010)

REFERENCES

Ahn, M. & Meeks, M. 2008. Building a Conducive Environment for Life Science Based Entrepreneurship and Industry Clusters. *Journal of Commercial Biotechnology*, 14: 20-30.

Aponte, J.J., Aide P., Renom, M., Mandomando, I., Bassat, Q., Sacarlal, I., Manaca, M.N., Renom, M., Mandomando, I.,Bassat, Q., Sacarlal, J., Manaca, M., Lafuente, S., Barbosa, A., Barbosa, A., Leach, A., Lievens, M., Lievens, M., Vekemans, J.L., Sigauque, B., Sigauque, B., Dubois, M.C., Demoitié, M.A., Sillman, M., Demoitié, M.A., Sillman, M., Savarese, B., Neil, J., Macete, E., McNeil, M.D., Macete, E., Balllou, W.R.,Cohen, J.I., Alonso, P.L. 2007. Safety of the RTS,S/ASO2D candidate malaria vaccine in infants living in a highly endemic area of Mozambique: a double blind randomized controlled phase I/IIb trial. *Lancet*. 370 (9598): 1543-1551.

Arrow K., Panosian C., Gelband, H. 2004. *Saving Lives, Buying Time: Economics of Malaria Drugs in an Age of Resistance*, Institute of Medicine of National Academies.

BioPharm Reports. 2008. *Cancer Drug Resistance*, http://www.researchandmarkets.com/reports/651804/cancer_drug_resistance, accessed May 13, 2009.

BioVentures for Global Health. 2009. *Priority Review Vouchers*, http://www.bvgh.org/resources/incentives/prv/default.asp, accessed May 20, 2009.

Castner, M., Hayes, J., Shankle, D. 2007. *The Global Pharmaceutical Industry International trade and contemporary trends*, http://www.duke.edu/web/soc142/team2/shifts.htm, accessed January 26, 2010.

Chernin P. 2009. Target: Malaria. *Wall Street Journal* (Eastern edition). New York, N.Y.: Apr 24: A.13

Cubist. 2009. *Cubist Pharmaceuticals to Acquire Calixa Therapeutics*, http://www.snl.com/irweblinkx/file.aspx?IID=4093793&FID=8747721, accessed January 26, 2010.

Wierenga, D.E. & Eaton, C.R. 2006. *Office of Research and Development, Pharmaceutical Manufacturers Association*, accessed January 7, 2010, www.phrma.org.

DesignMedix. 2008. *Company reports*. http://www.designmedix.com/management.htm, accessed January 25, 2010.

Goldman Sachs. 2007.*Pipeline Survey: Growth Accelerated with High Risk, Equity Research Report*, Accessed January 7, 2010, www.goldmansachs.com.

Hensley, S. 2004. New Antibiotic Could Boost Besieged Aventis, *Wall Street Journal*, Mar. 4: B.1.

Hensley, S. and Wysocki Jr., B. 2005. Missing Medicine—Shots in the Dark: As Industry Profits Elsewhere, U.S. Lacks Vaccines, Antibiotics; Incentives Are Low to Develop Some Public-Health Drugs, New Moves in Washington; A $200 Million Legal Fight, *Wall Street Journal*, November 8, 2005: A.1.

Infectious Diseases Society of America. 2009. *Facts About the Antibiotic R&D Pipeline: Why Antibiotics Require Special Treatment*, http://idsociety.org/Content.aspx?id=5652, accessed on May 2, 2009.

Jackson, D. 2003.The pharmaceutical market outlook to 2010: Essential analysis of key drivers of change. *Reuters Business Insight, Healthcare*. Datamonitor: New York.

Kalorama Information. 2003. Antibiotic market driven toward $30 billion. *Canadian Chemical News.http://www.highbeam.com/doc/1G1-103672630.html*, accessed February 1, 2010.

Kolchinsky P. 2008. *The Entrepreneur's Guide to a Biotech Start-up*. 4th Edition: Evelexa BioResources

Medicines for Malaria. 2009. *MMV Portfolio Q3 2009*, http://www.mmv.org/IMG/pdf/MMV_Portfolio__Q3_2009-_FINAL.pdf, accessed January 26, 2010.

Napodano, J. 2009. *Industry Outlook Pharmaceutical Industry*, zacks.com, http://www.zacks.com/stock/news/18795/Pharmaceutical+Industry, accessed May 19, 2009.

NIH (National Institutes of Health).2009. *Small Business Research Funding Opportunities*, accessed January 7, 2010, http://grants.nih.gov/grants/funding/sbir.htm

Opperdoes F, Hannaert V. 2002. *A Course on Tropical Parasitology.*http://www.icp.ucl.ac.be/~opperd/parasites/, accessed on December 15, 2009

Palzelt P., Brenner T. 2008.*Handbook of Bioentrepreneurship*. New York. Springer Science Business Media, LLC.

Perrone, M. 2008. *Global market expected to drive cancer drug growth*, usatoday.com, http://www.usatoday.com/money/economy/2008-05-15-2197938540_x.htm, accessed May 13, 2009.

PharmaInfo. 2010. Fluoroquinolones: An Overview, http://www.pharmainfo.net/reviews/fluoroquinolones-overview, accessed February 1, 2010.

The Roll Back Malaria Partnership. 2008. *The global malaria action plan for a malaria-free world*. http://www.rollbackmalaria.org/gmap, accessed January 26, 2010.

Roll Back Malaria Partnership. 2009. *Global Burden and Coverage Today.* http://www.rollbackmalaria.org/gmap/part1.pdf, accessed on December 15, 2009

Silver S. 2009. *Standard & Poor's Biotechnology Industry Report.* http://www.netadvantage.standardandpoors.com.proxy.lib.pdx.edu/NASApp/NetAdvantage/showIndustrySurvey.do?task=showIndustrySurvey&code=bio, accessed November 26, 2009

Stevenson, L. and Shotwell, S. 2009. *Personal interview,* April 28, 2009.

Webwire 2007. *Novartis makes offer for NeuTecPharma, acquisition adds two highly promising biotech drugs to portfolio for patients with severe infections,* http://www.webwire.com/ViewPressRel.asp?aId=14948, accessed January 26, 2010.

World Health Organization. 2008. *World Malaria Report 2008.* http://whqlibdoc.who.int/publications/2008/9789241563697_eng.pdf, accessed on November 20, 2009.

Oxigene: Realizing Value from Multiple Technology Platforms

MARK J. AHN[1], ANNE S. YORK[2], DAVID ACKERLEY[3] AND REBECCA BEDNAREK[3]

[1]*Atkinson Graduate School of Management, Willamette University;* [2]*College of Business, Creighton University; and* [3]*Victoria University of Wellington*

- **Key issue(s)/decision(s):** Oxigene is a development stage, biopharmaceutical company developing novel therapeutics to treat cancer and eye diseases. This case follows Dr. Richard Chin, CEO of Oxigene, as he faces the key strategic challenge of developing a technology platform with multiple product applications, each of which required significant resources. The case considers Chin and Oxigene's strategic options — continue to diversify from oncology to ophthalmology, spin off their ophthalmology product candidates into a new company, or license their ophthalmology to another company.

- **Companies/institutions:** Oxigene, Bristol-Myers Squibb Company, Arizona State University

- **Technology:** Vascular disrupting agents (VDA) are a new generation platform, which induce anti-angiogenesis to deprive the blood supply to tumors that occlude pre-existing tumor vessels.

- **Stage of development at time of issue/decision:** Zybrestat® is in advanced clinical trials in Phase III thyroid cancer and pre-clinical development for ophthalmology.

- **Indication/therapeutic area**: Zybrestat® is being developed for various cancers, including anaplastic thyroid cancer and is in preclinical development in ophthalmology, specifically in wet Age-related Macular Degeneration (wAMD).

- **Geography**: US, Sweden

- **Keywords**: angiogenesis, vascular disrupting agents, oncology, ophthalmology, technology transfer

---

* *This case was prepared as a basis for class discussion rather than to illustrate either effective or ineffective handling of an administrative situation.*

INTRODUCTION

D r. Richard Chin, CEO of Oxigene, had just completed a series of meetings with institutional mutual and hedge fund investors. He was walking at his usual fast pace across Times Square in New York City and reflecting on his key challenge—enhancing shareholder value of his development stage biotechnology company, Oxigene. At issue was that Oxigene had developed a technology platform with multiple product applications, each of which required significant resources. "Wall Street", however, solely valued the company on its lead oncology product, Zybrestat®, and seemingly did not impute any value to its other ophthalmology pipeline candidates. He was framing his strategic options—focus resources solely on that lead product, continue to diversify, or spin off other product candidates through business development—for his board of directors meeting the following day. As Chin looked up he saw a billboard for the Broadway musical "Jekyll & Hyde." He smiled ruefully: "at least somebody is getting value for being two things at once!"

A "RE-START-UP" BIOTECHNOLOGY COMPANY

OXIGENE'S SENSITIZERS

Oxigene was founded in New York in 1998 by Dr. Ron Pero, a DNA repair specialist, and Richard Brown, a career executive, after they observed an opportunity for the development of DNA research in the field of oncology. DNA research was relatively new in the late 1980s and represented a disruptive innovation with the potential to create multiple product opportunities. Initially, the company's focus was to inhibit, measure and stimulate the DNA repair process associated with cancers. For instance, by inhibiting the DNA repair process in cancerous tumor cells it was hoped that they would become more sensitive to traditional cancer therapies such as radiation and chemotherapy. Thus, Oxigene's first drug candidates were labelled 'sensitizers'. Between 1989 and 1992 the company developed its first sensitizer platform, Benzamide, and from this platform the company's first generation drug 'Sensamide' entered clinical trials (Gupta, 1994).

With its platform showing promise, the company embarked on an expansion plan. As a first step, Oxigene was floated on the NASDAQ in 1993, generating US $35 million for the company. Raising capital meant that Oxigene could increase its research and development capacity and, as a result, the company developed its second generation sensitizer, Neu-Sensamide, in 1995. In addition, Oxigene embarked on a geographic expansion, setting up a subsidiary in Stockholm in early 1994 to take advantage of European research and market opportunities. It was from the Swedish subsidiary that Oxigene hired its second CEO, Bjorn Nordenvall, who replaced co-founder Brown in 1995 (Gupta, 1993-1998).

Early in 1997 the company developed its third generation sensitizer declopromide. However, the following year disaster struck when the company's first and second generation benzamide agents failed their clinical trials with particularly negative results. In both cases a significant number of patients did not complete treatment due to central nervous system side effects (Scrip, 1998). While declopromide did not yield the same issues, questions were raised regarding the safety profile for all benzamide based agents: the trials showed that all the drugs the company had worked on since its inception were potentially toxic. Oxigene's share price tumbled and as a consequence the company was faced with the decision whether to close the business or develop an entirely new technology platform and product pipeline (Benoit, Brown, Gregan, & Wilson, 1999).

OXIGENE'S VASCULAR DISRUPTING AGENTS

Oxigene chose to continue and an opportunity came via a compound that was already known to the company – Combretastatin[1], for which the company had procured a worldwide license from Arizona State University in 1997 (Cookson, 1998). Combretastatin was completely independent from the benzamide based products: instead of being a sensitizer it was a vascular disrupting agent (VDA). Whereas a sensitizer is intended to disrupt the DNA repair process in cancer cells and thus make them more sensitive to radiation and chemotherapies, VDA aims to disrupt the blood supply to cancerous tumors, literally starving them and doing so in a way that healthy cells are not damaged. Based on the promise of this technology Nordenvall made the decision to move the company in a new direction and all benzamide-based products, apart from declopromide, were dropped and resources were channelled towards the development of Combretastatin.

The decision to focus on Combretastatin paid dividends almost immediately when an agreement was reached with Bristol-Myers Squibb Company, a large multinational pharmaceutical company, to develop Oxigene's first generation VDA, called CA4P (Scrip, 1999). The deal bolstered the company by providing it with up to $70 million in funding (The Wall Street Journal, 1999a). In addition, early clinical results for Combretastatin found that 18 patients showed signs of full remission (The Wall Street Journal, 1999b). With such promising results, more resources were sought to develop the VDA platform.

In 2000, Dr. Dai Chaplin, one of the original discoverers of Combretastatin, was appointed to the board as chief operating officer. This coincided with the departure of the company's second co-founder, Dr. Pero, who retired from the position of COO in early 2000. To complete a year of regeneration, the company headquarters migrated from New York to a larger facility in Watertown, Massachusetts near the Boston-based biotechnology cluster. According to CEO Nordenvall, "the transition was made to allow the company to raise its profile in the US".

Declopromide, the final Benzamide-based sensitizer, was finally dropped and all intellectual capital gathered during the development of the sensitizers was sold in 2001. This enabled the company to focus solely on developing a VDA platform. At the same time additional studies were revealing the potential of VDA's. These results coincided with positive news from the company's Phase I trial of CA4P (see appendix A for an overview of the US FDA clinical and regulatory process). Weighted against such good news, however, was the decision by Bristol Myers Squibb to end its licensing agreement with Oxigene concerning CA4P. The loss of the Bristol Myers Squibb deal sent the Oxigene share price into a tailspin, whereas the positive news from the clinical trials created a subsequent surge in the stock price. To gain further investment momentum, CEO Nordenvall decided to push the company to develop its VDA platform further by producing a second generation drug to CA4P. This resulted in the discovery of OXi 4503 later in the year (Bauman, 2001).

2002 signalled the beginning of a second period of expansion at Oxigene. The first phase I trial for CA4P (henceforth Zybrestatat®) concluded with a complete response from a patient with Anaplastic Thyroid Cancer (ATC).[2] With such encouraging results another trial was immediately initiated, and Zybrestat became the first VDA to enter a phase II trial. The results also sparked interest in the potential of VDA's for other disorders and additional trials were initiated (Cookson, 1998). A phase Ib trial began using a combination of Zybrestatat® and Carboplatin (a chemotherapy drug) after independent preclinical research found that Zybrestatat® synergized effectively with such chemotherapeutic agents in the treatment of solid tumours. In addition, preclinical studies concluded that OXi 4503 (Oxigene's

---

1   *A synthetic derivative of a natural product extracted from the African bush willow.*

2   *Anaplastic thyroid cancer (ATC) typically (over 90%) presents in patients as cervical metastasis (spread of the cancer to lymph nodes in the neck) at the time of diagnosis. The presence of lymph node metastasis in these cervical areas causes a higher recurrence rate and is predictive of a high mortality rate. The most common way these cancers are detected is by the rapid growth of a mass on the neck of a patient. ATC is very aggressive in spreading both adjacent tissues (e.g., lung) and metastasizing to distant parts of the body.*

second generation VDA) led to a complete regression of human breast cancer in mice. The discovery of OXi 6197 later in the year, a drug with a molecular structure distinct from Zybrestat®yet still a VDA, further expanded the drug pipeline and trial regimen.

The most significant strategic expansion, however, came as a result of an independent study by the University of Cambridge that found Zybrestatat® to be effective in the treatment of retinopathy (abnormal development of blood vessels in the eye). This enabled Zybrestat® to be viewed not only as an oncology drug, but as an ophthalmologic drug as well. As a result of the study, the Foundation for Fighting Blindness entered into a phase I/II research agreement with Oxigene, whereby Zybrestatat® was assessed for its effectiveness in combating wet Age-related Macular Degeneration (wAMD). It was the first time in the Foundation's history that funding had been directed towards the development of a therapeutic drug.

It was becoming clear that the VDA platform (consisting of the company's first and second generation VDA's), was garnering substantial interest. This drove an increase in share price and allowed the company to raise additional capital, in turn enabling the company to increase its drug pipeline and initiate more trials. However, there was another benefit for Oxigene: by focusing solely on the development of VDAs the company was significantly reducing the firm's cash burn rate. Between 2002 and 2003 net losses had plummeted by almost thirty percent, thereby extending the firm's operating runway.

With the company in a healthy state after surviving the Benzamide disaster and regenerating itself to the point of expansion, Nordenvall decided to leave the position of CEO and headed back to Sweden where he oversaw the closure of the Swedish subsidiary. Nordenvall was replaced as CEO by Fred Driscoll, a financial professional with over thirty years of experience in the biotechnology and biomedical industries. At the same time, the board of directors was strengthened with the arrival of its new chairman Joel Citron.

Driscoll chose to take Oxigene in a different direction from his predecessor. He consolidated Oxigene's expansion agenda and instead aimed to strengthen its existing drug platform, share price and company balance sheet. Patents were obtained for the drug platform in relation to the treatment of oncology and ophthalmologic disorders and the company sought to attain Orphan or Fast Track drug status for Zybrestat® in the treatment of ATC, which it gained in late 2003 (Forelle, 2003; The Wall Street Journal, 2003) (see Appendix A). In addition, three large institutional investors purchased $14 million in shares and the company completed a $24.2 million common stock offering resulting in a significantly strengthened balance sheet by the end of 2003.

Although the technology platform was consolidated, Oxigene *was* still expanding, albeit at a slower rate. A collaboration with Baylor University had uncovered a third second generation VDA, OXi 8007. Further, positive preclinical results demonstrated the effectiveness of Zybrestat® for the treatment of Myopic Macular Degeneration (MMD) induced Oxigene to expand its ophthalmology program. Thus, by the beginning of 2004, the company had six clinical trials in place, three second generation drug candidates, and four new clinical trials for Zybrestat®.

By the end of 2004, OXi 4503 became the first of the second generation drugs to enter a phase I trial. Subsequently, Zybrestat® entered a phase II trial for the treatment of MMD, together with entering advanced trials (four phase II and one phase III) for various oncology indications. By 2005, however, the effectiveness of Oxigene's strategy was beginning to wane. Despite the efforts to consolidate there were multiple trials now in progress, draining resources and decreasing the company's share price. New opportunities, but also more financial pressure, resulted from the opening of a research facility in Oxford, U.K. For 2005, Oxigene's net loss was $11.9 million, compared with a net loss of $10 million in 2004. While CEO Driscoll had been instrumental in the positioning of VDAs for success, it was time for a new phase to begin. In 2006, he was replaced by Dr. Richard Chin. Along with Chin, the company introduced seven new members to its board, many with significant financial experience (Business Week, 2005a; Business Week, 2005b).

Dr. Chin, a Harvard-trained, Board Certified internal medicine physician, brought to Oxigene extensive drug development experience in several clinical fields including ophthalmology, dermatology and cardiology, as well as senior level experience in the areas of clinical trials, regulatory requirements and corporate collaborations. He had overseen numerous drug candidates through development, approval and launch. Among the list of Food and Drug Administration (FDA) approved products which Dr. Chin had launched are: TNKase™ (for myocardial infarction), Raptiva® (for psoriasis), Xolair® (for asthma), Cathflo® (for catheter clearance) and Prialt® (for pain). He was also instrumental in developing Lucentis® (for age-related macular degeneration) at Genentech. Additionally, Dr. Chin brought extensive experience in managing multiple corporate partnerships, including relationships with Boehringer Ingelheim, Xoma, Roche, Biogen Idec, Aventis, and Cor (now Millenium), among others. Importantly, he played a key role in leading the development of Avastin, a potential complimentary drug to Zybrestat, at Genentech.

In nearly 20 years, Oxigene had gone through significant changes in management and technology which effectively framed the company as a "re-start-up" which needed focus to be attractive to institutional investors. Figure 1 provides an overview of Oxigene's product pipeline in 2007.

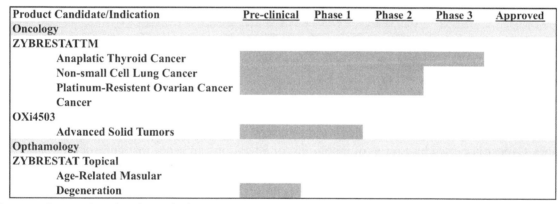

Figure 1: Oxigene's technology platform and current drug candidates.
Source: www.oxigene.com

THE BIOPHARMACEUTICAL INDUSTRY AND CANCER

The US FDA (Food and Drug Administration) approved the first biotechnology drug in 1982 and to date biopharmaceuticals have represented the largest area of biotechnology investment and revenues. In total the biopharmaceutical industry has had 254 drugs approved for 385 indications. In 2007, the global biotechnology industry posted revenues of $85 billion, invested over $32 billion in research and development, generated nearly 800 publicly listed companies growing at double digit rates across North America, Europe and Asia, employed over 200,000 knowledge workers addressing some of societies most pressing unmet needs, and raised $29.9 billion in investment capital. Despite this impressive growth, biotech industry profitability remains elusive with only a handful of firms achieving positive cash flow to date (BIO, 2006; Burrill & Company, 2007; Ernst & Young, 2007).

To differentiate from traditional multinational pharmaceutical companies, biotechnology companies have primarily focused on difficult to treat, unmet medical needs in complex diseases such as cancer, which have focused specialty physician populations and pricing power. The global cancer market was estimated at $50 billion in 2008 (Jackson, 2003) (see Figure 2). In addition to being a large market, cancer care is also highly concentrated which makes it ideal for a small company to commercialize. For instance, oncologists represent only 1% or 8,400 out of 635,000 total physicians in the US (often further concentrated in major metropolitan areas where specialists practice in teams). Thus, oncologists as a physician group can be promoted to by a specialty sales force (versus primary care therapeutics in areas such as cardiovascular which require thousands of sales representatives).

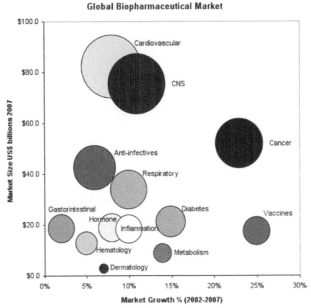

Figure 2: Global biopharmaceutical market
Source: IMS (2008)

Cancer is a group of diseases characterized by either the runaway growth of cells or the failure of cells to die normally. Often, cancer cells spread to distant parts of the body, where they can form new tumors. Cancer is caused by a series of mutations, or alterations, in genes that control cells' ability to grow and divide. Some mutations are inherited; others are promoted by environmental factors such as smoking or exposure to chemicals, radiation, or viruses that damage cells' DNA.

The economic cost of cancer to the healthcare system is significant. The National Institute of Health (NIH) estimates that the overall cost of cancer in 2004 was $189.8 billion. This cost includes $69.4 billion in direct medical expenses, $16.9 billion in indirect morbidity costs, and $103.5 billion in indirect mortality costs. Cancer is the second leading cause of death (after heart disease) in the United States where each year nearly 1.4 million new cases of cancer are diagnosed and one in four deaths are expected to be due to cancer. For all forms of cancer combined, the 5-year relative survival rate is 64% (American Cancer Society, 2006). Despite the fact that the cancer mortality rate in the U.S. has risen steadily for the past 50 years, scientific advances have perhaps begun to turn the tide. According to the National Center for Health Statistics, the cancer mortality rate increased for 50 years until in 2003, when it began a slight decline - the start of what researchers hope will be a long-term decline in cancer mortality.

Major treatments for cancer include surgery, radiotherapy, and chemotherapy. There are many different drugs used to treat cancer, including cytotoxics or antineoplastics, hormones, and biologics. Chemotherapy (anticancer drugs that destroy cancer cells by stopping them from multiplying), in particular, has dominated the market to date. Because cancerous cells are derived from healthy progenitors that have acquired only a very few key mutations, it is essentially impossible for a drug to kill cancer cells effectively without some degree of harm to healthy cells; and it is this harm to healthy cells that causes side effects. The undesired consequence of harming an organ not involved with cancer is referred to as a complication or a side effect that not only causes discomfort, but may also limit a patient's ability to achieve the best outcome from treatment by preventing the delivery of therapy at its optimal dose and time. Common side effects include anemia, fatigue, hair-loss, reduction in blood platelets and white and red blood cells, bone pain, and nausea and vomiting.

Targeted anticancer therapies have been developed as a result of biologic insights to create products with increasingly specific molecular targeting to enhance efficacy and reduce toxicity. Most of these targeted therapies must be used in combination with chemotherapy. Over 100 targeted anticancer agents are already on the market or in development, with the leading eight targeted therapies (Avastin, Rituxan, Herceptin, Erbitux, Gleevec, Tarceva, Sutent, and Nexavar) having estimated sales of more than $7.5 billion in 2006. Further, targeted therapies clearly dominate cancer pipelines with over 300 drugs in clinical development (Farmer & Pollack, 2006).

CANCER TREATMENT AND ANGIOGENESIS

Angiogenesis is a novel targeted technology being used to tackle tumors indirectly, aiming to slowly 'strangle' them by cutting off the networks of blood vessels that deliver small nutrients and oxygen. One way to achieve this is by inhibiting angiogenesis – hence the label of the new technology as 'anti'-angiogenesis. The 'father' of anti-angiogenesis is Judah Folkman, who nearly forty years ago came up with the idea that tumors induce the growth of blood vessels to obtain the nourishment they require to survive and grow (Cookson, 1998; Folkman, 1971). However, it was not until 2004 that the first clinically available angiogenesis inhibitor in the United States, Avastin (Genentech/Roche), was approved.

Avastin® reached nearly $2 billion in annual sales within two years after approval (Figure 3). Although, first approved for combination use with chemotherapy, Avastin was also approved for use in breast cancer in 2008. This was despite the fact that a panel of outside advisors voted against approval and expressed concerns that while the data showed reduction in tumor volumes and progression-free survival time, it did not show any increase in quality of life or prolonging of life. Clinical studies are currently being undertaken in areas including colon, breast, and ovarian cancers, and specific forms of prostate, pancreatic, and liver cancers.

Avastin® remains one of the most expensive and widely marketed drugs, and has been criticized for its high cost. For example, Avastin® extends the life of colorectal cancer patients by an average of 4.7 months at a cost of $42,800 to $55,000 (Mayer, 2004); the addition of Avastin® to standard treatment can cost around $100,000 a year, but only prolong life in breast and lung cancer cases by a few months (Berenson, 2006). Thus, while Avastin® has proven an important breakthrough in validating angiogenesis as a safe and effective approach to treating certain cancers, there is perhaps still room for expanding the therapeutic index and value of this technology.

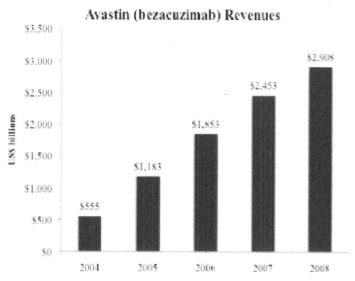

Figure 3: Avastin revenues 2004-2008
Source: Company reports

VDAS: A TARGETED CANCER TREATMENT

Leveraging the commercial success of Avastin, Vascular Disrupting Agents (VDA) has attracted interest in academia and industry for new uses in cancer, as well as novel areas such as ophthalmology. VDAs are another potential alternative to more traditional and invasive cancer therapies mentioned above (i.e., chemotherapy, surgery and radiation). Although the underlying biology is complex and not yet fully understood, VDAs used in cancer treatment essentially cause blood vessels inside a solid tumor to collapse, depriving the tumor of the nutrients and oxygen it requires for survival. A single blood vessel can support many tumors, so damaging one or two key blood vessels this way can have a major impact. Although similar in intent, VDAs differ from anti-angiogenesis drugs in that their primary targets are pre-existing vessels that already feed tumors, whereas anti-angiogenesis drugs prevent new blood vessels from developing (Lippert, 2007). Achieving a common goal through quite different mechanisms may potentially make VDAs and anti-angiogenesis agents a powerful one-two combination punch against solid tumors. Furthermore, VDAs have an advantage over conventional disease treatments such as radiation and chemotherapy because they target newly formed abnormal blood vessels in preference to the established blood vessels in healthy tissue, causing fewer side effects. Finally, they are designed to address the complete spectrum of solid tumors, whereas approaches that directly target tumor cells require different drugs for different tumors.

With the commercial success of Avastin, the pace and intensity of competition in the pro- and anti-angiogenesis arena has been demonstrably increased, with over 750 on-going clinical trials with existing and investigational drugs in this area. Companies involved in the development of VDAs include large multinationals such as AstraZeneca and Sanofi-Aventis, as well as smaller biotech firms such as Antisoma, Nereus and MediciNova. Many of these companies are also utilizing their multinational-sized budgets to test new cancer therapies beyond anti-angiogenesis, and to find ways of increasing the effectiveness of existing therapies. Oxigene's objective is to become a leader in the VDA technology platform and product development arena.

OXIGENE VDA TECHNOLOGY PLATFORM

Oxigene has rebuilt itself as a VDA technology platform company, with Zybrestat® (also known as combretastatin CA4P and fosbretabulin) as their lead product. Zybrestat® was discovered at Arizona State University (ASU), from whom Oxigene obtained an exclusive, worldwide, royalty bearing license in 1997. Oxigene believes its VDA product candidates may offer advantages over current anti-angiogenic drugs, including superior efficacy and reduced side-effects. Furthermore with Zybrestat® there is potential for Oxigene to be first to market with a VDA drug.

Zybrestat® is an agent that targets structures that play a critical role in the function of cells. A major consequence of this activity is disruption of the scaffold within the cells that form the inner lining of blood vessels (endothelial cytoskeleton). The swelling of these cells can cause blood vessels to become blocked or collapse, shutting down blood circulation (Figure 4). This effect is particularly pronounced within tumor blood vessels, which typically have a variety of abnormalities that make them more susceptible to endothelial cell swelling. Indeed, so pronounced is this effect that the anti-vasculature activity of Zybrestat® is apparent at well below its maximum tolerated dose in a wide variety of established tumor models (Tozer, Kanthou & Baguley, 2005). This is very important, as it suggests that Zybrestat® will have a wide therapeutic window (i.e., the ability to exert maximum anti-tumor activity with minimal side-effects).

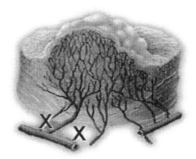

Figure 4: VDA Mechanism of action. VDAs induce anti-angiogenesis to deprive the blood supply to tumor which causes the collapse and occlude pre-existing tumor vessels
Source: Company reports

In solid tumor clinical studies, Zybrestat® has demonstrated potent and selective activity against tumor vasculature, as well as clinical activity against Anaplastic Thyroid Cancer (ATC), ovarian cancer, and various other solid tumors. Zybrestat® is currently being evaluated in a pivotal registration study in Anaplastic Thyroid Cancer (ATC) under a Special Protocol Assessment agreement with the U.S. Food and Drug Administration (FDA). Anaplastic Thyroid Cancer (ATC) is an aggressive and lethal malignancy. The disease is the least common (0.5 to 1.5 % of cases) and most deadly of all thyroid cancers, with only 10% of patients surviving 3 years after diagnosis even with the current treatments (including specific forms of radiation, chemotherapy and in rare cases surgery). As ATC affects a relatively small amount of people (between 1,000 and 2,000 throughout Europe and the U.S.), it is difficult to recruit large numbers of trial participants. Consequently, the U.S. Food and Drug Administration (FDA) granted Zybrestat® both "Fast-Track" and "Orphan Drug" status (see Appendix A), which has been integral to Oxigene's goal of being first to market with a VDA.

Beyond ATC, Zybrestat® has been shown to have broad therapeutic utility across a wide range of different solid tumor types including ovarian and lung tumors. Zybrestat® can also potentially be combined with mainstay oncology treatment modalities: chemotherapy, radiation therapy, and newer "molecularly-targeted therapies" such as tumor angiogenesis inhibitors. Preclinical studies have demonstrated that Zybrestat® has synergistic or additive effects when incorporated in various combination regimens with all of these different treatment modalities. In particular, there is a strong scientific rationale for combining Zybrestat® with tumor angiogenesis inhibiting drugs, and Zybrestat® is the first VDA to be tested in humans in combination with Avastin.

Another newer generation VDA technology being developed by Oxigene for the oncology market is OXi4503. They believe this is the first in a new class of compounds known as ortho-quinone prodrugs (OQPs): a unique and highly potent dual-mechanism VDA. Like its structural analog, Zybrestat®, OXi4503 blocks and destroys tumor vasculature, resulting in extensive tumor cell death and necrosis.[3] Preclinical studies have demonstrated that OXi4503 has single-agent activity against a range of tumor models, as well as synergistic or additive effects when incorporated in various combination regimens (chemotherapy, molecularly-targeted therapies and radiation therapy). OXi4503 is currently being evaluated as single agent therapy in a Phase II dose-escalation study in patients with advanced solid tumors.

Moving into a different market, Oxigene has also begun trials on the applicability of Zybrestat® in ophthalmology, specifically in wet Age-related Macular Degeneration (wAMD). In eye diseases such

---

3   In addition, preclinical data demonstrated that OXi4503 is metabolized by oxidative enzymes (e.g., tyrosinase and peroxidases) whose levels are elevated in many solid tumors and tumor infiltrates, yielding an orthoquinone chemical species that is directly toxic to tumor cells, preventing their reproduction or growth. Thus, OQPs appear to attack not only the blood vessels that feed solid tumors, but also the tumors themselves.

as wAMD, characterized by abnormal blood vessel growth, Zybrestat® has been shown in preclinical studies to suppress development and induce regression of these abnormal blood vessels. Although several angiogenesis-inhibiting drugs have recently been approved for the treatment of wAMD, they must be injected on a regular basis directly into the eye (intravitreal injection) and can cause side-effects. Oxigene believes that a patient-friendly, topical formulation of Zybrestat® (e.g., an eye-drop) is feasible and would have significant advantages for patients with wAMD when administered either as a monotherapy or in combination with current intravitreally-injected products. In clinical studies in patients with forms of macular degeneration, intravenously-administered Zybrestat® demonstrated encouraging clinical activity, and Oxigene is working to develop a convenient and patient-friendly topical formulation of Zybrestat for ophthalmological indications.[4]

To summarize, Oxigene uses its VDA platform in two ways. On the one hand, and reflecting its traditional focus, the company is developing oncology therapies, namely drug candidates that combat solid tumor growth. Yet, the company is also developing products with an ophthalmological focus, whereby drug candidates previously developed as cancer therapies are being tested for their applicability to certain eye diseases. The combined Oxigene pipeline of oncology and ophthalmology drug candidates has brought new breadth and depth to the company. However, this breadth of focus has also brought attendant concerns about research and development costs for the early-stage biotechnology firm.

STRATEGIC OPTIONS AT OXIGENE

As CEO Richard Chin met with his board of directors, and reflected on the advice of current and potential institutional investors, he outlined his current thinking around the strategic options facing Oxigene. He began: "A key decision currently facing Oxigene is the management of our twin foci on oncology and ophthalmology, as well as our imminent need to raise capital and the valuation we can obtain. Thus, the central question is whether Oxigene should narrow its resources to focus solely on oncology drug development."

One of the board members then cautioned: "I think we need to be conscious of the danger of investors valuing Oxigene predominantly on the Zybrestat Phase III trials in thyroid cancer. This is precisely the situation we find ourselves facing; meaning that one setback on Zybrestat® would put the entire company in jeopardy again. Also, I'm not sure we could get appropriate value if we licensed or spun out the ophthalmology products at their early stage of development."

Chin replied: "I agree that focusing solely on Zybrestat® increases our risk-return profile, but Wall Street is fairly unanimous in that they are valuing our shares solely on Zybrestat and are frowning on the way we are using capital to diversify. Because we need to raise money from them, the investment decision and the financing decision are not separated as in the usual text book scenario!"

"While we can certainly choose to stay the course, we have also identified two options for our ophthalmology drugs," Chin continued. "One option is to spin off our ophthalmology products to a syndicate of venture capital investors and in the process gain the cash required to fund our oncology platform. The issue is that we will lose control over the drugs and diversification if we have a setback with Zybrestat."

"The other option is to license the ophthalmology drugs to a large pharmaceutical company with the resources and capabilities to rapidly develop it. This is essentially the model that Antisoma followed for their own VDA drug when they licensed it to Roche." The board agreed that there was certainly precedent within the industry for such a move. Large pharmaceutical firms were increasingly turning

4   *It is the ability of VDAs to selectively target newly formed or abnormal blood vessels that makes them well-suited for certain ocular diseases, such as age-related macular degeneration, in which the formation of abnormal blood vessels in the eye plays a key role in disease. Emphasizing the way in which the oncology and ophthalmic markets are potentially closely interrelated for companies pursuing VDA and anti-angiogenesis technology, applications of Avastin in the ophthalmic domain are also being pursued.*

to small biotechnology companies to augment their pipelines. This was due to a persistent lack of internal research productivity and decreasing importance of scale in bioscience discovery research. It has been estimated that 30-50% of new molecular entities (NMEs) came from in-licensing versus internal development in the last five years. As a result, the number of pharma-biotech alliances has risen from just 69 in 1993 to 502 in 2004 (Credit Suisse, 2008; Roner, 2005).

After discussing these options, another board member noted, "Whether we attempt to develop all of our drug candidates by ourselves, spin-out opthamology into a new company, or seek to license some of them we need to raise money on acceptable terms to stabilize our share price and extend our operating runway."

APPENDICES

APPENDIX A: BIOPHARMACEUTICAL DRUG DEVELOPMENT

The average development costs per product are over $1.0 billion and 12 years from research to approval. Only five in 5,000 compounds that enter preclinical testing make it to human testing, with oncology drug candidates twice as unlikely to obtain marketing approval due to the complexity of the disease and setting. The table below provides an outline of the drug development process, success rate of drugs and the length of time each step takes (Wierenga and Eaton, 2006).

	Clinical Trials									
	Preclinical testing		Phase I	Phase II	Phase III		FDA		Phase IV
Years	3.5	File IND at FDA	1	2	3	File NDA at FDA	2.5	12 Total	Additional Post marketing testing required by FDA
Test population	Laboratory and animal studies		20 to 80 healthy volunteers	100 to 300 patient volunteers	1000 to 3000 patient volunteers		Review process / Approval		
Purpose	Assess safety and biological activity		Determine safety and dosage	Evaluate effectiveness, look for side effects	Verify effectiveness, monitor adverse reactions from long-term use				
Success rate	5,000 compounds evaluated		5 enter trials				1 approved		

Figure 5: Drug development

New biopharmaceutical products generally progress through the following steps: (1) pre-clinical testing to establish biological activity against the targeted disease, (2) Investigational New Drug Application (IND) filing to allow human clinical trials, (3) Phase I, II and III clinical trials to establish statistically significant safety and efficacy, and (4) New Drug Application (NDA) for approval for a specific type and stage of disease.

Further, under the Food and Drug Administration Modernization Act of 1997 (FDAMA), the FDA has established a number of processes—Fast Track, Priority Review, and Accelerated Approval—to accelerate the review of medicines, which treat life threatening unmet medical needs such as cancer. Fast Track review refers to a process for scheduling meetings to seek FDA input into development plans, option of submitting a New Drug Application in sections rather than all components simultaneously, and the option of requesting evaluation of studies using surrogate endpoints. Priority Review is a designation for an application that accelerates the review period to 6 months for FDA action versus the standard review period of 10 months (e.g., Ethyol (amifostine) by US Biosciences to reduce post-radiation xerostomia for head and neck cancer where the radiation port includes a substantial portion of the parotid glands). Accelerated Approval or Subpart H Approval is a program which allows the FDA evaluation to be performed on the basis of a surrogate marker (a measurement intended to substitute for the clinical measurement of interest, usually prolongation of survival) that is considered likely to predict patient benefit (e.g. Velcade (bortezomib) for the treatment of multiple myeloma patients who have received at least two prior therapies and have demonstrated disease progression on the last therapy by Millennium Pharmaceutical) (U.S. FDA, 2007).

APPENDIX B: SMALL PUBLIC ONCOLOGY COMPANIES

Small Cap Oncology Companies

Company	Stock Symbol	Stock Price as at 6th Nov 2007 ($US)	Stock Price as at 3rd Jan 2006 ($US)	% Change	90 Day Trading Volume as % of Float	Shares Out (M)	Market Capitalisation ($USM) - As at 6th Nov 2007	Cash & Equivalent ($USM)	Enterprise Value	Retained Earnings (Deficit) ($USM)	Number of Employees	Institutional Ownership (%)	Top Ten Institutional Ownership Development (%)	Enterprise as % of Mkt Cap	Cash as % of Mkt Cap
Oxigene	OXGN	$3.05	$4.12	-26%	0.31	28.5	$86.9	$34.8	$52.1	-$126.7	22	13%	10%	60%	40%
Telik	TELK	$3.74	$16.75	-78%	1.24	52.6	$196.6	$137.4	$59.1	-$392.9	118	95%	78%	30%	70%
Kosan	KOSN	$4.86	$4.46	9%	0.52	42.5	$206.7	$65.2	$141.5	-$160.3	82	88%	60%	68%	32%
MGI Phrama	MOGN	$32.54	$17.18	89%	1.52	80.6	$2,623.0	$299.7	$2,323.3	-$479.6	540	84%	60%	89%	11%
Pharmion	PHRM	$51.98	$17.71	194%	3.73	36.9	$1,917.5	$203.0	$1,714.6	-$226.8	417	84%	38%	89%	11%
Exelixis	EXEL	$9.62	$9.74	-1%	0.82	104.5	$1,005.5	$262.3	$743.2	-$705.3	651	83%	46%	74%	26%
Seattle Genetics	SGEN	$11.24	$4.74	137%	0.93	66.6	$748.9	$85.4	$663.5	-$179.6	151	58%	48%	89%	11%
Ariad	ARIA	$5.08	$5.88	-14%	1.04	69.2	$351.3	$41.6	$309.6	-$309.0	103	56%	37%	88%	12%
Altos	ALTH	$6.11	$2.30	166%	0.54	66.3	$404.9	$35.7	$369.2	-$208.6	66	53%	36%	91%	9%
Poniard	PARD	$4.70	$5.04	-7%	0.34	34.7	$162.9	$54.5	$108.4	-$279.6	42	52%	36%	67%	33%
Supergen	SUPG	$4.16	$5.03	-17%	1.59	57.5	$239.1	$71.7	$167.4	-$365.3	89	52%	30%	70%	30%
SGX	SGXP	$6.56	$5.75	14%	0.39	15.4	$100.7	$39.0	$61.7	-$163.7	120	44%	44%	61%	39%
Sunesis	SNSS	$2.13	$5.14	-59%	1.15	34.3	$73.1	$64.2	$8.9	-$240.3	138	36%	34%	12%	88%
Biocryst	BCRX	$7.82	$17.21	-55%	2.38	29.5	$231.0	$45.8	$185.2	-$195.5	85	35%	26%	80%	20%
Sonus	SNUS	$0.50	$5.13	-90%	3.73	36.9	$18.5	$66.9	-$48.4	-$111.7	61	34%	27%	-262%	362%
Threshold	THLD	$0.58	$14.90	-96%	0.57	37.3	$21.7	$53.4	-$31.7	-$134.7	45	26%	24%	-147%	247%
YM Bio	YMI	$1.29	$3.14	-59%	0.25	55.8	$72.0	$72.0	$0.0	-$114.8	37	18%	15%	0%	100%
Vion	VION	$0.64	$1.59	-60%	0.48	72.9	$46.7	$31.2	$15.5	-$175.2	39	18%	15%	33%	67%
Adventrx	ANX	$0.57	$3.30	-83%	0.89	89.8	$51.2	$52.3	-$1.2	-$89.3	24	16%	14%	-2%	102%
Cougar	CGRB	$25.60	$17.50	46%	0.39	17.5	$447.2	$32.0	$415.2	-$23.2	11	16%	16%	93%	7%
Novacea	NOVC	$2.86	$6.50	-56%	2.99	23.5	$67.3	$65.7	$1.6	-$91.4	59	14%	11%	2%	98%
Entremed	ENMD	$1.53	$1.99	-23%	0.33	85.1	$130.2	$54.9	$75.3	-$311.6	56	13%	11%	58%	42%
Cell Therapeutics	CTIC	$3.10	$2.26	37%	0.21	50.0	$154.9	$64.5	$90.4	-$961.1	197	7%	6%	58%	42%
Average		$8.27	$7.71	-1%	1.15	51.6	$406.9	$84.1	$322.8	-$262.9	137	43%	31%	35%	65%
Minimum		$0.50	$1.59	-96%	0.21	15.4	$18.5	$31.2	-$48.4	-$961.1	11	7%	6%	-262%	7%
Maximum		$51.98	$17.71	194%	3.73	104.5	$2,623.0	$299.7	$2,323.3	-$23.2	651	95%	78%	93%	362%
Std Dev		$12.33	$5.84	81%	1.07	24.4	$643.4	$72.8	$582.3	$213.6	168	28%	19%	83%	83%

Figure 6: Competitors' share prices/company value
Source: www.yahoo.com

APPENDIX C: OXIGENE'S FINANCIAL POSITION

Profit & Loss	2001	2002	2003	2004	2005	2006
Revenue	9.9	0.03	0	0	0	0
Gross Profit	8.4	0.3	0	N/A	N/A	N/A
SG&A	5.5	7.4	5.3	4.5	6	7.1
R&D	6.1	5.1	3.9	5.9	7.1	10.8
Other	0.3	0.1	0.1	0.1	0	0
Operating Income	-3.5	-12.3	-9.3	-10.5	-13.1	-17.9
Net Income	-4.1	-11	-8.4	-10	-11.9	-15.5

Balance Sheet	2001	2002	2003	2004	2005	2006
Cash & Equivalents	19.0	3.8	0.9	16.0	32.3	15.7
Short Term Investments	0.0	8.1	17.7	8.0	23.4	29.7
Cash and Short Term Investments	19.0	11.8	18.6	24.0	55.70	45.35
Prepaid Expenses	0.46	0.02	0.03	0.06	0.08	0.27
Other Current Assets, Total	0.01	0.02	0.39	0.05	0.17	0.37
Total Current Assets	19.5	11.87	18.98	24.08	55.95	45.99
Property/Plant/Equipment	0.87	0.87	0.91	0.95	1.05	1.25
Goodwill, Net	-	-	-	-	-	-
Intangibles, Net	1.94	1.17	1.07	0.97	0.87	0.78
Long Term Investments	-	-	-	6.53	3.16	0.49
Other Long Term Assets, Total	0.09	0.07	0.11	0.11	0.15	0.14
Total Assets	22.15	13.60	20.20	31.76	60.27	47.64
Accounts Payable	1.14	1.42	1.55	0.49	0.69	0.68
Accrued Expenses	1.78	1.72	2.03	2.13	3.04	3.54
Notes Payable/Short Term Debt	0.00	0.00	0.00	0.00	0.00	0.00
Other Current liabilities, Total	0.27	0.29	0.15	0.00	-	-
Total Current Liabilities	3.19	3.42	3.73	2.62	3.73	4.22
Total Long Term Debt	0.00	0.00	0.00	0.00	0.00	0.00
Total Debt	0.00	0.00	0.00	0.00	0.00	0.00
Other Liabilities, Total	0.44	0.15	0.00	0.00	-	-
Total Liabilities	3.63	3.58	3.73	2.62	3.73	4.22
Common Stock, Total	0.11	0.13	0.14	0.17	0.28	0.28
Additional Paid-In Capital	82.39	83.47	97.67	119.53	160.88	160.57
Retained Earnings (Accumulated Deficit)	-60.64	-71.65	-80.02	-90.05	-101.95	-117.41
Other Equity, Total	-3.34	-1.92	-1.32	-0.51	-2.68	-0.02
Total Equity	18.52	10.02	16.47	29.14	56.53	43.42
Total Liabilities & Shareholders' Equity	22.15	13.60	20.20	31.76	60.27	47.64
Total Common Shares Outstanding	11.4	12.7	14.0	16.7	28.0	28.2

Figure 7: Oxigene's financials
Source: www.yahoo.com

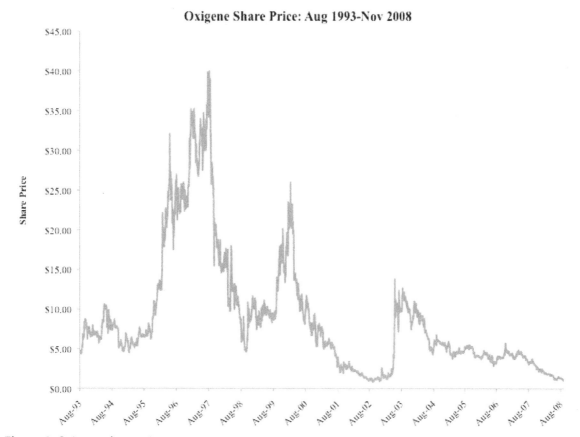

Figure 8: Oxigene share price
Source: www.yahoo.com

REFERENCES

American Cancer Society. 2006. *Cancer Statistics 2006.* www.cancer.org, accessed December 11, 2009.

Bauman, L. 2001. Oxigene and WebMethods surge, as tech issues Lift Russell 2000. *The Wall Street Journal,* *www.wsj.com,* accessed February 21, 2010.

Benoit, B., Brown, J.E., Gregan, P.E. & Wilson, N. 1999. Tax on hold for ECB Meeting *The Wall Street Journal,* *www.wsj.com,* accessed February 21, 2010.

Berenson, A. 2006. A cancer drug shows promise, at a price that many can't pay. *The New York Times,* February 15, 2006.www.nytimes.com, accessed January 14, 2009.

BIO (Biotechnology Industry Organization). 2006. *BIO 2005-2006: Guide to Biotechnology* Washington DC: BIO.

Biocentury. 2007. www.biocentury.com, accessed Dec 19, 2009.

Burrill & Company. 2007. *Biotech 2007 Life Sciences: A Global Transformation.* San Francisco: Burrill Life Sciences Media Group.

Business Week. 2005a. Oxigene, Inc.; Biopharmaceutical compound developer announces U.K. and U.S. facility expansion. *Managed Care Business Week,* 70.

Business Week. 2005b. Oxigene; Biopharma compound developer names new board member. *Lab Business Week*: 170.

Cookson, C. 1998. New weapons in the war against cancer: Health cancer research, *Financial Times,* May 14, 25.

Credit Suisse. 2008. Pfizer, Wyeth, Amgen, *Equity Research: Major pharmaceuticals.*

Ernst & Young. 2007. *Beyond borders: Global Biotechnology Report.* EYGM Limited. www.ey.com. Accessed May 15, 2007.

Farmer, G. & Pollack, F. 2006. *Targeted Cancer Therapeutics: Application of Evolving Themes.* Wachovia Capital Markets, LLC.

Folkman, J. 1971. Tumor angiogenesis: therapeutic implications, *New England Journal of Medicine,* 285: 1182-1186.

Forelle, C. 2003. Oxigene receives FDA 'Fast Track' for thyroid drug. *The Wall Street Journal* D.4, www.wsj.com, accessed February 21, 2010.

Goldman Sachs. 2007. *Pipeline Survey: Growth Accelerated with High Risk, Equity Research Report.*

Gupta, U. 1994. Medicine & health: The bottom line – Experience pays: There's more money for biotech firms; But start-ups need not apply," *Wall Street Journal, www.wsj.com,* accessed February 21, 2010.

Jackson, D. 2003. The pharmaceutical market outlook to 2010: Essential analysis of key drivers of change. *Reuters Business Insight, Healthcare* & *Datamonitor.*

Lehman Brothers. 2007. *Global Pharmaceuticals: Key Metrics, Equity Research.* www.lehman.com, accessed October 24, 2007.

Lippert, J.W. 2007. Vascular disrupting agents. *Bioorganic* & *Medical Chemistry,* **15: 605-615.**

Mayer, R.J., 2004. Two steps forward in the treatment of colorectal cancer. *New England Journal of Medicine,* 23(350): 2406-2408.

Roner, L. 2005. Forging strategic alliances with biotech partners. *Eyeforpharma Briefing.*

Scrip, 1998. "OXiGENE's Sensamide disappoints," *Scrip,* 2376: 11.

Scrip. 1999. BMS in talks to develop combretastatin, 2494: 11.

The New York Times. 2009. F.D.A. approves drug's use for breast cancer. February 22, 2008.www.nytimes.com, accessed January 14, 2009.

The Wall Street Journal. 1999a. Bristol-Myers, Oxigene sign pact. *www.wsj.com,* accessed February 21, 2010.

The Wall Street Journal. 1999b. Oxigene says a drug for tumors produced full-remission case. B2, *www.wsj.com,* accessed February 21, 2010.

The Wall Street Journal. 2003. Oxigene cancer drug gets 'Orphan' Status. A10,*www.wsj.com,*accessed February 21, 2010.

Tozer, G.M., Kanthou, C. & Baguley, B.C. 2005. Disrupting tumour blood vessels. *Nature Reviews,* 5(6): 423-435.

U.S. Food and Drug Administration. 2007. *Center for Drug Evaluation and Research.* www.accessdata.fda.gov, accessed January 17, 2010.

Wierenga, D.E. & Eaton, C.R. 2006. *Office of Research and Development, Pharmaceutical Manufacturers Association.* www.phrma.org, accessed January 7, 2010.

PART II: LAWS, REGULATIONS, POLITICS

Tysabri Re-launch Decision: Promise and Perils of Addressing Unmet Needs

MARK J. AHN & LAURA UEKI

Atkinson Graduate School of Management, Willamette University

- **Summary and key issue/decision:** In November 2004, Biogen-Idec and Élan, two well-established biotechnology companies based in Cambridge, Massachusetts and Dublin, Ireland, respectively, were granted permission from the FDA to market their drug, Tysabri (natalizumab) for the treatment of multiple sclerosis (MS). The FDA approval was fast-tracked because of Tysabri's efficacy in late stage clinical trials, in which the drug, when accompanied with the drug Avonex (beta interferon; Biogen-Idec), reduced probability of relapses in individuals with MS by 54% versus treatment with Avonex alone. Unfortunately, within four months, two cases—one fatal—of the rare and usually terminal disease, progressive multifocal leukoencephalopathy (PML), had been confirmed in patients participating in the trials. Accordingly, in February 2005, Biogen-Idec and Élan voluntarily suspended Tysabri from the market. Because of the drug's effectiveness and the dearth of other effective treatments for MS, however, there was widespread call for its return. On June 6, 2006, after examining two years worth of clinical trial data and hearing from an advisory committee, the FDA re-approved natalizumab for the treatment of relapsing MS, with the stipulation that prescribers, pharmacies, infusion centers, and patients of the drug are tracked. By the end of the month, the EC approved natalizumab for the European market with their own conditions, and by late July, Biogen-Idec and Élan re-launched Tysabri in the U.S., Germany, Ireland, Sweden and the U.K. This case analyzes the companies' decision to re-launch their drug despite its association with PML.

- **Companies/institutions:** Biogen-Idec and Élan

- **Technology:** Tysabri (Natalizumab), a humanized monoclonal antibody, inhibitor of the adhesion molecule α4-integrin.

- **Stage of development at time of issue/decision:** The results from two Phase III clinical trials demonstrated that Tysabri use reduced the occurrence of disability progression by 42% (p>0.001), and caused a 68% reduction (p>0.001) in patients' annual relapse rates compared to placebo. When re-launched, three confirmed cases of PML in patients taking Tysabri had been reported.

- **Indication/therapeutic area:** Multiple sclerosis occurs when inflammatory cells pass through the blood-brain barrier into the central nervous system (CNS) and demyelinate the body's own nerve cells, impairing their ability to function. Natalizumab works to prevent this by inhibiting the receptors on endothelial cells responsible for allowing inflammatory cells to pass into the CNS.

- **Geography:** US and Europe

- **Keywords:** Tysabri (natalizumab), multiple sclerosis monoclonal antibody, autoimmune disease, progressive multifocal leukoencephalopathy

* *This case was prepared as a basis for class discussion rather than to illustrate either effective or ineffective handling of an administrative situation.*

INTRODUCTION

In November 2004, when Biogen-Idec and Elan Corporation launched Tysabri, a monoclonal antibody therapeutic for relapse-remitting multiple sclerosis, expectations were sky high. Two year data from Phase III clinical trials showed a 42% decrease in the probability of disability progression and a 68% reduction in annual rate of relapse after two years (Polman *et al.*, 2006). More effective than any other treatment available, Tysabri was predicted to become the leading multiple sclerosis drug on the market. With approximately 300,000 multiple sclerosis patients in the United States alone, sales were expected to exceed $1 billion (Maggos, 2005). As Morgan Stanley analyst Marc Goodman stated: "The [FDA] approval is very good news. We expect this drug to be a paradigm shift in MS therapy as we believe its therapeutic profile is so much better than existing therapy" (Pritchard & Linnane, 2004).

Unfortunately, this is where the success story ends. On February 28, 2005, just four months after approval, Biogen-Idec and Elan voluntarily withdrew Tysabri when two cases—one fatal—of progressive multifocal leukoencephalopathy (PML) were reported in trial patients (Biogen-Idec, 2005). Though PML is estimated to occur in only one in 1,000 patients treated with Tysabri (Yousry *et al.*, 2006), the problem is that multiple sclerosis is not usually a fatal disease but PML is. Most doctors find it hard to justify a potentially life-threatening treatment for a non-fatal disease. In this case, however, with Tysabri's effectiveness hard to ignore and few other treatment options available, many patients are willing to take that risk. As Aaron Miller, medical director of the Corinne Goldsmith Dickenson Center for Multiple Sclerosis, and chairman of the medical advisory board of the National Multiple Sclerosis Society, noted: "I personally think it may very well be reasonable to allow the drug back on the market to be used in patients who are seriously ill and are not responding to other therapies." (Krasner, 2005)

As a result of physician and patient demand for Tysabri, there was a widespread call for Tysabri's re-release to the market. The reintroduction of Tysabri could prove to be a disaster for the company if toxicities increased further. Moreover, the exact connection between Tysabri and PML was still unknown. Jim Mullen, CEO, Biogen-Idec had to carefully weigh his company's options.

COMPANY HISTORY AND FINANCIALS

Biogen-Idec is a Cambridge, Massachusetts-based biopharmaceutical company that specializes in oncology, immunology, and neurology. It was formed from the merger of two well-established biopharmaceutical companies, Biogen and IDEC Pharmaceuticals. Biogen was founded in 1978 in Geneva, Switzerland by a group of distinguished biologists including the Nobel Prize winners, Walter Gilbert of Harvard University and Phillip Sharp of MIT. Biogen's first line of research was beta interferon, and in 1986, their first product, Intron A (interferon alfa-2b), was released for the treatment of hairy cell leukemia. Ten years later, the FDA approved their multiple sclerosis drug, Avonex (Interferon beta-1a), which was later approved for rheumatoid arthritis treatment in 2006. IDEC Pharmaceuticals was founded in 1985 in San Francisco, California with the goal of developing "anti-idiotype" monoclonal antibodies for the treatment of non-Hodgkin's lymphoma. In 1997, their drug, Rituxan (rituximab), became the first monoclonal antibody approved for the treatment of cancer (Biogen-Idec, 2009).

Biogen and Idec merged in November 2003 and formed a fully integrated firm with global capabilities in the discovery, development, manufacturing, and commercialization of therapeutics. Biogen-Idec's international headquarters are located in Zug, Switzerland, and it has manufacturing plants in Cambridge, Research Triangle Park, North Carolina, and Hillerod, Denmark. Currently the company has four drugs on the market, Avonex, Rituxan, Tysabri, and Fumaderm, which generated a total of $3.96 billion in 2008 revenues. The company also received $116.2 million last year in royalty revenues form licensed technologies and products. With research facilities in Cambridge and San Diego, Biogen-Idec spent $1,072.1 million on research and development last year, and currently has 22 products in development that are in Phase II trials or beyond (Biogen-Idec, 2009) (Appendix A).

In 2000, Biogen-Idec partnered with Elan Corporation to develop a promising new drug for multiple sclerosis and Crohn's disease called Tysabri (natalizumab) (Biogen-Idec, 2009). Elan Corporation is a biotechnology company headquartered in Dublin, Ireland, with a focus on neurology (Elan, 2009).[1]

TECHNOLOGY, PRODUCTS, AND PIPELINE

Biogen-Idec has four approved products on the market, the newest of which is Tysabri (nalituzimab), a humanized monoclonal antibody originally approved in November 2004 for the treatment of relapsing MS to reduce the frequency of clinical relapses.

MULTIPLE SCLEROSIS PREVALENCE

Multiple sclerosis (MS) is a chronic autoimmune disease of the central nervous system. A frequently mild but potentially disabling disease, there is no cure for MS and its true etiology has yet to be determined. There are approximately 300,000 diagnosed cases of MS in the United States (NINDS, 2009), with approximately 2.5 million people affected worldwide (Compston, Lassmann, & McDonald 2006). In the United States, this means about 200 new cases are diagnosed each week and approximately 1 in every 1,000 people has the disease (Campagnolo et al., 2009). In the United States, multiple sclerosis costs about $47,000 per patient per year, and nationwide, total costs exceed $14 billion each year (Ransohoff, 2007).

MS affects people of all ethnic backgrounds all over the world, but is most prevalent among white women of northern European descent (Campagnolo et al., 2009). In the US, Caucasians are more than twice as likely as people of other races to develop the disease, and it is about twice as common in women as in men (NINDS, 2009). Studies indicating a relationship between certain climates and geographical locations and incidence of MS suggest that environment may play a role in development of the disease, but this role has yet to be elucidated (NINDS, 2009).

DISEASE COURSE

One of the most concerning aspects of MS is that the disease course and symptoms vary widely and are largely unpredictable. Clinical variables, laboratory tests, and imaging techniques offer no more than modest predictions of the individual patient's prognosis. There is, however, a correlation between age at onset of the disease and time to disability; the older the patient is at disease onset, the shorter the time to disability (Confavreux & Compston, 2006).[2]

The average onset of MS is 30 years of age, and the disease then takes one of four courses of progression. Relapse-remitting MS is characterized by episodes of acute onset of symptoms, separated by

---

1   Elan was founded in 1969, and has facilities in Ireland, Bermuda, and the United States. Elan Corporation is divided into two units: Biopharmaceuticals and Elan Drug Technologies. Biopharmaceuticals encompasses research and development of drugs for neurodegenerative and autoimmune diseases including Alzheimer's disease, Parkinson's disease, multiple sclerosis and Crohn's disease. Elan Drug Technologies is a drug delivery business; it offers assistance with formulation development, clinical trials, product registration, and commercial manufacturing for pharmaceuticals developed by other companies (Elan, 2009).

2   MS affects the central nervous system, which is comprised of the brain and spinal cord and which coordinates the activities of the entire nervous system by receiving sensory impulses and transmitting motor impulses. The disease is characterized by damage to myelin in nerve cells of the CNS. Myelin is the fatty substance that insulates the nerve fibers—called axons—that connect nerve cells. Myelin facilitates fast, smooth transmission of signals from cell to cell, and this "demyelination" impairs cells' ability to communicate. Demyelination is caused by attacks of inflammation, which result in patchy CNS scars called plaques, or lesions. The initial cause of the disease has not been identified, but one theory holds that in a genetically susceptible host, exposure to any one of many common agents, including the Epstein–Barr virus, can activate or dysregulate T cells that recognize myelin protein antigens (Compston, Lassmann, and Smith, 2006).

periods of remission in which symptoms lessen or disappear entirely. Approximately 85% of patients are initially diagnosed with relapse-remitting MS (Confavreux & Compston, 2006), but approximately 50% of these patients will develop secondary-progressive MS within twenty years, and that number increases to 80% within 35 to 40 years (Vukusic & Confavreux, 2003).

Primary-progressive MS is marked by a gradual clinical decline with no distinct remissions. Progressive-relapsing MS takes a progressive course punctuated by acute attacks. And finally, second-ary-progressive MS begins with a relapsing-remitting course followed by a progressive course. The progressive form of the disease is defined as a situation of increasing, irreversible disability over time. Factors that are correlated with a faster onset of secondary progression include being male, an older age at onset of the disease, and shorter intervals between relapses (Confavreux & Compston, 2006).

Not all patients develop irreversible disabilities, but about 50% of MS patients will reach the point of being unable to maintain a household or employment within ten years of diagnosis, and will require assistance walking within fifteen years (Ransohoff, 2007). The disease itself is not considered fatal; the median time from onset of MS to death of the patient is 31 years, and though average life expectancy is reduced 5 to 10 years compared to the general population, this is generally attributed to the bed-ridden state of disabled patients and the complications that arise thereof. Patients with a severe form of the disease may pass away from an acute attack of symptoms, but these instances are rare (Confavreux & Compston, 2006).

SYMPTOMS AND DIAGNOSIS

Symptoms of MS include mild to severe sensory, motor, and cognitive impairment. Severity and duration of symptoms varies according to what part of the nervous system is affected. Vision impairment is common and includes blurred or double vision, red-green color distortion, or even blindness in one eye, called optic neuritis, which occurs in 20% of patients at initial onset of the disease. Sensory symptoms such as paresthesias—temporary numbness or prickling sensations—occur in over eighty percent of patients. Other common symptoms include weakness of one or more limbs, overall fatigue, ataxia, bladder problems, cramps, poor memory, and facial pain. In addition to its physical symptoms, the emotional toll of the disease is immeasurable for both patients and loved ones. Unexpected onset of symptoms can render patients temporarily or permanently unable to work, while at the same time incurring expensive medical bills. Not surprisingly, 17% of MS patients experience depression, which is thought to be a reaction to living with the disease rather than a direct symptom (McDonald & Compston, 2006).

There is no single test for diagnosing multiple sclerosis; a doctor, often a neurologist, must look at a combination of clinical signs and laboratory tests before making a diagnosis. Since there are many nervous system disorders with similar symptoms, practitioners must decide if their patients' symptoms are caused by the multiple demyelinating lesions characteristic of the disease. The most basic criteria for positive diagnosis include two or more episodes suggestive of demyelination, at least thirty days apart, creating lesions on two or more separate sites in the central nervous system. Recently, however, new criteria developed by an International Panel allow for a positive diagnosis after only one episode or when only one lesion is present if there is MRI evidence for the spread of demyelination in time and space (Miller, McDonald, & Smith, 2006).[3]

---

3   *MRI evidence for new or worsening of lesions is often not associated with clinical manifestation of symptoms. In fact, studies show that as few as ten percent of episodes of demyelination detected by MRI actually result in clinical symptoms. This can prove stressful for patients experiencing a remission in symptoms but a worsening of subclinical MRI activity. Most practitioners choose to guide treatment based on symptoms rather than MRI activity (Miller et al. 2006).*

TREATMENT

Patients experiencing a relapse are first instructed to rest, as strained impulse activity can promote axonal damage. Often practitioners will prescribe a high dose of the intravenous corticosteroid, methylprednisolone, a treatment that has been shown to be effective in two recent studies. There is also the option for more convenient oral corticosteroids, but there is unconfirmed evidence that they cause adverse side effects not worth the ease of use (Noseworthy, Confavreux, & Compston, 2006). Since MS manifests in many different forms, patients require a wide variety and combination of symptom-specific treatments, based on the symptoms they experience. Such treatments include Lioresal, tizanidine, or a benzodiazepine to treat muscle spasms, cholinergic medications for urinary problems, amantadine for fatigue, and antidepressants for psychological symptoms (Noseworthy, Miller, & Compston, 2006b).

Though there is not yet a cure for multiple sclerosis, there are many disease-altering treatments that may slow its progression. For patients with relapsing MS, there are five immunological therapeutics on the market, all of which alter the functioning of the immune system, aiming to reduce inflammation in the CNS. Beyond Tysabri, three other products—including one of Biogen-Idec's other products, Avonex—are beta-interferons. Avonex and Refib are different formulations of interferon beta-1a, and Betaseron—Betaferon in Europe—is interferon beta-1b. The last is Copaxone (glatiramer acetate), a non-interferon, non-steroidal immunosuppressant (Noseworthy, Miller, & Compston, 2006a). While all of these medications have been shown to decrease the number of relapse episodes and slow onset of disability to some degree, Tysabri is currently the most effective treatment option (Polman *et al.*, 2006).

Biogen-Idec's Avonex generated revenues of $1.87 billion in 2007. Rebif, co-promoted by EMD Serono and Pfizer, sold $1.7 billion in 2007. Copaxone, which is sold by Teva in the United States, and both Teva and Sanofi-Aventis in Europe, also had worldwide sales of $1.7 billion in 2007. Betaseron/Betaferon, from Bayer Schering Pharma AG, had revenues of $1.4 billion in 2007.

TYSABRI: A NOVEL TREATMENT FOR MS

The immune system functions by distinguishing between "self" and "non-self," protecting the body by rejecting harmful "non-self" substances. When the immune system fails to tolerate and actually attacks "self" molecules, what results is an autoimmune disease (Tortora, Funke, & Case, 2003; Campbell & Reece, 2007). MS is considered an autoimmune disease as it is thought that demyelination occurs when inactive T cells that recognize "self" myelin protein antigens—"naive autoreactive T cells"—somehow become activated and promote an immune response, including inflammation, against the myelin in the CNS (Wekerle & Lassman, 2006).[4] It is thought that naive autoreactive T cells are initially activated in the blood by antigens that resemble myelin proteins and then are reactivated in the CNS by myelin antigens presented by microglial cells. Upon reactivation, these cells may secrete cytokines that cause inflammation and directly damage myelin, and they may also activate B cells to produce myelin

---

4   *Substances that are recognized by the immune system and provoke a specific response are called antigens. Examples of foreign antigens are invading microbes, pollen, or foreign RBCs. Antigens are taken up by antigen-presenting cells (APCs) which, as implied by their name, present a portion of the antigen to white blood cells called lymphocytes, which are divided into T and B lymphocytes, or T and B cells. Each of these cells has surface receptors that recognize and bind to only one specific antigenic determinant, (an external portion of an antigen that is specific to that antigen). There are millions of different T and B cells in the body that recognize different antigens. When inactive, or naive, T cells bind to antigens on APCs, they become activated and undergo cell division. Some replicated T cells then activate other lymphocytes, including B cells, and/or may secrete compounds called cytokines. Some replicated B cells become plasma cells which secrete antibodies. Antibodies recognize and bind to the original antigen and can help neutralize or destroy it (Tortora et al., 2003; Campbell & Reece, 2007).*

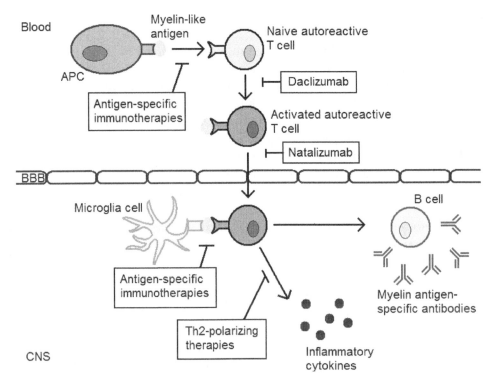

Figure 1: The pathogenesis of MS and points of therapeutic intervention
Source: Adapted from Expert Reviews in Molecular Medicine (2005)

antigen-specific antibodies (Gran & Rostami, 2001). To enter the central nervous system and cause damage T cells must cross the blood-brain barrier, a membrane of endothelial cells that regulate passage of molecules between the blood and the CNS, functioning to protect the CNS from infection. For this to occur receptors on the exterior of T cells, called integrins, must interact with complementary receptors on endothelial cells. In MS, the interaction between the Alpha-4, Beta-1 integrin on T cells and complementary receptors on the endothelial cells of the blood brain barrier is what allows the T cells to enter the CNS and cause damage. Tysabri (natalizumab) is a recombinant, humanized monoclonal antibody specific for Alpha-4 integrin subunits.[5] Natalizumab recognizes and binds to Alpha-4, blocking the bonding of T cells to endothelial cells and preventing T lymphocytes from entering the CNS (Ransohoff, 2007).

Tysabri's patents in the US, which cover the protein, the DNA encoding the protein, the manufacturing methods and pharmaceutical compositions, as well as various methods of treatment using the product, expire between 2014 and 2020.

TYSABRI CLINICAL STUDIES

The efficacy of natalizumab as treatment for multiple sclerosis has been evaluated in two Phase III clinical trials. In both trials, the drug was administered only to patients with the relapsing–remitting form of the disease. The first study, a two-year, Phase III monotherapy trial, called AFFIRM, randomly assigned 942 MS patients to receive either natalizumab or placebo every four weeks for the course of the study. A neurologist with no knowledge of the specific assignments evaluated duration and severity

5   *Monoclonal antibodies (MAb) are highly selective fused protein molecules that recognize and bind to specific targets (antigens) which have been developed as research tools, diagnostics, and therapeutics. MAbs are a highly versatile biotechnology therapeutic platform which can be used alone or to deliver drugs, toxins, or radioactive materials. Their ability to directly target specific diseased cell populations make MAbs an ideal technology platform while avoiding healthy normal cells (Brekke & Sandlie, 2003).*

of relapses and the progression of disability with the Expanded Disability Status Scale. MRI brain imaging was administered at the start of the trial, at one year, and at two years. Their results showed that natalizumab treatment reduced the probability of sustained disability progression to 17% (P<0.001), compared to 29% for those taking placebo, and decreased the average annual rate of relapse by 68%. Furthermore, patients taking natalizumab had 92% fewer lesions than those taking placebo at both one and two years (P<0.001)" (Polman *et al.*, 2006).

The second study, a similar two-year Phase III trial called SENTINEL, added natalizumab or placebo to interferon beta-1a (Avonex) for patients who had been treated with interferon beta-1a for twelve previous, consecutive months. The study included 1171 patients and found treatment with both drugs to be more effective than interferon beta treatment alone. After one year, patients on combination therapy showed a 23% chance of developing sustained progression, compared to 29% for those who received beta interferon alone. Those given both drugs also had a 54% reduction in annual rate of relapse after one year compared to the control group, and showed an 83% reduction in new or enlarging lesions. Unfortunately, this study was ended a month early because two patients receiving combination treatment developed progressive multifocal leukoencephalopathy (PML), a devastatingly fatal viral disease affecting the white matter of the brain (Rudick, *et al.* 2006).

PROGRESSIVE MULTIFOCAL LEUKOENCEPHALOPATHY (PML)

PML is a rare and progressive demyelinating disease of the central nervous system that has no effective treatment and typically causes permanent disability or death. It is caused by activation of the JC virus, which occurs in its harmless, latent form in about 65-80% of the normal adult population.[6] The virus infects and destroys brain cells called oligodendrocytes, which causes demyelination and leads to degeneration of the central nervous system. PML predominately occurs in immunocompromised patients, whose immune systems have been weakened by certain diseases or medications (Tornatore & Clifford, 2009).

PEOPLE AND CULTURE

As a pioneer in the biopharmaceutical industry, Biogen-Idec possessed a talented and experienced team of industry veterans led by James C. Mullen, chief executive officer and president, as well as board member, since the merger of Biogen, Inc. and IDEC Pharmaceuticals Corporation in November 2003. Mr. Mullen joined Biogen in 1989 as director, facilities and engineering, then served as vice president-operations, vice president-international, and CEO and chairman of the board of directors. From 1984 to 1988, Mr. Mullen held various positions at SmithKline Beckman Corporation (now GlaxoSmithKline plc). Mr. Mullen is a member of the board of directors and executive committee of the Biotechnology Industry Organization, or BIO, and is a former chairman of the board of BIO. Mr. Mullen is also a director of PerkinElmer, Inc.

Biogen-Idec's management team at the time of the Tysabri challenge is indicated in Figure 2 and Biogen-Idec's vision, mission, and core values are shown in Figure 3.

Questions:
HR   201-930-0100   ext. 70420

---

6   *The JC virus is a type of human polyomavirus (formerly known as papovavirus) and is genetically similar to BK virus and SV40. It was discovered in 1971 and named after the two initials of a PML patient from whose autopsy it was isolated from (Tornatore & Clifford, 2009). The virus causes PML and other diseases only in cases of immunodeficiency, as in AIDS or during treatment with drugs intended to induce a state of immunosuppression (e.g. organ transplant patients).*

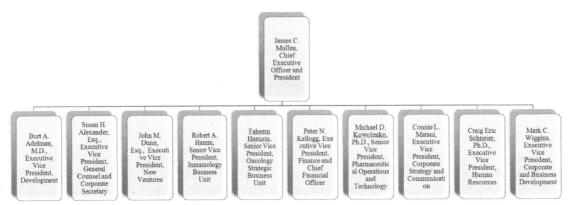

Figure 2: Biogen-Idec management structure
Source: Company reports

Vision: With passion, purpose and partnerships, we transform scientific discoveries into advances in human healthcare.

Mission: We create new standards of care in neurology, oncology and immunology through our pioneering research, our global development, manufacturing and commercial capabilities.

Core Values

Courageous Innovation: We apply our knowledge, talent and resources to yield new insights and bold ideas. We confront challenge and uncertainty with zeal, tenacity and vision and seize opportunities to excel.

Quality, Integrity, Honesty: Our products are of the highest quality. Our personal and corporate actions are rooted in mutual trust and responsibility. We are truthful, respectful and objective in conducting business and in building relationships.

Team as a Source of Strength: Our company is strong because our employees are diverse, skillful and collaborative. We pursue our fullest potential as individual contributors, team members and team leaders.

Commitment to Those We Serve: We measure our success by how well we enable people to achieve and to thrive. Patients, caregivers, shareholders and colleagues deserve our best.

Growth, Transformation and Renewal: Consistent with our core values, we as individuals and as a corporation are dedicated to creative and constructive growth, transformation and renewal as a source of inspiration and vitality.

Figure 3: Biogen-Idec's vision, mission, and core values
Source: www.biogenidec.com

CONCLUSION

Jim Mullen, CEO of Biogen-Idec, met with his management team to review the situation and generate options to review at his upcoming board of directors meeting. Jim Mullen reflected on the months of challenging discussions with the FDA, shareholders, patients, and physicians and asked, "What are our options?"

First, Dr. Gordon Francis, senior vice president, clinical development summarized a recent meeting with the FDA. "We had a thorough and rigorous review of the Tysabri data with the FDA. Though the two Tysabri trial patients who acquired PML did not have a history of immmunosuppresion, they were taking steroids, which are immunosuppressants, but not at a dosage that would be expected to be associated with increased PML risk. A retrospective analysis of more than 3,700 subjects failed to reveal any additional cases of PML. However, a third case of PML had been reported in a Tysabri trial for Crohn's disease.

Overall, the estimated risk of PML was one case per 1,000 patients treated with Tysabri for an average of 17.9 months. This is much higher than PML's incidence in the general population of one in 200,000, thus indicating a definite association between the disease and Tysabri use. There was speculation that use of the drug along with Avonex and other immunosuppresants increased the risk of developing PML, but the FDA's advisory committee that investigated the matter concluded that "there is a treatment-associated risk of PML even when [Tysabri] is given as monotherapy."

Peter Kellogg, the chief financial officer stated: "re-launching Tysabri induces significant financial and litigation exposure to the company. We have other products and a robust pipeline, including other drugs for MS. One option may be to permanently withdraw Tysabri form the market."

Kellogg continued, "another option could be to give it back to Elan and let them re-launch the drug. We could negotiate a royalty deal that would allow us to participate in the sales, but limit our legal exposure."

Connie Matsui, executive vice president, corporate strategy and communication, noted, "We're getting calls and letters daily from patients and their physicians in desperate need. We can't let those calls for care go unanswered."

Francis added, "our discussions with the FDA provided another option. The FDA has indicated that re-launching Tysabri with a black box warning on the product label noting the risks of taking Tysabri treatment is possible.[7] We could also establish a monitoring system of registering infusion centers and every patient to monitor their progress. This would allow us to monitor trends and establish an 'early warning system' to watch for any emerging issues and warn physicians about any concerns."

Jim Mullen reflected, "our core values include courageous innovation and measuring our success by how well we enable people to achieve and to thrive. Patients, caregivers, shareholders and colleagues deserve our best. How do we balance serving patients and appropriately managing risk?"

---

7   In the United States, a black box warning (also sometimes called a black label warning or boxed warning) is a type of warning that appears on the package insert for prescription drugs that may cause serious adverse effects. It is so named for the black border that usually surrounds the text of the warning. A black box warning means that medical studies indicate that the drug carries a significant risk of serious or even life-threatening adverse effects. The U.S. Food and Drug Administration (FDA) can require a pharmaceutical company to place a black box warning on the labeling of a prescription drug, or in literature describing it. It is the strongest warning that the FDA requires.

APPENDICES

APPENDIX A: BIOGEN-IDEC—SELECTED FINANCIAL DATA

Years Ended December 31
(In thousands, except per share amounts)

	2005	2004	2003	2002	2001
Product revenues	$ 1,617,004	$ 1,486,344	$ 171,561	$ 13,711	$ -
Revenue from unconsolidated joint business	$ 708,881	$ 615,743	$ 493,049	$ 385,809	$ 251,428
Royalties	$ 93,193	$ 98,945	$ 12,010		
Corporate partner revenue	$ 3,422	$ 10,530	$ 2,563	$ 4,702	$ 21,249
Total revenues	$ 2,422,500	$ 2,211,562	$ 679,183	$ 404,222	$ 272,677
Total costs and expenses	$ 2,186,460	$ 2,168,146	$ 1,548,852	$ 190,346	$ 141,540
Income (loss) before income taxes (benefit)	$ 256,195	$ 64,093	$ (880,624)	$ 231,522	$ 161,604
Net income (loss)	$ 160,711	$ 25,086	$ (875,097)	$ 148,090	$ 101,659
Diluted earnings (loss) per share	$ 0.47	$ 0.07	$ (4.92)	$ 0.85	$ 0.58
Shares used in calculating diluted earnings (loss) per share	$ 346,163	$ 343,475	$ 177,982	$ 176,805	$ 178,117
Cash, cash equivalents and marketable securities available-for-sale	$ 2,055,131	$ 2,167,566	$ 2,338,286	$ 1,447,865	$ 866,607
Total assets	$ 8,366,947	$ 9,165,758	$ 9,503,945	$ 2,059,689	$ 1,141,216
Notes payable, less current portion	$ 43,444	$ 101,879	$ 887,270	$ 866,205	$ 135,977
Shareholders equity	$ 6,905,876	$ 6,826,401	$ 7,053,328	$ 1,109,690	$ 956,479

Source: Company reports

APPENDIX B: BIOGEN-IDEC SHARE PRICE V. NASDAQ INDEX V. BIOTECHNOLOGY INDUSTRY INDEX (BTK)

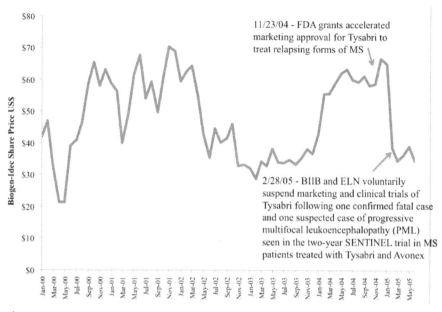

Source: www.yahoo.com

REFERENCES

Biogen Idec. 2005 . Press Release: Biogen Idec and Elan Announce Voluntary Suspension of Tysabri®. http://investor.biogenidec.com/phoenix.zhtml?c =148682&p=irol-newsArticle&ID=1324695&highlight, accessed November 10, 2009.

Biogen Idec. 2009. Annual Report on Form 10-K For the Year Ended December 31, 2008. Washington, DC: United States Securities and Exchange Commission.

Biogen-Idec. 2009. Company History. http://www.biogenidec.com/site/history.html, accessed October 9, 2009.

Campagnolo, D., Vollmer, T., Shi, F., Scott, D., & Williamson, S. 2009. Multiple Sclerosis. *eMedicine*. http://emedicine.medscape.com/article/310965-overview, accessed November 10, 2009.

Campbell, N., & Reece, J. 2007. *Biology, Eighth edition*. San Francisco: Benjamin Cummings.

Compston, A., Lassmann, H., & McDonald, I. 2006. The story of multiple sclerosis. In *McAlpine's Multiple Sclerosis*: 3-68. Philadelphia: Elsevier.

Compston, A., Lassman, H., Smith, K. 2006. The neurobiology of multiple sclerosis. In *McAlpine's Multiple Sclerosis*: 449-490. Philadelphia: Elsevier.

Confavreux, C., Compston, A. 2006. The natural history of multiple sclerosis. In *McAlpine's Multiple Sclerosis*: 183-272. Philadelphia: Elsevier.

Elan. 2009. *Company: About Us*. http://www.elan.com/company, accessed October 9, 2009.

Gran, B., & Rostami, A. 2001. T-cells, cytokines, and autoantigens in multiple sclerosis. *Current Neurology and Neuroscience Reports*, 1:263–270.

Koralnik, I. 2004. New insights into progressive multifocal leukoencephalopathy. *Current Opinion Neurology*. 17: 365-370.

Krasner, J. 2005. Safety review boosts hopes for return of MS drug: No more cases of brain disease. *The Boston Globe*. August 10: Section A.

Maggos, C. 2005. Deconstructing Tysabri. *BioCentury, The Bernstein Report*. http://www.biocentury.com, accessed November 10, 2009.

McDonald, I.,& Compston, A. 2006. The Symptoms snd Signs of multiple sclerosis. In *McAlpine's Multiple Sclerosis*: 287-346. Philadelphia: Elsevier.

Miller, D., McDonald, I., & Smith, K. 2006. The diagnosis of multiple sclerosis. In *McAlpine's Multiple Sclerosis*: 347-388. Philadelphia: Elsevier.

National Institute of Neurological Disorders and Stroke (NINDS). 2009. Multiple Sclerosis: Hope Through Research. *National Institutes of Health*. http://www.ninds.nih.gov/disorders/multiple_sclerosis/detail_multiple_sclerosis.html, accessed October 5, 2009.

Noseworthy, J., Miller, D., & Compston, A. 2006. Disease modifying treatments in multiple sclerosis. In *McAlpine's Multiple Sclerosis*: 729-802. Philadelphia: Elsevier.

Noseworthy, J., Miller, D., & Compston, A. 2006. The treatment of symptoms in multiple sclerosis and the role of rehabilitation. In *McAlpine's Multiple Sclerosis*: 701-728. Philadelphia: Elsevier.

Noseworthy, J., Confavreux, C., & Compston, A. 2006. Treatment of the acute relapse. In *McAlpine's Multiple Sclerosis*: 683-699. Philadelphia: Elsevier.

Polman, C., O'Connor, P., Havrdova E, Hutchinson, M, Kappos, L., Miller, H, Phillips, J. T., Lublin, F., Giovannoni, G., Wajgt, A., Toal, M., Lynn, F., Panzara, M. A., & Sandrock, A. W. 2006. A randomized, placebo-controlled trial of natalizumab for relapsing multiple sclerosis. *New England Journal of Medicine*, 354 (9):899-910.

Pritchard, C., & Linnane, C. 2004. Biogen, Elan shares move on FDA OK. *Market Watch*. http://www.marketwatch.com/story/biogen-elan-shares-move-on-fda-ok-of-ms-drug?siteid=google&dist=google%2C, accessed November 10, 2009.

Ransohoff, R. 2007. Natalizumab for Multiple Sclerosis. *New England Journal of Medicine*, 356(25): 2622-2629.

Rudick R., Stuart W., Calabresi P.,Christian Confavreux, C., Galetta, S., Radue, E., Lublin, F., Weinstock-Guttman, B., Wynn, D. R., Lynn, F., Panzara, M. A., & Sandrock, A. W. 2006. Natalizumab plus interferon beta-1a for relapsing multiple sclerosis. *New England Journal of Medicine*, 354 (9):911-923.

Tornatore, C., & Clifford, D. 2009. Clinical vigilance for progressive multifocal leukoencephalopathy in the context of natalizumab use. *Multiple Sclerosis*, 15(S4):S16-S25.

Tortora, G., Funke, B., & Case, C. 2003. *Microbiology: An Introduction, Eighth edition*. San Francisco: Benjamin Cummings.

Vukusic, S., & Confavreux, C. 2003. Prognostic factors for progression of disability in the secondary progressive phase of multiple sclerosis. *Journal of Neurological Science*, 206: 135–7.

Wekerle, H., & Lassman, H. 2006. The immunology of inflammatory demyelinating disease. In *McAlpine's Multiple Sclerosis*: 491-556. Philadelphia: Elsevier.

Yousry, T., Major, E., Ryschkewitsch, C., Fahle, G., Fischer, S., Hou, J., Curfman, B., Miszkiel, K., Mueller-Lenke, N., Sanchez, E., Barkhof, F., Radue, E., Jäger, H., & Clifford, D. 2006. Evaluation of Patients Treated with Natalizumab for Progressive Multifocal Leukoencephalopathy. *New England Journal of Medicine*, 354:924-933.

XDx: Navigating Regulatory & Reimbursement Challenges

LAURA ELIAS AND MICHAEL ALVAREZ
Stanford University

- **Summary and key issue/decision:** This case focuses on XDx, a molecular diagnostics company, that had introduced their first product, AlloMap˚, to the market. When the United States Food and Drug Administration (FDA) and the Centers for Medicare and Medicaid Services (CMS) announced policy changes that had the potential to adversely affect AlloMap's adoption in the marketplace, the XDx management team was forced to decide whether to proceed with a passive strategy of waiting for the uncertainties to unfold before taking action or to undertake a very costly business model reformulation. XDx chose to rebuild their business model with the hope of maximizing market adoption of their novel diagnostic and preserving company value but in the process incurred significant costs.

- **Companies/institutions**: XDx, Centers for Medicare and Medicaid Services (CMS), US Food and Drug Administration (FDA), Division of Laboratory Services

- **Technology**: AlloMap, a molecular diagnostic test that aids in the identification of heart transplant patients with a low probability acute cellular rejection

- **Stage of development at time of issue/decision:** AlloMap was on the market, and XDx's laboratory was certified under the Clinical Laboratory Improvement Amendment Act (CLIA). The California Medicare contractor had granted reimbursement for AlloMap. XDx was anticipating an Initial Public Offering (IPO).

- **Indication/therapeutic area:** Heart transplantation monitoring

- **Geography:** US

- **Keywords:** XDx, Initial Public Offering (IPO), Laboratory Developed Test (LDT), Clinical Laboratory Improvement Amendment (CLIA), *In Vitro* Diagnostic Multivariate Index Assay (IVDMIA), 510(k), molecular diagnostics, genomics, heart transplant

Acknowledgements: We would like to acknowledge Pierre Cassigneul and Tod Klingler for participating in interviews and follow-up clarifications. In addition, we would like to thank Hannah Valantine for participating in an interview. Finally, we would like to thank Guillermo Elias and Tiara Kawahara for discussions about the case and feedback on drafts.

* *This case was prepared as a basis for class discussion rather than to illustrate either effective or ineffective handling of an administrative situation..*

BUILDING A BUSINESS AT THE MOLECULAR DIAGNOSTIC FRONTIER

"Today the world is joining us here in the East Room to celebrate the completion of the first survey of the entire human genome. Without a doubt, this is the most important, most wondrous map ever produced by humankind."
–U.S. President Bill Clinton (2000)

The world was watching as genomics was poised to revolutionize the prevention, diagnosis, and treatment of disease through the development of personalized medicine. In 2001, back-to-back papers in premier scientific journals reported the completion of a draft sequence of the human genome (International Human Genome Sequencing Consortium *et al.*, 2001; Venter *et al.*, 2001). In 2003, the scientific community marked the 50-year anniversary of the discovery of the double-helical structure of DNA (Watson & Crick, 1953) with a commemoration of the remarkable advances that had been made in genomics, including a complete sequence of the human genome (Jasney & Roberts, 2003; Collins, Green, Guttmacher & Guyer 2003). Silicon Valley was awash with excitement and expectations; the leaders at XDx, a molecular diagnostics company, were no exception (see Appendix A for molecular diagnostics market projections).

XDx was one of the pioneers of the personalized medicine era, playing a leading role in the development and commercialization of a new class of medical tests called molecular diagnostics. The idea behind molecular diagnostics is to elucidate or predict specific states of health or disease through measurements of DNA, RNA, or protein. Unlike traditional diagnostics, which rely on convenience, efficiency, and low cost to maintain market position (e.g., testing for glucose levels in diabetes), companies pursuing molecular diagnostic tests are often attempting to apply the latest technologies to assess a complicated biological state that informs a critical clinical decision (Klingler, 2010). It is ultimately the utility of the information provided that determines a molecular diagnostic's potential adoption in the marketplace and value to the customer, although many other practical matters influence whether a product can actually reach its potential.

XDx focused on developing a molecular diagnostic to monitor organ rejection in heart transplant patients (XDx, 2007). The company developed a first-in-class test that was based on solid scientific and clinical observations and had an apparent clinical utility; the test had the potential to become a non-invasive alternative to the traditional surgical test for transplant rejection. XDx's new test, AlloMap, was leading the company to success, and XDx anticipated an Initial Public Offering (IPO) in 2008 (XDx, 2007).

However, as XDx was busy commercializing its new diagnostic, the regulatory and reimbursement agencies were still in the process of formulating their policies around the use of molecular diagnostic tests. XDx encountered regulatory and reimbursement policy changes that had the potential to threaten the viability of the company's business model. Pierre Cassigneul, president and CEO, and the XDx team were faced with the decision of how to manage their business in this unpredictable environment. While XDx considered a variety of complex strategic options, this continuum can be represented by the binary choice between a passive approach of waiting for the uncertainties to resolve before taking action and a very costly proactive business model reformulation. Successful companies have taken both approaches, and when facing this decision, each company needs to carefully weigh the costs, risks, and benefits to determine how best to preserve their company's value in the midst of uncertainties.

XDx COMPANY BACKGROUND

The founding CEO and president of XDx was Peter Altman, Ph.D., who was the CEO of BioCardia. At the time, BioCardia was focused on developing products aiding in the delivery of therapeutic agents to the heart. In 2000, the excitement about the future of genomics inspired Dr. Altman together with

Tom Quertermous and Jay Wohlgemuth from Stanford to consider how genomics might be applied to cardiology (Klingler, 2010). In 2002, Expression Diagnostics was spun out of BioCarda with the initial goal of applying genomic technologies to monitor patients with immune-mediated conditions, including heart transplant patients. In July 2007, Expression Diagnostics changed its name to XDx.

In 2003, Pierre Cassigneul was brought in as the president and chief executive officer because of his extensive experience in developing and launching diagnostic products (Cassigneul, 2010). Prior to XDx, Cassigneul was an operating partner at Stone Bridge Management, a consulting firm. He also had industry experience, serving as vice president of diabetes management at Becton Dickinson & Co., a medical technology company.

From the outset, Cassigneul was driven by the company's mission to, "Improve patient care by developing molecular diagnostics that translate an individual's immune status into clinically actionable information." For XDx, this meant creating value around developing, verifying, and commercializing non-invasive molecular diagnostic tests to monitor the immune system and thus provide clinically actionable information to optimize medical treatment and/or replace more invasive, costly procedures (XDx, 2007). Rather than focusing on autoimmune diseases (arising from an overactive immune response to one's own cells or tissues) that represented a larger market opportunity, XDx first focused on the alloimmunity response (immune response to foreign cells or tissues) in organ transplant patients (see Figure 1 for product opportunities).

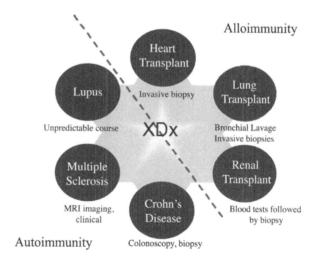

Figure 1: XDx Product Opportunities in Immune Mediated Inflammatory Diseases
Source: Company reports

Transplant based immune responses had a known cause and time of onset thus providing a clearer path to a successful development and validation program. While the transplant market is relatively small, the potential for repeat monitoring during the remainder of a patient's life made the business model viable (Klingler, 2010). In 2007, an estimated 18,800 heart transplant recipients were living in the US; each year approximately 2,100 new heart transplants are performed (XDx, 2007; Organ Procurement and Transplantation Network, 2010). An estimated 40,000 biopsies are performed each year in the United States (Klingler, 2010; see Appendix B for market projections).

HEART TRANSPLANT MANAGEMENT

The first heart transplant was performed in 1967 in Cape Town, South Africa by Dr. Christiaan Barnard; the patient survived 18 days. The following year, 100 heart transplants were performed. However, immunological rejection of the foreign tissue led to low survival rates and physicians became reluctant to perform the surgery. In 1970 the number of transplants dropped to only 18. Since these early years, immunosuppressant therapies have advanced tremendously and today rejection of organ transplants can be managed for long-term patient survival (Columbia University Medical Center, 2010).

The prevailing immunosuppressive regimen for post-transplant patients includes antimetabolites (e.g., Azathioprine), antiproliferatives (e.g., Cyclosporine), and steroids (e.g., Prednisone), which together inhibit the immune system from making cells that modulate rejection. The dosage of immunosuppressive medications is modulated based on the rejection state of the patient as measured by endomyocardial biopsy (EMB) along with signs and symptoms and cardiac functional tests (Columbia University Medical Center, 2010). Long-term side effects of immunosuppressant therapy include development of cancer and infections. The EMB procedure is an outpatient surgery performed under local anesthesia in which an intravenous line is placed in the patient's neck and the bioptome is directed to the heart where small pieces of tissue are collected for histological analysis in the lab (Columbia University Medical Center, 2010; XDx, 2007; see Figure 2 for EMB procedure).

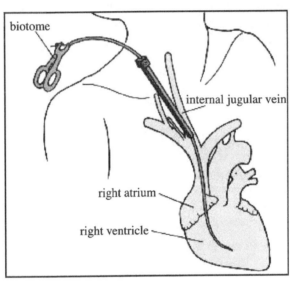

Figure 2: The endomyocardial biopsy (EMB) procedure
Source: Focosi, 2010

Side effects of EMBs include damage of the tricuspid valve, cardiac perforation, infection, and bleeding. Typically, EMBs are performed once a month for the first year, three to four times each year from year two to five, and annually after year five (XDx, 2007; Valantine, 2010). Approximately 50% of heart transplant patients survive 10 years (International Society for Heart and Lung Transplantation, 2009). Thus a patient that survives 10 years would have more than 20 surgical EMBs during the course of their post-transplant care. XDx's vision was to develop a product that would require only a blood sample to test for rejection status, thereby significantly improving post-transplant care for patients by reducing the need for biopsies.

ALLOMAP'S TECHNOLOGY

AlloMap, a non-invasive diagnostic test to monitor acute cellular rejection in heart transplant patients, was developed and validated during the course of a study initiated in 2001, the Cardiac Allograft Rejection Gene expression Observational (CARGO). CARGO was a large clinical trial that included nine transplant centers, 700 subjects, and 6,000 samples over a four year period (XDx, 2007).

To develop the test, XDx first created a custom microarray to assay the gene expression levels of

8,000 candidate genes known to be expressed in leukocytes, white blood cells that are part of the immune system. XDx's goal was to discover genes whose expression level in the patient's peripheral blood mononuclear cells (PBMCs) was correlated with the presence of rejection as measured by the patient's corresponding biopsy grade. From the 8,000 genes tested, XDx identified 262 candidate markers of which 68 were confirmed using quantitative real-time polymerase chain reaction, a sensitive and reproducible technique to measure RNA expression levels.

From these 68 marker genes, AlloMap was derived. AlloMap measures 11 rejection marker genes along with 9 genes used for normalization and quality control. These measurements are combined using an algorithm that produces a score between 0 and 40 (Klingler, 2010). For patients more than 6 months post-transplant with a score less than 34 there is a 98.9 percent chance that the heart transplant patient is not experiencing acute cellular rejection and a biopsy is not recommended. If the test is above 34, follow-up is recommended to confirm the state of rejection. In 2005, the Invasive Monitoring Attenuation through Gene Expression (IMAGE) study was initiated to compare the effectiveness of managing transplant patients with AlloMap versus standard EMBs with respect to patient outcomes including death and graft dysfunction (XDx, 2007); the results were due to be announced in 2010.

ALLOMAP AND A PROMISING PIPELINE

AlloMap was XDx's first commercialized product and entered the market in 2005 (XDx, 2005; XDx, 2007). In addition, at the time of filing for IPO, XDx was developing a test to monitor rejection in lung transplant patients (AlloMap LTx) and was in the process of designing a study to develop a diagnostic test to monitor disease flares in systemic lupus erythematosus patients (XDx, 2007). However, the success of the IPO was resting largely on the company's one commercialized test, AlloMap.

Beginning in 2005, XDx began to generate revenues based on sales of the AlloMap test, which grew in 2006 and 2007, although and it was clear that the company would not be profitable for some time (XDx, 2007; see Appendix C for company financials). As of September 30th 2007, 6,700 AlloMap tests had been processed, the list price for each test was $2,950, and the test was being used at 50 of the 190 US heart transplant management centers (XDx 2007).

FINANCING GROWTH

The first two financing rounds (A and B) at XDx were lead by angel investors, some of whom were family and friends of the founders (Klingler, 2010). In 2003, Series C financing was led by Kleiner Perkins Caufield & Byers and Texas Pacific Group, both prestigious venture capital firms, giving XDx a new level of financial stability and credibility in the world of Silicon Valley biotechnology startups.

In December 2004, XDx completed a $20 million Series D financing round led by the Sprout Group (New Leaf Venture Partners) (XDx, 2004). Additional new investors included JP Morgan's Bay Area Equity Fund, Integral Capital Partners, and Burrill & Company. In January 2006, XDx acquired an additional $26.5 million in Series E financing (XDx, 2006a). The round was led by Duff, Ackerman & Goodrich and had participation from all the previous investors in addition to Intel Capital. In May 2007, XDx completed a $26.5 million dollar series F financing round led by Intel Capital (Cassigneul, 2010).

Based on the successful development and early sales of AlloMap, XDx was ready for an IPO and filed an S-1 with the US Securities and Exchange Commission on October 24th 2007 to list the company on the public stock market (XDx, 2007). As seen by the recent success of companies like Genomic Health and Myriad Genetics, the market was clearly prepared to begin backing molecular diagnostic companies. As of September 30th 2007, XDx had accumulated a deficit of $78.4 million and expected to incur net losses for at least the next couple of years (XDx, 2007). Through the IPO, XDx hoped to raise up to $86.25 million in new capital to support the expansion of AlloMap testing and the development and commercialization of new diagnostics in transplantation, such as for lung transplant, and in autoimmunity, such as for lupus (XDx, 2007).

REGULATORY AND REIMBURSEMENT CHALLENGES SURFACE

XDx was facing a number of critical risks in late 2007 when the company filed their S-1 (XDx, 2007). Interestingly, the main risks did not concern the science or the fundamentals of the business, which were very strong, but rather the external policies on regulation and reimbursement. XDx was not the only pioneering molecular diagnostic company impacted by the evolving regulatory and reimbursement environment. A research group at McKinsey & Company analyzing the microeconomics of personalized medicine determined that, of a number of risks facing diagnostic companies, time to regulatory approval and payer reimbursement have the largest effect on company profitability (Davis *et al.*, 2009; see Figure 3 for sensitivity analysis).

The Davis *et al.* model predicted a 10-year net present value of $15 million for an average diagnostic test; a one-year delay in regulatory approval reduced the value to $10 million, and delayed reimbursement further diminished the diagnostic's value (2009). Interestingly, new drugs are typically reimbursed within the first year on the market, while diagnostics can take six years (Davis *et al.*, 2009).

The development of clear and efficient regulatory and reimbursement polices would allow molecular diagnostic companies to maximize their potential value, but XDx launched AlloMap at a time when such standards were still far off and delays were inevitable. XDx's challenge was to preserve AlloMap's and the company's value in the face of regulatory and reimbursement uncertainties. XDx realized that dealing with these challenges appropriately would critically impact AlloMap's commercial success and the company's future financing options (see Figure 4 for timeline of events).

REGULATION OF LABORATORY DEVELOPED TESTS

AlloMap was developed and launched in the midst of a controversial regulatory environment that was still catching up to the scientific advances of the genomics era. At the time, the laboratories that processed Laboratory Developed Tests (LDTs) were subject to regulation under the Clinical Laboratory Improvement Amendment (CLIA) Act. The objective of CLIA, operated by the Division of Laboratory Services at the Centers for Medicare and Medicaid Services (CMS), was to ensure the quality of laboratory testing, rather than to regulate the interpretation or validation of the test (CMS, 2010), which was left to clinical experts who would utilize these tests. Interestingly, the FDA did regulate the interpretation and validation of test kits (to be used by the provider, consumer, or independent laboratory) while

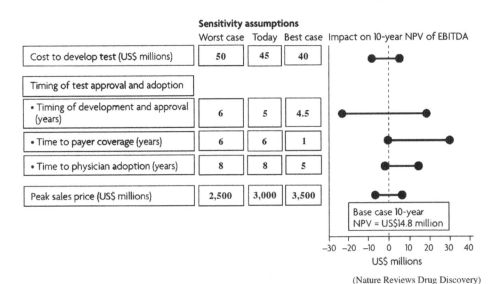

(Nature Reviews Drug Discovery)

Figure 3: Sensitivity analysis of factors affecting the Net Present Value (NPV) of a molecular diagnostic
Source: Davis et al., 2009

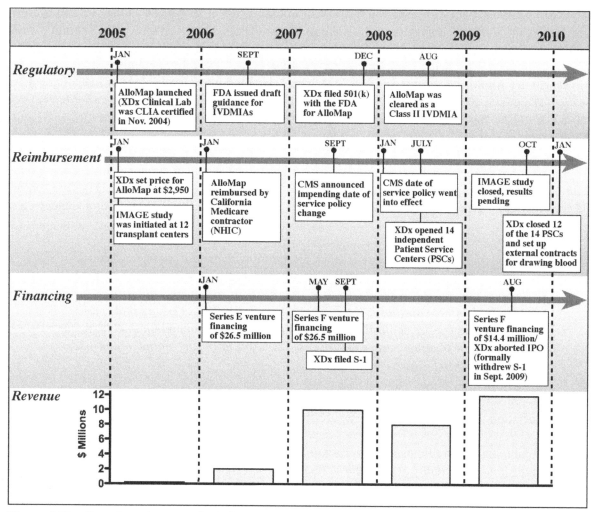

Figure 4: Timeline of regulatory and reimbursement events and their impact on company financing and AlloMap revenue. *Reported revenue recognized under the generally accepted accounting principles (cash and contractual agreements with a payment history)*

there was no official regulation in place for LDT services (performed in a company laboratory). Many felt that these inconsistencies needed to be reconciled.

Beginning in the 1990s a debate was ensuing over whether LDT services, like testing kits, should fall under FDA regulatory authority (Eaton, 2004). In 1994, the Institute of Medicine published a book outlining the risks associated with the initiation of the human genome project and inevitable advancement of genetic testing. One of the risks identified was the lack of quality control and FDA regulation of LDT services (Andrews, Fullarton, Holtzman, & Motulsky, 1994: 116-145). In 1997, members of the Task Force on Genetic Testing including Niel Holtzman, the chair, once again called attention to the undeveloped regulatory environment in an article published in *Science*.

> *"Under CLIA, there are no requirements for demonstrating clinical validity... the acquisition of sufficient data to warrant the transition of predictive genetic testing into health care cannot be ensured." (Holtzman, Murphy, Watson, & Barr, 1997: 602)*

Dr. Holtzman and the taskforce went on to recommend additional regulatory requirements for ensuring the validity of newly developed genetic tests with the objective of easing the providers' burden to protect the patient. However, when AlloMap was launched in 2005 (XDx, 2005; XDx, 2007), no amendments to the regulatory procedures were in place. XDx had received CLIA certification for its South San Francisco, California laboratory (later moved to Brisbane, California), enabling the company to commercialize their diagnostic test (XDx, 2004; XDx, 2007). In September of 2006, almost two years after AlloMap was launched, the FDA issued draft guidance for a subset of LDTs known as *In Vitro* Diagnostic Multivariate Index Assays (IVDMIAs), which included tests like AlloMap (FDA, 2007; XDx, 2007). Based mostly on risk to the patient, tests would be classified as Class I, II, or III and thereby require registration with the FDA, 510(k) clearance, or Pre Market Approval (PMA) respectively.

The issuance of the guidelines was met with both resistance as well as support. Some leading biotechnology companies such as Genentech supported the FDA's move to regulate LDTs. The senior vice president and general council at Genentech filed a petition with the FDA outlining Genentech's stance.

> "Given the potential risks to patient safety associated with the use of diagnostic tests that make unsubstantiated claims intended to guide specific drug or biologic therapeutic decision making, it is imperative that FDA exercise its regulatory authority consistently over all in vitro diagnostic test products, including LDTs, to ensure that claims made for the tests are scientifically proven. Failure to do so could present safety risks for patients as more treatment decisions are based in whole or in part on the claims made by such test makers." (Jhonston, 2008:3)

Others quickly responded to Genentech's statement expressing their disagreement. The president of the Association for Molecular Pathology published the following statement.

> "Adoption of the recommendations contained in Genentech's petition would significantly hinder the ability of laboratories to provide access to many tests critical to patient care. Moreover, this would curtail the innovative impact that laboratory developed tests have on the advancement of personalized medicine, a concept that the FDA and numerous other federal agencies strongly support." (Nowak, 2009: 4)

In addition, other biotechnology groups became engaged in the discussion by forming alliances such as The Coalition for 21st Century Medicine. In sum, the issuance of the draft guidelines by the FDA to regulate IVDMIAs stimulated a conversation between the regulators and the innovators, centering on balancing the benefits of patient protection with the importance of fostering innovation. In the beginning of 2010, the FDA had not yet finalized the IVDMIA regulation; indeed, the process of establishing final regulatory guidelines for a novel area of biomedicine is long and complex and fraught with uncertainties.

WEATHERING THE STORM OF REGULATORY UNCERTAINTIES

When XDx filed its S-1 in October 2007, the company was facing a regulatory risk. In the worst-case scenario, the FDA could have issued final guidelines for IVDMIAs forcing XDx to stop testing in their current CLIA certified laboratory pending regulatory clearance of AlloMap. While the company's management did not see this as a likely or reasonable possibility, there were no guarantees. XDx's S-1 filing explained these risks to potential investors.

"We believe that recently issued draft [FDA] guidance will, once finalized, require us to seek FDA clearance or approval for our AlloMap test and any future tests in order to market such tests in the United States, which we may be unable to obtain in a timely and cost-effective manner." (XDx, 2007: 4)

What should a company do upon learning of the FDA's intention to institute new regulatory standards? In this case, a variety of molecular diagnostic companies took very different approaches and no one had the final authority on the best course of action. Some companies waited for the uncertainties to resolve and in the meantime joined in the movement to resist the FDA's proposed guidelines claiming that CLIA provided sufficient regulation and the involvement of the FDA would stifle innovation; other companies proactively cooperated with the FDA hoping that communication would lead to reasonable policies that would both ensure safety and promote innovation.

The XDx executive team determined that the best strategy for the company was to work side by side with the FDA. This collaborative attitude grew out of Pierre Cassigneul's previous experiences in companies that had resisted the FDA and gotten into trouble precisely because of their lack of cooperation. In August 2006, just prior to the issuance of the draft guidelines, the management at XDx visited the FDA headquarters to share information about their science and business, so that the FDA could make better choices on how to regulate this novel class of diagnostic tests. Tod Klingler, Ph.D., the vice president of information sciences, recounted XDx's approach to FDA regulation, stating that, "If something happens, you want to be a part of it; [You want to] have influence to make sure whatever comes out is as mutually beneficial as possible" (Klingler, 2010). However, XDx's strategy of proactive engagement with respect to regulation wasn't without controversy within the company or with its investors.

Complying with the FDA had its own risks as AlloMap might not have been cleared, it might have been classified as a class III and not a class II test (which would have more than doubled the money and time required for approval), or the use of the test might have been limited by the issuance of a narrow label. Furthermore, the policy was not as of yet written in a clear and compelling manner and "could be improved to facilitate innovation and help protect the public" (Cassigneul, 2010). In addition, there was no guarantee that the regulations would be finalized. In spite of these potential risks and disagreements about the policy language, XDx submitted its 510(k) application in December 2007, shortly after the S-1 filing. During the subsequent period of back and forth, XDx worked very closely with the FDA to ensure that their test would meet the FDA's requirements. Nine months from the submission of the application, at a cost of approximately three years of one full-time employee's work, the AlloMap test became the third test to receive clearance as a Class II IVDMIA (XDx, 2008; see Appendix D for FDA label).

The process of receiving clearance from the FDA was certainly an additional commitment of time and money that many of XDx's peers chose to hold off on, as it was not yet required. However, when XDx filed their S-1, Pierre Cassigneul recounted that, "[The team] felt confident that we would be able to meet the FDA requirements" (Cassigneul, 2010). In his mind, there was very little risk of not receiving clearance because they were openly working with the FDA and had the regulatory experience to understand what the FDA would require. In fact, Cassigneul surmised that receiving clearance after going public would serve as a nice indicator to the investors that the company was heading in the right direction (Cassigneul, 2010). When the clearance was announced, Pierre Cassigneul stated in a press release "The FDA clearance further demonstrates the clinical relevance and benefit of AlloMap testing in assessing the potential risk of rejection in heart transplant patients" (XDx, 2008). The management at XDx had led the charge in gaining FDA clearance and hoped their hard work would eventually encourage provider adoption, payer reimbursement, and help them work with the FDA to continue to build the future of molecular diagnostics.

REIMBURSEMENT OF A NOVEL DIAGNOSTIC

Decisions that affect pricing and reimbursement of a new product are critical for the economic success of a biotechnology company. At the same time, the importance of cost savings in the medical system is rapidly becoming one of the most prominent factors in health care decision making. Thus, as biotechnology companies introduce novel products to the market, they not only have to consider provider and patient adoption but also the insurance companies that ultimately inform their profit margins.

Companies introducing new molecular diagnostic tests to the market have a number of challenges including coming up with an appropriate pricing strategy (see Figure 3). Medicare and other insurance companies would of course prefer a cost-based pricing scheme in which they get to determine a margin based on the cost to run the test which is likely to be less than the procedure it is replacing and thus could represent savings for the insurance company. The pioneering biotechnology company, on the other hand, wants to maximize the value of their product to recoup the tremendous sunk research and development costs. XDx sought to determine the commercial value of AlloMap based on the current standard of care, EMB, which has been estimated to cost approximately $4,000 (Evans *et al.*, 2005; XDx, 2007). From CARGO data, XDx estimated that in approximately 75% of tests the score would be below the threshold and no further biopsy would be required. In the other 25% of cases, an additional biopsy to check for rejection could be necessary. Taking these statistics into account, the break-even point for the test came to $3,000[1], and XDx priced their test at $2,950, leading to approximately $50 of savings per test or a total savings of more than $1,000 over an average patient's lifetime (Klingler, 2010). Independent analysis confirmed that AlloMap had the potential to be cost cutting for payers (Evans *et al.*, 2005; Davis *et al.*, 2009; see Appendix E for cost analysis).

Next came the challenge of defining a reimbursement strategy. Reimbursement is complicated by the multiple organizations, businesses, and decision makers involved in the process. Medicare can make national determinations or leave the decisions up to regional contractors. Adding to the complexity, both large and small private insurance companies around the country make independent determinations on reimbursement. Furthermore, the billing responsibilities can be assumed by the company itself or handed over to the providers.

From the outset, XDx adopted the strategy that until favorable reimbursement policies were widespread, they would bear the financial reimbursement risk instead of passing it along to the providers (Klingler, 2010). Prior to any major coverage decisions, claims could be submitted for reimbursement on a case-by-case basis. XDx wanted to ensure that providers were not discouraged from adopting the test due to the auxiliary paperwork and resources involved in submitting the claims and appealing when reimbursement was declined. On the other hand, this was a labor and time intensive undertaking that was certainly costly for XDx. Ultimately, XDx needed to gain favorable reimbursement decisions form insurers around the country, and stated in their S-1 filing that, "Coverage and reimbursement of AlloMap testing by third-party payers is essential to our commercial success" (XDx, 2007: 36).

In keeping with XDx's commitment to engaging the regulatory bodies, XDx sent a team to Washington, DC to meet with Medicare and discuss reimbursement for AlloMap. Upon arriving at the meeting, XDx was greeted by about 30 people including many of Medicare's medical directors, who expressed a sincere interest in XDx's test and the new diagnostic technologies it represented (Klingler, 2010). After an engaging discussion, the national body of Medicare recommended that XDx work with the regional California Medicare contractor, National Heritage Insurance Company (NHIC), because the service was provided solely in California where the tests were performed (Klingler, 2010). As of January 1st, 2006, AlloMap was approved for reimbursement by NHIC (XDx, 2006b; XDx, 2007). This was a big step forward for XDx and represented an independent confirmation of the clinical validity and cost effectiveness of the test.

Gaining reimbursement approval from private insurers proved to be a more difficult task. AlloMap

---

1   *The calculation of the AlloMap break-even point is as follows: Price of EMB = $4,000; Probability of follow-up biopsy = 0.25; x = break-even point; $4,000 = x + 0.25*$4,000; x = $3,000.*

coverage decisions were made by select private insurers including Kaiser and some smaller insurance companies (Klingler, 2010). Larger insurers, such as Blue Cross Blue Shield, required outcome studies (Klingler, 2010), which XDx had initiated but not yet completed (the IMAGE study). Such an outcomes study would be necessary to prove the non-inferiority of monitoring patients with AlloMap versus traditional EMB. Thus, XDx continued to bill these larger private insurers on a case-by-case basis.

Overall, the use of AlloMap was rising, and reimbursement was proceeding as fast as could be expected for a novel diagnostic test. Everyone at the company was excited at the prospect of an IPO, and it seemed that the pieces were in place to make the offering successful. This all changed when the CMS announced a seemingly small policy change that would, in fact, drastically alter XDx's current reimbursement strategy and put the value of the company in jeopardy.

UNPREDICTABLE REIMBURSEMENT HEADWINDS HALT THE IPO

In late 2007, CMS announced a change in reimbursement policy that would go into effect on January 1st 2008. This change was with regards to the date of service of a test (XDx, 2007). Previously, the date of service of a test like AlloMap was considered to be the day that the test was run in the laboratory, which allowed XDx to bill the California Medicare contractor directly. However, for bundling purposes, the new policy stated that if the blood was drawn in a hospital while the patient was an outpatient, the test could no longer be billed separately, but rather had to be bundled with the other services provided on that date (XDx, 2007). Medicare instituted this policy so that they could understand the total cost of outpatient services and eventually institute a fixed bundled rate for outpatient visits similar to their bundling policy for inpatient services (Cassigneul, 2010). The risk that this policy change presented to XDx was noted in the company's S-1 filing.

> *"Effective January 1, 2008, we will no longer bill directly for a test performed on a Medicare patient who has been registered as an outpatient, which could adversely affect the adoption of our test, our success in obtaining adequate reimbursement and our results of operations." (XDx, 2007: 4)*

At the time of the S-1 filing, XDx was clearly aware of the proposed change in policy, but they were not yet certain of the implications. XDx was taking responsibility for the billing of all tests at the time of the S-1 filing (Klingler, 2010). Under the new CMS policy, if the hospital continued to draw the blood used for AlloMap testing, the hospital would have to bundle the test with the patient's other medical services and take responsibility for all of the billing. Furthermore, the fact that XDx had only dealt with the California Medicare contractor as advised by Medicare, and not proceeded with a national coverage decision, meant that hospitals in states other than California would not be guaranteed reimbursement by the local Medicare contractor.

In the final months of 2007, XDx was facing a critical decision. Cassigneul and the XDx team had two choices: the first option was to hold off on making any changes to the business model until it was clear whether hospitals would take on the responsibility for AlloMap billing once reimbursement became more widespread; the second option was to proactively adapt their business model and set up mechanisms to draw blood outside of hospitals and continue to take responsibility for the billing. Potential strategies for instituting blood draw services outside of hospitals included setting up XDx operated centers that would collect the samples and initiating contracts with partners that would provide this service. Cassigneul had to carefully consider the costs and risks associated with this critical decision and do his best to select the path that would most likely enable XDx to retain its value.

XDx spent the last six weeks of 2007 scrambling to set up their own Patient Service Centers (PSCs), which could collect blood samples outside of hospitals, so that their testing would not be interrupted by the change in Medicare policy rapidly approaching on January 1st. The XDx team was racing under pressure to preserve the value of their test and ultimately their company. While XDx mobilized their managers and sales force as well as outside consultancy to speed the process, it was a daunting challenge, and XDx was forced to prioritize its efforts, leaving some PSCs for later (Klingler, 2010). Furthermore, XDx debated whether inconveniencing the patient to travel to a new PSC would negatively impact their sales or whether the patient and their clinicians would willingly take the extra steps to preserve their new heart and avoid invasive biopsies. The answer to this debate became clear in January when AlloMap sales dropped (Cassigneul, 2010). Dr. Klingler recounted an experience visiting a center in the Philadelphia area. The PSC was just across the street from the hospital, but it was an extremely busy street full of potholes. One of the patients was in a wheelchair and during the trip across the street had a negative experience, prompting her to request not to continue with her AlloMap testing (Klingler, 2010). Thus, coordinating and integrating the PSCs became a central challenge for XDx.

In the short term, XDx continued to focus on preserving its business by setting up *de-novo* PSCs that were conveniently located in close proximity to the major transplant hospitals using AlloMap. By July 2008, fourteen AlloMap PSCs were open. Pierre Cassigneul lamented on the unsustainable nature of this solution.

> "When you run a business you want to try to have as little fixed cost as possible… The patient service centers were an extremely expensive solution, but a necessary one to stay in business. That also gave us time to identify partners such as Consentra, Health Resources, E-Lab Test, and finally Lab Corp to draw the blood for us on a variable cost basis so we only pay per draw." (Cassigneul, 2010)

Setting up the PSCs gave XDx the time they needed to transition from a short-term fix to a long-term business model that would support the growth and success of their novel diagnostic. The expense of the private PSCs was significant and reduced XDx's margin. XDx began focusing on setting up contracts with a broader network of labs that could collect and process the samples before sending them to XDx for testing. As of January 2010, XDx had closed down 12 of its own PSCs and had contracts set up with large independent networks (Cassigneul, 2010; see Table 1 for AlloMap blood draw costs).

Table 1: Approximate costs to XDx per AlloMap blood draw at hospitals, patient service centers (PSCs) or through external contracts

	Hospitals	PSCs	Contracts
Service Fee	$50-80	--	$100
Supplies	$20	--	--
Fixed Costs*	--	$3,000	--
Total Cost	$70-100	$3,000	$100

Source: Cassigneul, 2010
* Fixed costs are spread over the number of draws performed and reported per draw.

Furthermore, the management at XDx once again took a proactive approach to making the most of the policy change. In this case, they turned to the old marketing adage, "If you can't fix it, feature it" (Cassigneul, 2010). XDx's new contracts allowed them to collect the patient's blood in the days prior to their appointment in closer proximity to the patient's home. This increased the convenience for the patient and gave the doctor the advantage of having the test score in hand when deciding whether the patient would need a biopsy, a change that might allow for increased frequency of testing and add significant value to the test from both the patient's and the doctor's perspective (see Appendix B for market projections).

While XDx's proactive and rapid response allowed them to minimize damage to the company's value, the timing of the reimbursement policy change resulted in XDx withdrawing their S-1 as it became clear that the company would need more time to rebuild its reimbursement strategy. Going public before addressing the logistical issues of setting up draw centers and reformulating their business model would have most likely resulted in missed financial targets and disappointed investors, a situation from which it would have been hard to recover (Klingler, 2010). In addition, the company was low on cash, and the financial crisis that would hit later in 2008 was brewing (Cassigneul, 2010). Pierre Cassigneul was forced to abort the IPO and restructure the company.

> *"There was no way we would get into an IPO where we were running out of cash, having a major business model problem, and in the midst of a financial crisis that made it look like the end of the world." (Cassigneul, 2010)*

It took XDx almost two years to rebuild its business strategy to meet the new reimbursement stipulations (Cassigneul, 2010). Medicare, focusing on their primary intent to inform future outpatient reimbursement bundling, presumably did not fully consider the profound impact of this new policy on small growing companies like XDx.

CONCLUSION: *QUO VADIS?*

Cassigneul knew that in sailing when you run into headwinds you can just sit there letting your sails flap in the wind going nowhere, or you can build momentum and advance by setting a new course. XDx's ultimate goal of becoming a successful company rested on the adoption of AlloMap by heart transplant centers as a means to manage their patients. To make this happen, XDx needed to communicate the clinical utility of the test to providers and eliminate the financial risk of reimbursement for the hospitals. When XDx was faced with regulatory and reimbursement challenges, their goal of provider adoption helped guide the company's strategic positioning.

Pierre Cassigneul's proactive attitude towards addressing issues with the FDA was certainly a strong influence on the company, but not everyone agreed with this approach, and the investors were concerned about the potential risks of approaching the FDA when it was not yet mandated. However as Cassigneul recounted, when considering a new technology, "Investors always ask who will win and who will lose because of it" (Cassigneul, 2010). In the case of XDx, the old technology, EMB, was the big loser, and in some cases the physicians, if they were private practitioners, would have a financial incentive to perform an EMB rather than use AlloMap. Cassigneul thought that it was necessary to provide compelling evidence to help physicians make the decision to use AlloMap and convinced the investors that gaining FDA clearance would bolster the company's and AlloMap's credibility overall.

> *"When that private practitioner makes the decision to use AlloMap he or she also makes the decision not to make money because he or she is not going to do an EMB. Because of that we felt that we had to be at the highest standard and have the most compelling evidence. Having an FDA clearance when it was not mandated was one way to create this compelling evidence. The second part was the IMAGE comparative effectiveness research where we are saying that the test has the same outcome as biopsy but on top of that it is cheaper, it is non-invasive, and it is objective. Basically, we are providing overwhelming evidence to those hard-to-convince physicians." (Cassigneul, 2010)*

XDx was committed to demonstrating overwhelming evidence of the test's effectiveness. In fact, cardiologists at Stanford including Dr. Hannah Valantine approached XDx with a proposal for what became the IMAGE study. Motivated by the clear clinical benefits of avoiding biopsies, the doctors sought to determine the safety and effectiveness of AlloMap in managing their patients with a state of the art study that compared AlloMap with the more traditional EMB (Valantine, 2010).

Dr. Valantine, the principal investigator and chair of the IMAGE steering committee, commented that, "This is a wonderful example of full collaboration between a biotech company and clinical investigators in coming up with a clinical trial… There is a huge unmet medical need in this area of monitoring patients for acute rejection. The current gold standard is the endomyocardial biopsy. It is invasive. Patients live in fear of having a heart biopsy" (Valantine, 2010). Dr. Valantine clearly felt that this type of comparative effectiveness study was absolutely essential for a big paradigm shift in transplant management such as AlloMap. At the same time she conceded that such studies are hugely expensive and XDx's willingness to fund the study was highly unusual (Valantine, 2010). According to Dr. Valantine, the whole transplant community is waiting to hear the results of the IMAGE study to see if they can safely transition to using AlloMap as the primary means of monitoring transplant patients (Valantine, 2010). If the doctors are convinced that AlloMap is as effective as EMB, the path should finally be clear for XDx's success in addressing this unmet medical need and building a sustainable company.

While the future potential of XDx looks bright, delaying the IPO was very disappointing and more difficulties followed as the company cut the lung transplant research program and shrank from a total of 129 to only 80 employees in November of 2008 (Cassigneul, 2010). Despite the difficulties that the company endured, by 2010 Pierre Cassigneul believed that the company was a much more rational company aiming to become profitable and self-sustaining much sooner than it otherwise would have (Cassigneul, 2010). To get there, XDx had to pioneer a novel genomic-based technology, keep pace with the developing regulatory environment, and initiate a state of the art comparative effectiveness study to gain the trust and respect of the cardiologist community while at the same time reformulating their business model to adapt to the changing reimbursement policies. In contemplating the difficulties that the company faced and strategic options ahead, Pierre Cassigneul highlighted the importance of fundamentals in science and medicine as well as business in navigating the many uncertainties inherent in building an innovative biotechnology company.

"You need to have a product that makes a difference from a healthcare standpoint to the patient, to the provider, or to the payer, but you also need to have a business that makes sense on its own." (Cassigneul, 2010)

APPENDICES

APPENDIX A: PROJECTIONS FOR THE MOLECULAR DIAGNOSTICS MARKET (US$ BILLIONS)

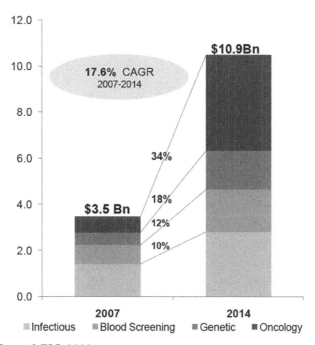

Source: *The Sharma Group & TSG, 2009*

APPENDIX B: ALLOMAP POTENTIAL FUTURE MARKET ANALYSIS*

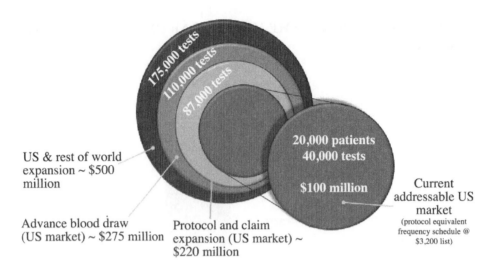

US & rest of world
expansion ~ $500
million

Advance blood draw
(US market) ~ $275 million

Protocol and claim
expansion (US market) ~
$220 million

20,000 patients
40,000 tests

$100 million

Current
addressable US
market
(protocol equivalent
frequency schedule @
$3,200 list)

Source: Company reports

* 40,000 tests - current use; 87,000 tests - assuming an expanded protocol to reduce immunosuppressant levels;
110, 000 tests - assuming increased monitoring if the tests can be performed close to the patients home;
175,000 tests - worldwide market

APPENDIX C: XDX FINANCIALS—INCOME STATEMENT

	Year Ended December 31,			Six Months Ended June 30,	
	2004	2005	2006	2006	2007
				(Unaudited)	
Revenues:					
Testing revenue	$ —	$ 179	$ 2,922	$ 229	$ 3,972
Contract revenue	—	148	20	—	—
Total revenues	—	327	2,942	229	3,972
Operating costs and expenses:					
Cost of testing	—	1,650	2,770	1,225	2,152
Research and development	9,526	8,848	12,261	6,081	7,646
Selling and marketing	1,692	3,737	5,067	2,408	3,752
General and administrative	1,799	2,349	3,049	1,493	2,273
Total operating costs and expenses	13,017	16,584	23,147	11,207	15,823
Loss from operations	(13,017)	(16,257)	(20,205)	(10,978)	(11,851)
Interest expense	(78)	(104)	(412)	(27)	(704)
Interest income	57	197	832	440	424
Other income (expense), net	(2)	(241)	(83)	(18)	297
Net loss before cumulative effect of change in accounting principle	(13,040)	(16,405)	(19,868)	(10,583)	(11,834)
Cumulative effect of change in accounting principle	—	(238)	—	—	—
Net loss	$ (13,040)	$ (16,643)	$ (19,868)	$ (10,583)	$ (11,834)

Source: XDx, 2007

XDX FINANCIALS—BALANCE SHEET

	December 31,		June 30, 2007	Pro Forma June 30, 2007
	2005	2006		
			(Unaudited)	
ASSETS				
Current assets:				
Cash and cash equivalents	$ 24,831	$ 7,939	$ 8,800	$ 8,800
Short-term investments	—	7,973	19,207	19,207
Accounts receivable	—	384	530	530
Inventory	443	303	342	342
Prepaid expenses	208	348	282	282
Other current assets	—	35	498	498
Total current assets	25,482	16,982	29,659	29,659
Property and equipment, net	1,082	5,616	5,730	5,730
Restricted cash	266	393	248	248
Deposits	82	86	43	43
Other non-current assets	—	61	39	39
Total assets	$ 26,912	$ 23,138	$ 35,719	$ 35,719
LIABILITIES, CONVERTIBLE PREFERRED STOCK AND STOCKHOLDERS' EQUITY (DEFICIT)				
Current liabilities:				
Accounts payable	$ 675	$ 1,990	$ 2,270	$ 2,270
Accrued liabilities	1,206	1,748	1,886	1,886
Current portion of capital lease obligations	161	95	72	72
Current portion of long-term debt	—	3,877	4,080	4,080
Liability for early exercise of stock options	321	170	184	184
Total current liabilities	2,363	7,880	8,492	8,492
Deferred rent, net of current portion	43	3,400	3,558	3,558
Capital lease obligations, net of current portion	122	74	47	47
Long-term debt, net of current portion	—	6,171	4,247	4,247
Convertible preferred stock warrant liability	2,023	2,638	2,324	—
Total liabilities	4,551	20,163	18,668	16,344

Source: XDx, 2007

APPENDIX D. FDA LABEL FOR ALLOMAP
Source: FDA, 2008:1

Indication(s) for use:

AlloMap Molecular Expression Testing is an *In Vitro* Diagnostic Multivariate Index assay (IVDMIA) test service, performed in a single laboratory, assessing the gene expression profile of RNA isolated from peripheral blood mononuclear cells (PBMC). AlloMap Testing is intended to aid in the identification of heart transplant recipients with stable allograft function who have a low probability of moderate/severe acute cellular rejection (ACR) at the time of testing in conjunction with standard clinical assessment.

APPENDIX E. ALLOMAP IS COST SAVING FOR PAYERS

a Companion diagnostics			b Procedure-focused diagnostics	c Genetic risk markers		
HER2	BCR–ABL	Warfarin	AlloMap	BRCA1	Familial BRCA1	KIF6 (statin)
Savings from changed decision (US$ thousands)						
40	80	2	4	25	25	-3.5
× probability that diagnostic changes treatment decision (percentage)						
70	5	35	75	2	20	50
= savings per test (US$ thousands)						
28	4	0.7	3.1	0.5	5	<0
Cost of test (US$ thousands)						
0.1	1	0.3	3	2-3	2-3	NA
Cost saving for the payers?						
✓	✓	✓	✓	✗	✓	✗

(Nature Reviews Drug Discovery)

Source: Davis et al., 2009

REFERENCES

Andrews, L. B., Fullarton, J.E., Holtzman, N. A., & Motulsky, A.G. (Eds.). 1994. *Assessing genetic risks: Implications for health and social policy.* Institute of Medicine. Washington, DC: National Academy Press, http://www.nap.edu/openbook.php?record_id=2057&page=2

Cassigneul, P. 2010. Personal Interview, Jan 19th. (Follow-up calls on Feb 9th and March 4th)

CMS. 2010. Clinical Laboratory Improvement Amendments (CLIA) > Overview. http://www.cms.hhs.gov/clia/, first accessed Jan 4th.

Clinton, B. 2000. Completion of a draft of the human genome. http://www.dnalc.org/view/15073-Completion-of-a-draft-of-the-human-genome-Bill-Clinton.html, June 26th.

Collins, F. C., Green, E. D., Guttmacher, A. E. & Guyer, M. S. 2003. A vision for the future of genomics research. *Nature,* 422: 835-847.

Columbia University Medical Center. 2010. Heart Transplant. http://www.columbiasurgery.org/pat/hearttx/, first accessed Jan 4th.

Davis, J. C., Furstenthal, L., Desai, A. A., Norris, T., Sutaria, S., Fleming, E., & Ma, P. 2009. The microeconomics of personalized medicine: today's challenge and tomorrow's promise. *Nature Reviews Drug Discovery,* 8: 279-286.

Eaton, M. 2004. Anticipating and Managing Postmarket Problems. In, *Ethics and the Business of Bioscience:* 329-376. Stanford, CA: Stanford University Press.

Evans, R. W., Williams, G. E., Baron, H. M. , Deng, M. C., Eisen, H. J. , Hunt, S. A., Khan, M. M., Kobashigawa, J. A. , Marton, E. N. , Mehra, M. R. & Mital, S. R. 2005. The Economic Implications of Noninvasive Molecular Testing for Cardiac Allograft Rejection. *American Journal of Transplantation,* 5(6): 1553-1558.

FDA. 2007. Draft Guidance for Industry, Clinical Laboratories, and FDA Staff - *In Vitro* Diagnostic Multivariate Index Assays. http://www.fda.gov/MedicalDevices/DeviceRegulationandGuidance/GuidanceDocuments/ucm079148.htm, July 26th.

FDA, 2008. 510(k) Substantial Equivalence Determination Decision Summary. K073482. XDx. http://www.accessdata.fda.gov/cdrh_docs/reviews/K073482.pdf, Aug 27th.

Focosi, D. 2010. Endomyocardial Biopsy. http://www6.ufrgs.br/favet/imunovet/molecular_immunology/endomyocardial_biopsy.jpg, accessed Feb 28th.

Holtzman, N. A, Murphy, P. D., Watson, M. S., & Barr, P. A. 1997. Predictive Genetic Testing: From Basic Research to Clinical Practice. *Science,* 278: 602-605.

International Human Genome Sequencing Consortium *et al.* 2001. Initial sequencing and analysis of the human genome. *Nature* , 409: 860-921.

International Society for Heart and Lung Transplantation. 2009. Registries - Heart/Lung Registries > Overall Heart and Adult Heart Transplantation Statistics http://www.ishlt.org/registries/slides.asp?slides=heartLungRegistry, Accessed Feb 3rd, 2010.

Johnston, S. A. 2008. Genentech, Inc. Citizen Petition: Regulation of *In Vitro* Diagnostic Tests. FDA-2008-P-0638-0001/CP. December 5th.

Jasney, B.R. & Roberts, L. 2003. Building on the DNA Revolution. *Science,* 300(5617): 277.

Klingler, T. 2010. Personal Interview, Jan 8th.

Nowak, J. 2009. Re: Comments to Docket No. FDA-2008-P-0638-0001/CP. American Association for Molecular Pathology. http://www.amp.org/Gov/ResponsetoGenentechPetition_042309.pdf, April 23rd

Organ Procurement and Transplantation Network. 2010. http://optn.transplant.hrsa.gov/latestData/rptData.asp, accessed January 3rd.

The Sharma Group & TSG Partners. 2009. Strategic Opportunities & Directions in Molecular Diagnostics: Sector Review & Analysis by TSG. https://chidb.com/newsarticles/MDx_StrategicReview.pdf

Valantine, H. 2010. Personal Interview, Jan 22nd.

Venter, C.J. *et al.* 2001. The Sequence of the Human Genome. *Science,* 16(291): 1304-1351.

Watson, J.D. & Crick, F.H.C. 1953. A Structure for Deoxyribose Nucleic Acid. *Nature,* 171: 737-738.

XDx. 2004. Press Release. XDx Receives CLIA Certification Enabling Launch of AlloMap Testing. http://www.xdx.com/news-press.html, Dec 1st.

XDx. 2005. Press Release. XDx Launches Noninvasive Test to Enable Better Management of Heart Transplant Patients. http://www.xdx.com/news-press.html, April 6th.

XDx. 2006a. Press Release. XDx Completes $26.5 Million Private Equity Financing. http://www.xdx.com/

news-press.html, Jan 5th.

XDx. 2006b. Gene Expression Test from XDx Now Indicated for Earlier Use and Prediction of Acute Cellular Rejection. http://www.xdx.com/news-press.html, Oct 31st.

XDx. 2007. Form S-1. United States Securities and Exchange Commission. http://www.secinfo.com/dVut2.u7W7.htm, Oct 23rd.

XDx. 2008. XDx's AlloMap® Gene Expression Test Cleared by U.S. Food and Drug Administration. http://www.xdx.com/news-press.html, Aug 27th.

Dyadic International: From Doom to Dawn— What's Next?

POLLY S. RIZOVA, ADELAIDA PATRASC LUNGU, AND MARK J. AHN
Atkinson Graduate School of Management, Willamette University

- **Summary and key issue/decision:** Dyadic International, Inc. is an early-stage biotechnology company based in Jupiter, Florida dedicated to research, development and manufacturing of products which enable the development of biofuel, biotherapeutics, and industrial enzymes from its patented C-1 fungus technology platform. The company was founded in 1979 by Mark Emalfarb, Dyadic's chairman of the board and CEO, and became publicly traded in 2004. Simultaneously pursuing an aggressive growth strategy and transitioning to a public entity, Dyadic faced a series of events which resulted in what appeared to be an insurmountable calamity for both the organization and its founder. A combination of financial misconduct of Dyadic's subsidiary in China, combined with internal legal maneuvering and organizational politics, culminated in the board of directors ousting Emalfarb as CEO in 2007. He sued, won, and was reinstated a year later through a shareholder board election. The case catches up with Emalfarb shortly after he was reinstated, as he faced a series of critical governance decisions for considering Dyadic's future: How to regain the confidence of the important stakeholders' groups? How to pull the organization from the brink of the abyss it was staring at and on the path to growth?

- **Companies/institutions:** Dyadic International Inc., Ernst & Young, Puridet

- **Technology:** C-1 fungus technology, a fungal system for gene discovery and expression and the production of proteins

- **Geography:** US, China

- **Keywords:** enzymes, bioenergy, biopharma, biotechnology

---

* *This case was prepared as a basis for class discussion rather than to illustrate either effective or ineffective handling of an administrative situation.*

QUO VADIS?[1]

A s Steve Emerson, a major Dyadic shareholder, met for coffee with founder and CEO Mark Emalfarb, he shook his head in wonderment. Somehow, Mark Emalfarb never doubted this day. Somehow, the thought that his and Dyadic's days were numbered never crossed his mind. His passion and belief that the time when biomass can replace most petroleum products was drawing nearer and his resolve that the company he started almost three decades ago fresh from college would have a lot to do with this gigantic transformation were unwavering. However, a glance at the gauntlet of challenges which confronted Dyadic would suggest that it might take more than loads of passion, tenacity, leadership, and luck to just stay afloat in the days ahead. As one major shareholder put it, it might take a "miracle."

On a sweltering hot day in Jupiter, Florida in June 2008, Mark Emalfarb, CEO of Dyadic entered his office to find his world turned upside down. He had not been there since 2007 when he was summarily fired by the board of directors (BOD), due to financial and accounting irregularities at Puridet—Dyadic's recently acquired subsidiary in China. While he "was away," Dyadic was reduced to 31 employees from 125; its cash dwindled by a factor of 9, from $27 million to just about $3 million; the value of its stock fell from $5.30 to a mere 50 cents. Furthermore, instead of proceeding with a $10 million investment deal in Dyadic's stock to support research in large-scale fungus's production, the Spanish Abengoa Bioenergy New Technologies, one of the worlds' largest alternative energy organizations, was now suing the company. The plans to build new headquarters and a biotech incubator next to the prestigious Scripps's laboratory had to be abandoned, as the prime piece of real estate which Dyadic bought for that purpose was sold at way below market value to pay legal bills, and a likely research grant from the Department of Energy (DOE) was never even applied for in Emalfarb's absence.

As if this list of calamities was not long enough, almost the entire BOD had to be replaced; key managerial positions were vacant; an investor class-action lawsuit was pending; financial reports had not been filed with the Securities and Exchange Commission (SEC) for two years since 2007; and to make matters even more unpredictable, the company no longer had an accounting firm. It is to point out the obvious that the reputation of both Emalfarb and Dyadic in the industry were badly damaged; in the words of a key investor, "demolished." How did things go so terribly wrong? *Quo Vadis?* These were the questions on Mark's mind. He did regain his authority, but how was he going to regain the confidence of stakeholders and begin rebuilding? Where was he to start from? Was it even worth trying? Mark never even considered the last question.

DYADIC COMPANY HISTORY AND FINANCIALS

Dyadic International, Inc., a small, publicly traded industrial biotechnology company founded in 1979, evolved within its three decades from a clothing company into an industrial bioscience organization. Its ambitious goal, according to its founder, "is not just to make cellulosic ethanol, it is to create enzymes to produce glucose from biomass in order to replace not just ethanol, but all petroleum produces" (telephone interview, 2009). To these ends, the company is dedicated to applying its patented and proprietary technologies to the discovery, development and manufacturing of products which are to facilitate the finding of solutions for the bioenergy, industrial enzyme, and pharmaceutical industries.

Dyadic's roots, however, are in the stonewashing industry, where Emalfarb began his adventure as an entrepreneur two years after he graduated from the University of Iowa with a B.A. degree in journalism. Emalfarb successfully navigated Dyadic's progression from its leadership position as a supplier of components for the stonewashing of blue jeans; to the discovery, development, manufacturing and commercialization of enzymes with an industrial application in the textiles, food and beverage, animal feed, deter-

---

1   In Latin, "Where are you going?"

gent, and paper and pulp industries; to its current biotechnology endeavors. His *Business Week* executive profile reads: "He is a named inventor on over 25 United States, foreign and international biotechnology patents and patent applications, resulting from various inventions related to Dyadic's proprietary C-1 microorganism. He has been instrumental in forging strategic research and development and marketing alliances between Dyadic and global organizations located in the United States, China, Ecuador, Iceland, the Netherlands, Russia, Switzerland, and Poland" (*Business Week*, 2010). It is probably his experience as a wrestler on the university team, though, that offers a deeper and richer insight into his go-getter attitude and perseverance with sticking with the fight until emerging as a victor.

Dyadic pioneered the stonewash industry by supplying Icelandic, Ecuadorian, American and other pumice[2] for stonewashing jeans to iconic companies such as Guess, Lee, Levi's, and Wrangler. With a first mover's advantage, Dyadic enjoyed rapid revenue and volume growth for several years. In the mid-1980's, however, textile companies introduced a superior alternative technology to achieve the look the customers were craving for—enzymes (proteins that catalyze chemical reactions). It was a Danish biotech company, Novozymes, that launched an enzyme to replace pumice in the process of stonewashing. Emalfarb seized the moment and quickly moved to become a distributor for Novozyme's products. As more distributors entered the market, though, Dyadic's sales volumes started to decrease. Emalfarb's competitive drive did not sit well with playing a second fiddle and he made the bold decision to get into the business of developing and manufacturing his own proprietary enzymes from fungus. As he put it plainly, "I thought it was my birth right to sell to these laundries, whatever it was. Whether it was pumice or enzymes, they were going to buy it from *me*." (Obermueller, 2009(a):1).

He had a compelling and relentless vision for Dyadic—not only will the company compete in the $3-billion industry of industrial enzymes, but it will become "a star." As one analyst noted, "What makes him a great CEO is a combination of his vision, his passion for it, and an insatiable competitive drive. It's almost hard to believe that Emalfarb did not set out to change the world. But he is going to." (Obermueller, 2009a:1).

Thus, Mark began his company's transition "from jeans to genes," as he himself labeled it. He scanned the globe for "enzyme producing" talent; from Stanford to MIT, across the Atlantic to the Netherlands and further east to Russia in search of a scientific team that would help him to realize his vision. He set up a "research arm" in Jupiter, Holland and North Carolina and began working with a group of dedicated scientists from the enzymology laboratory at the Moscow State University who had been studying enzymes and making a solid progress since the 1970's oil crisis. Taking advantage of the world-altering events in 1989 amidst the fall of the Berlin Wall, Dyadic secured the aid of two dozen world class Russian scientists at low cost. To complete the cycle, he set up production facilities in Poland, as well as two distribution centers: one in Florida and one in China.

The first promising results came in 1994 when Dyadic launched its first commercial cellulase product based on a fungus called *trichoderma*; the same fungus that destroyed the cellulose in the cotton tents of World War II soldiers (Obermueller, 2009a). While trying to cultivate a better enzyme for stonewashing by breeding fungus, the Russian researchers had a lucky break. By serendipity, they came to a mutation which afforded the production of a larger number of enzymes at a lower cost. The new fungus, called C-1, was isolated from alkaline soil in Eastern Russia and had almost twice the number of biofuel genes in its genome in comparison with *trichoderma* (Sandra, Hinz, Pouvreau, Joosten, Bartels, Jonathan, Wery & Schols, 2009; Obermueller, 2009b) (Figure 1). C-1's main use came from breaking down the cellulose in the cell wall, thus releasing the sugar that can be fermented into ethanol; the latter accounts for a small share on the gasoline market which is expected though to rise to 7.5% of the market in 2017 (Westcott, 2007).[3] The discovery was also welcomed by the textile, brewing,

---

2 *Pumice: a textural term for a volcanic rock that originates from solidified lava.*

3 *According to Westcott (2007), "In 2006, ethanol (by volume) represented about 3.5 percent of motor vehicle gasoline supplies in the United States. By the middle of the next decade, ethanol production (by volume) is expected to represent less than 8 percent of annual gasoline use in the United States."*

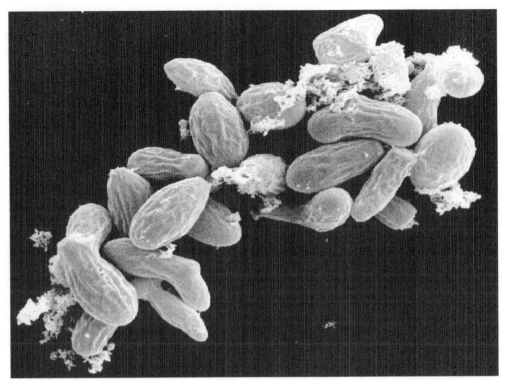

Figure 1: Dyadic platform technology. Strains of fungal microorganism *Chrysosporium lucknowense* (C-1), which through a random serendipitous mutation form novel propagules, are at the heart of Dyadic's C1 Platform Technology. The latter can be used as a host organism for manufacturing novel products of both eukaryotic and prokaryotic genes. The C-1 gene expression system, consisting of a robotic screening system and protein expression system, can be used for gene discovery, product development, and a large scale product manufacturing in a wide variety of industrial and biopharmaceutical applications.

paper, and animal feed industries. As Emalfarb described:

> *"We hired Russian scientists to look for something better [to soften jeans with], and we found this organism called C-1. It seemed to do the job on cotton better than we needed it to do. As we bred this fungus to make more and more of what we wanted, we had a serendipitous mutation that allowed it to be able to be produced on a large scale at low cost. [C-1] seems to have almost twice the biofuel genes in its genome than trichoderma. Those genes and codes for enzymes seem to operate at broader temperatures and pH ranges so that it has applicability for the microbial world in the broader sense. It can be used with bacteria and yeast to make ethanol from sugar. We patented it and we've developed it for over 15 years now... So we're steering it toward biofuels and biopharma, as well as industrial enzymes, cosmetics, and nutraceuticals. It's basically a platform technology that's scalable and very productive and can produce things at low cost in large volume."*

The invention of C-1, which Dyadic patented in 1994, was one of the most important events in the company's shift "to genes"; it shaped its decision-making in terms of growth, strategy and markets. While the company's success in scientific research and development was solid and had a positive impact on its performance, internal changes, external events, and some of the resultant decisions made between 2003 and 2007 had the opposite effect.

Confident in the promise of C-1, and to keep up with shifts in the textile industry which were taking place after NAFTA was originally signed into law by President Clinton in 1993, Dyadic decided to aggressively expand its operations. The textile industry was going to Asia, and Dyadic established operations in the region to be closer to its customers. Between 1998 and 2006 it acquired a Hong Kong company—Puridet, which manufactures and distributes enzymes through its wholly-owned operation in mainland China. This took place in three stages: in 1998 Dyadic purchased a majority interest in Puridet; in 2003, it made Puridet a subsidiary; and in 2006, acting upon the strong business and accounting recommendation, counsel, and assistance of the company's lawyer Robert Schwimmer and of accountant Ernst & Young, Dyadic became a 100% owner of Puridet—not knowing that its small foreign subsidiary would later open Pandora's proverbial box.

Mark Emalfarb was confident that Dyadic carefully structured its operations and governance with Puridet to appropriately coordinate and manage potential market risk in China. Dyadic's aim was to set up a structure to allow them to operate in a foreign country as seamlessly as possible. Initially, Dyadic was using two different auditing companies for its US and overseas operations. In the post-Enron era, after Arthur Andersen ceased to practice public accounting, Ernst & Young (E&Y) "actively solicited Dyadic to become a client." In 2002, Dyadic retained E&Y to serve its domestic auditing and accounting needs while it kept using a local company for its operations in Hong Kong and China. In 2003, E&Y assured Emalfarb that it would be in Dyadic's best interest if it used one auditing organization, namely E&Y, to "do both sides of the equation" and as a result, to maximize efficiencies and strengthen auditing. Emalfarb agreed and retained the services of E&Y's branch in Hong Kong, (EY-HK) "to specifically monitor and deal with its Chinese subsidiary in Hong Kong, Puridet."

At this point, Emalfarb felt Dyadic was in a very good position. In addition to having secured the services of two of the world's leading auditing firms, he also retained a local accounting firm in Hong Kong which had the explicit responsibility to provide Dyadic with reports on a monthly basis and to ensure compliance. "I don't know what else we could have done on the accounting side differently," he said in a personal phone interview. "We did more than anyone would have done... more than companies with better resources than us to protect ourselves." What is more, he put two people "on the ground" in China and Hong Kong to keep an eye on what was going on and to make sure that business was conducted properly and legally. He also visited Puridet with his controller, as well as ensured a steady stream of meetings with US personnel, to keep communications.

Emalfarb was already considering taking the company public as a means of accessing the capital markets and gaining liquidity. He also felt that being public would increase the exposure of Dyadic's research and development, including C-1, which could accelerate business development and strategic alliances—to speed up scaling the company. In addition, going public would lead to less dilution for shareholders, he reasoned. "So, by 2003, we had the infrastructure to go public," felt Emalfarb. However, by that time he had been at the helm of a privately-held organization for about a quarter of a century and was well aware that he did not have the necessary knowledge to take the company public himself. Notwithstanding, neither the board or advisors considered hiring a new CEO either, as research on change in the organization's life cycle may suggest (Daft, 2010).

Recognizing that he had no formal business training or experience running a public company, Emalfarb "sought to identify and hire a qualified team of experienced professionals upon whom [he and the company] could heavily rely to ensure that all of the requirements of becoming and maintaining a public company were satisfied." (Delaware Court Case, 2009: 9). Emalfarb took several major steps in preparation for the transition. First, Dyadic hired a top tier law firm, Jenkens & Gilchrist (J&G), and started working with a lawyer Bob Schwimmer and two of his partners. J&G portrayed themselves as experienced experts in transitioning organizations from private to public, including SEC filings, registrations, and REG FD disclosures. Around that time, E&Y approached Dyadic once again and, in addition to their auditing services, proposed that Dyadic could also benefit greatly from its equally high-quality service and expertise as business consultants. Mindful of his lack of experience with both—going through the transition and managing a public company—Emalfarb thought that although costly, E&Y's consultancy services would be positive for Dyadic and ensure a swift transition. Thus, by the summer of 2004, with their network of lawyers, accountants, auditors and consultants in place, Dyadic proceeded to obtain a public stock exchange listing via a reverse merger.[4] Dyadic effected a reverse merger in October 2004, filed its registration statement with the SEC in December of 2004, and was ultimately listed on American Stock Exchange (Amex) in 2005. Simultaneously, Dyadic completed a financing of $24.5 million in a common stock offering.

PANDORA'S BOX OPENS

In December 2003, while still privately held but in the midst of preparing to go public, Emalfarb received an anonymous e-mail from an employee in China regarding suspected financial improprieties taking place at Puridet. The e-mail alleged that Puridet had set up a dummy company through which it was boosting revenues. Emalfarb asked for receipts or any other forms of evidence and after several months had elapsed he received another rambling e-mail in which the allegations were repeated along with some "ranting" about "George Bush, Saddam Hussein, Osama Bin Laden, Israel, September 11th and more." In the absence of specific guidance and a policy for whistle blowers (as a private company, Dyadic did not have one in place), Emalfarb forwarded the e-mails to his Controller at the time and notified Robert Smeaton, an Australian who ran Puridet, of the allegations. In addition, he contacted his two "people on the ground" and asked them to get more information about the situation. He also informed E&Y about the whistle blowing e-mail during a routine audit and, given his inexperience in accounting, he fully expected the auditors to appropriately investigate the matter. In the end, it turned out that there was indeed a dummy company and Emalfarb ordered Puridet to cease all transactions with the bogus entity.

At the end of 2004, the now-public Dyadic, was advised by E&Y on the need to address an important aspect of its governance; namely, to identify the critical skills necessary to enhance transparency and controls in managing the company. E&Y's view was that Dyadic needed people for internal controls, and therefore advised to begin searching for a chief financial officer (CFO). Emalfarb, based on the enthusiastic recommendation of E&Y, Schwimmer, and other board members chose Joel Wayne Moor, among several candidates, who was hired and started his new job as Dyadic's CFO in January 2005. Moor brought significant financial and internal controls experience, particularly with publicly-listed firms.

The next order of business and a critical component to the success of the transition from private to public was putting together an independent board of directors. Thinking of what was best for the business and the company, Emalfarb opted to put together an independent BOD of outsiders rather than inviting people he already knew (something which later, upon reflection, he felt he should have done). He wanted to recruit board members who had experience being on and running public company; who would open doors for Dyadic; and would convey trust and stability to investors. Hence, by soliciting recommendations from numerous sources he managed to enlist Stephen Warner, a managing director of a venture capital firm in Florida; Robert Shapiro, the former CEO of Monsanto; and Harry Rosengart, who had solid connections on Wall Street and in the biotech industry.

All board members were vetted by Bob Schwimmer who also directed the search for a lead director of the board. After a consultation with Fred Frank, the vice chairman of Lehman Brothers, Richard Berman—a CEO of a small public biotech company—was selected for the position. Berman had already been trusted to play the same role by seven other public companies. Emalfarb was rather pleased with the way things were going: "From a Wall Street perspective, we built a first class board. We did the same

4   *A reverse merger, also known as a reverse takeover, occurs when a private company merges with a public company after which the private company has controlling ownership. Reverse mergers are generally pursued to take advantage of the greater financing options available to public companies and achieve liquidity. As such, the reverse merger is an alternative to the traditional Initial Public Offering (IPO) as a method for going public.*

with the scientific advisory board (SAB). In fact, we had Richard Lerner, the president of Scripps, to chair the board and a whole bunch of first class scientists. So, we felt like we had a great board, great SAB, outstanding scientists… Everything was going great." Indeed, Dyadic was gaining significant momentum. It raised money in the capital markets, scientists were making solid progress, and strategic alliance opportunities seemed endless. Dyadic expected to move from Amex to a full NASDAQ stock exchange listing in Q2 of 2007.

In April 2007, however, Emalfarb received another anonymous whistle-blower e-mail about Puridet's financial doings which he immediately forwarded to his CFO, Wayne Moor, who happened to be in Hong Kong at the time. He also told him about a similar e-mail he had received in 2003 and mistakenly insisted that he had e-mailed the first e-mail to Moor. This would have been impossible as Moor joined the company a little over a year after the fact. In a paranoid post-Enron atmosphere of endless accounting scandals though, Moor took the statement as a direct accusation of ignoring evidence of financial improprieties. According to Stephen Warner, a board member and investor, "That was the end of Wayne's regard for Mark"; he felt as if he were handed a rope to hang himself. Realizing his mistake, Emalfarb offered his apology the next day but it came a day too late for his CFO who stopped speaking to him (Florida Trend, 2009: 56).

From this point on, events unfolded rather swiftly. Moor, without the knowledge of Emalfarb, contacted Schwimmer, E&Y, and Dyadic's Audit Committee about the whistle-blower e-mail and the alleged improprieties at Puridet. He also stipulated wrongdoing by Mark. Consequently, Schwimmer suggested that Dyadic needed to conduct a formal internal inquiry and hired a Miami law firm, Moscowitz & Moscowitz, to conduct an independent investigation. Schwimmer and Moor also impelled the board to advise Emalfarb to take a voluntary leave of absence based on the belief, instilled by the legal counsel, that he "withheld material information regarding the events at Puridet." On his part, Emalfarb agreed to step down after he received written and verbal assurances from Schwimmer that "proper corporate procedures would be implemented in the conduct of the investigation and that the appropriate parties would be subject to the investigation" (Delaware Court Case, 2009: 25).

These assurances, however, were not followed through and the parties with the relevant professional knowledge, expertise and authority—Moor, E&Y, and Schwimmer, effectively turned the investigation away from themselves while pointing it on Emalfarb exclusively. What is more, Moor was promoted to interim CEO and within a month became a member of the board. E&Y, in the meantime, insisted that it would not review and sign off on Dyadic's first quarter filings or assist in correcting its own mistakes in prior financial statements without assurances that E&Y will not become a part of the investigation. In May 2007, Moscowitz & Moscowitz, "inexplicably entered into an agreement with E&Y that, as part of its independent investigation, it would not opine on the quality of E&Y's accounting or audit work—or whether the audits were performed in accordance with applicable professional standards in return for E&Y agreeing to surrender certain of its work papers and make personnel available for interviews." (Delaware Court Case, 2009: 30). Despite receiving such requested assurances, E&Y refused to cooperate and as a direct result, Dyadic was prevented from trading shares publicly and was delisted from AMEX in early 2008. Not only did E&Y not review accounting and audit reports, but somehow both E&Y and Schwimmer & Greenberg "forgot" to mention that they had a conflict of interest since at the time of the investigation of Dyadic was taking place E&Y was a client of Schwimmer & Greenberg regarding a separate legal mater.

Meanwhile, Schwimmer advised Dyadic to temporarily hold trading in the company's stock. Upon the counsel of Moscowitz & Moscowitz, Dyadic also hired Bilzin Sumberg Price & Lxelrod LLP—yet another law firm—to review the report which Moscowitz & Moscowitz were finishing up and to advise the Audit Committee on further courses of action. In August 2007, Moscowitz & Moscowitz provided their "independent" conclusions to Dyadic. With Schwimmer, Moor, E&Y and EY-HK effectively shielding themselves from the investigation, the report put all the blame for Puridet's financial troubles on Mark Emalfarb alone. Consequently, Bilzin recommended for the BOD to terminate Emalfarb. The members of the board, with one exception—Stephen Warner—agreed and Mark Emalfarb, Dyadic's founder, chief executive, "principal architect," visionary, and largest shareholder, was fired for cause in November 2007. In the aftermath of Enron, the CFO, the company's lawyers and E&Y aligned and blamed him for the debacle with the Hong Kong subsidiary, Puridet. Although Emalfarb was the largest shareholder with 35% ownership, he found himself on the outside looking in and still waiting for the "guaranteed" thorough investigation of the matter.

Throughout this entire time, the board members met only via teleconference and never face-to-face; Emalfarb was kept in the dark and not even provided a copy of the independent report. Furthermore, not only was the report from the special investigation delivered in writing, as opposed to the standard practice of orally reporting to the board, but it was made public and sent to SEC—significantly raising the probability that Dyadic would become the perfect candidate for a class action lawsuit.

COUNTERATTACK

Emalfarb, the former wrestler, was "the classic entrepreneur, the guy who lives, eats and sleeps his business." He did not even consider throwing away almost three decades of his life and the future of the company; at least not without a fight. Technically, he still held a board seat. Using his own money, Emalfarb sued in a Florida court and filed eleven counts of professional negligence, breach of fiduciary duty, and constructive fraud against E&Y, EY-HK, Jenkens and Schwimmer, Greenberg and Schwimmer, Moscowitz & Moscowitz P.A., Norman Moscowitz and Jane Moscowitz, and Bilzin; and one count for "civil conspiracy" against Greenberg, Schwimmer and E&Y (Delaware Court Case, 2009: 36-70).

Against the odds, Emalfarb won and the court ordered a shareholder meeting in June 2008. Large shareholders, who together with Emalfarb held 52% of the company's stock, threw their support behind him and Emalfarb was back in his old office "by 1 pm on the day of the meeting" (South Florida Business Journal, 2009). In the same month E&Y resigned as Dyadic's auditor.

Bruised but not beaten, Emalfarb provided a steadfast vision of Dyadic's potential and promise as it embarked on a new phase of growth:

> *"We could become the next ExxonMobil. A barrel of sugar can replace a barrel of petroleum, and in a much more environmentally-friendly way. It has a better environmental footprint for the planet, and it's renewable. And the beautiful thing is you can grow it in Texas; you can grow it in China. You can grow it in India. You don't have to import it from hostile foreign nations. From a national defense standpoint and a jobs-creation standpoint, growing energy crops or taking biomass and turning it into sugar is absolutely the answer I think the world needs. It's a matter of time now. We can for sure make sugar, we just have to drive the cost down. Through the advent of modern biology and the sequencing of the human genome, and other genomics and protein technology like Dyadic C-1, it's 'he who can make proteins in large tanks affordable is going to rule the world.'"*

As he left his office after a year-long hiatus, Emalfarb drove down Ocean Boulevard and reflected on the ups and downs of taking Dyadic public. He was determined to seize the new opportunities in front of his battered company, which continued to make scientific progress despite the tumultuous period. Emalfarb looked up at a billboard advertising "Wrestlemania," an upcoming professional wrestling tournament featuring a menacing, heavily muscled, and masked figure performing the "Kung Fu Chop". They should call it the "Kung Fu Chop II" he chuckled, "I already got Kung Fu chopped."

APPENDICES

APPENDIX A: DYADIC COMPANY FINANCIALS

INCOME STATEMENT

	2008	2007
Revenue:		
Product Related Revenue, Net	$ 9,156,529	$ 9,676,581
Research and Development Revenue	3,958,546	3,603,373
Total Revenue, Net	13,115,075	13,279,954
Cost of Goods Sold:	9,112,945	8,521,251
Gross Profit	4,002,130	4,758,703
Expenses:		
General and Administrative	9,733,989	8,247,588
Sales and Marketing	2,344,736	3,381,430
Research and Development	4,102,516	5,008,615
Foreign Currency Exchange Losses, Net	24,153	23,947
Impairment Loss	-	3,993,276
Total Expenses	16,205,394	20,654,856
Loss from Operations	(12,203,264)	(15,896,153)
Other Income (Expense)		
Interest Income	145,765	1,092,778
Interest Expense	(388,980)	(272,252)
Liquidated Damages	-	(1,103,906)
Gain on Sale of Land	-	200,000
Gain on Reversal of Related Party Payable	-	475,393
Other	80,081	120
Total Other Income (Expense), Net	(163,134)	392,133
Net Loss	$ (12,366,398)	$ (15,504,020)
Basic and Diluted Net Loss Per Common Share	$ (0.41)	$ (0.52)
Weighted Average Common Shares Used in Calculating Net Loss Per Share:		
Basic and Diluted	29,990,675	29,961,563

Source: Company reports

BALANCE SHEET

	2008	2007
ASSETS		
Current Assets:		
Cash and Cash Equivalents	$ 2,826,542	$ 15,953,984
Restricted Cash	344,355	349,193
Accounts Receivable, Net	1,504,200	1,411,070
Inventory, Net	3,775,750	6,122,758
Prepaid Expenses and Other Current Assets	637,202	794,318
Total Current Assets	9,088,049	24,631,323
Fixed Assets, Net	1,039,458	1,186,617
Intangible Assets, Net	162,420	90,337
Other Assets	137,502	182,409
	$ 10,427,429	$ 26,090,686
LIABILITIES AND STOCKHOLDERS' EQUITY		
Current Liabilities:		
Accounts Payable	$ 2,678,794	$ 1,941,754
Accrued Expenses	325,634	1,521,174
Accrued Interest Payable to Stockholder	194,260	48,896
Deferred Research and Development Obligation	3,332,863	3,332,500
Note Payable to Stockholder	2,424,941	2,404,742
Income Taxes Payable	8,658	6,456
Total Current Liabilities	8,965,150	9,255,522
Long-Term Liabilities:		
Deferred Research and Development Obligation	-	3,332,863
	8,965,150	12,588,385
Commitments and Contingencies		
Stockholders' Equity:		
Preferred Stock, $.0001 Par Value:		
Authorized Shares – 5,000,000; None Issued and Outstanding	-	-
Common Stock, $.001 Par Value,		
Authorized Shares – 100,000,000; Issued and Outstanding –		
29,990,675	29,991	29,991
Additional Paid-In Capital	75,843,581	75,517,205
Accumulated Deficit	(74,411,293)	(62,044,895)
	1,462,279	13,502,301
	$ 10,427,429	$ 26,090,686

Source: Company reports

APPENDIX B: DYADIC SHARE PRICE

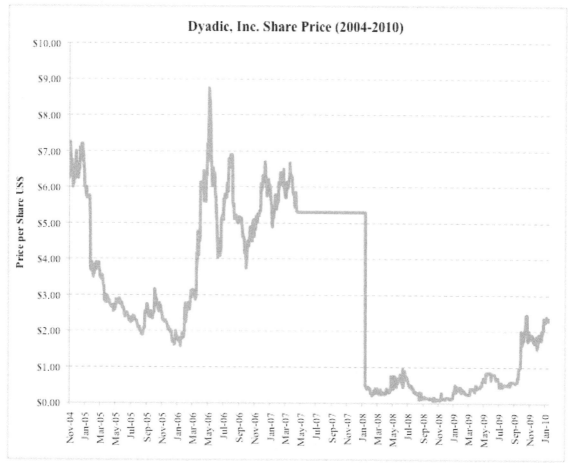

Source: Yahoo Finance

APPENDIX C: BIOS OF BOARD OF DIRECTORS AND KEY MANAGEMENT AT THE TIME OF THE CRISIS IN 2007
Source: Dyadic, Inc. 10-K 2007

- **Mark A. Emalfarb: Chairman, President and Chief Executive Officer** since October 2004 and has been the president, chief executive officer and chairman of the board of directors of wholly-owned subsidiary Dyadic International (USA), Inc., a Florida corporation ("Dyadic-Florida"), since its inception. Since founding Dyadic-Florida in 1979, Mr. Emalfarb has successfully led and managed the evolution of Dyadic-Florida—from its origins as a pioneer and leader in providing ingredients used in stone-washing of blue jeans—to the discovery, development, manufacturing and commercialization of specialty enzymes used in various industrial applications and the development of the Dyadic Platform Technology and our other proprietary technologies. Mr. Emalfarb is an inventor of over 25 U.S. and foreign biotechnology patents and patent applications resulting from discoveries related to Dyadic-Florida's C1 organism, and has been the architect behind its formation of several strategic research and development, manufacturing and marketing relationships with U.S. and international partners. Mr. Emalfarb earned a B.A. degree from the University of Iowa.

- **Glenn E. Nedwin, Ph.D.: Board of Directors and our Chief Scientific Officer and Executive Vice President and President of BioPharma Business** since March 2006. Before joining Dyadic, Dr. Nedwin co-founded and served as president of Novozymes, Inc. since 1991. At Novozymes, Inc., a Davis, California-based research & development subsidiary of Novozymes A/S (CSE:NZYMb.CO), Denmark, a global leader in enzymes and microorganisms with over $1 billion in worldwide revenues, Dr. Nedwin was responsible for all scientific, financial and administrative activities, and was a member of Novozymes A/S global R&D management team and its biosolutions strategy group, and was involved in technology/product licensing. From 1989 to 1991, Dr. Nedwin served as vice president of corporate development, Xoma Corporation (NASDAQ:XOMA), a biotechnology company based in Berkeley, California. Earlier, he was vice president, business development and co-founder of Ideon Corporation, Redwood City, California, and senior research scientist and co-founder of Molecular Therapeutic, Inc. (now Bayer Pharmaceuticals Corporation), West Haven, Connecticut. Dr. Nedwin received his Bachelor of Science degree in biochemistry from the State University of New York at Buffalo and his Ph.D. in biochemistry from the University of California, Riverside. Dr. Nedwin did his postdoctoral fellowship in molecular biology at Genentech, Inc. Dr. Nedwin also holds an M.S. in the management of technology from the Massachusetts Institute of Technology and is currently a co-editor of the *Industrial Biotechnology Journal*.

- **Wayne Moor: Chief Financial Officer and Vice President** since January 2005. During the past five years Mr. Moor has served as a chief financial officer of several public companies, and has over 25 years experience in real estate and hotel operations, asset management, debt restructurings, recapitalizations and developing strategic turnaround plans. From October 2002 through December 2004, Mr. Moor served as the senior vice president, treasurer and chief financial officer of Boca Resorts, Inc, a hospitality company. From October 2001 to October 2002, Mr. Moor was senior vice president and chief financial officer for ANC, the parent company of Alamo and National rental car companies. In November 2001, following the terrorist attacks of September 11, 2001, ANC and its U.S. operating subsidiaries voluntarily filed petitions for reorganization under Chapter 11 of the U.S. bankruptcy code in Wilmington, Delaware. From September 2000 to October 2001, Mr. Moor was senior vice president and chief financial officer for Gerald Stevens, Inc., a floral products marketer and retailer. In April 2001, Gerald Stevens, Inc. and certain operating subsidiaries voluntarily filed petitions for reorganization under Chapter 11 of the U.S. bankruptcy code in Miami, Florida. From June 1997 to January 2000, Mr. Moor was executive vice president and chief financial officer for US Diagnostic, Inc., an operator of outpatient medical diagnostic imaging and related facilities. Prior to that, Mr. Moor held the position of chief financial officer or executive vice president for a number of privately and publicly held financial institutions and real estate operating companies. He began his career in public accounting.

- **Kent M. Sproat: Executive Vice President, Manufacturing and Special Projects** responsible for all manufacturing functions of our enzymes business activities since January 2007. Prior to that he served as executive vice president, enzyme business from April 2005 through December 2006. Mr. Sproat served as our vice president, manufacturing from 1997 through March 2005. Mr. Sproat joined Dyadic-Florida in 1997 from Genencor International, where since 1996 he served as its elkhart site manager. From 1990 to 1996, Mr. Sproat was vice president, manufacturing, of Solvay Enzymes, Inc. From 1989 to 1990, he was director of international manufacturing of the enzyme division of Miles, Inc. Between 1981 and 1990, he served as plant manager of Miles' Elkhart, Indiana and Clifton, New Jersey-based enzyme plants. Between 1973 and 1981, Mr. Sproat was a production superintendent at Miles' citric acid division; start up manager of Miles' citric acid facility in Brazil; and production supervisor and project engineer. Mr. Sproat is the recipient of a patent for his design in the purification process of amylases. Mr. Sproat is a chemical engineer with a B.S. degree from Purdue University.

- **Alexander (Sasha) Bondar: Vice President, Strategy & Corporate Development**, with primary responsibility for corporate development, organization planning, merger & acquisition opportunities, fund-raising activities and investor and public relations, and secondarily, when requested, for assisting in business development for the company's biopharma and enzyme businesses, since April 2005. Mr. Bondar served as executive director, business development from May 2003 through March 2005. Mr. Bondar joined Dyadic-Florida in May 2003 from The Aurora Funds, a venture capital firm based in Research Triangle Park, North Carolina, where he was focused on investing in early stage life sciences companies. Prior to that, from 1996 to 2001, Mr. Bondar served in a variety of management roles at Incyte Genomics, now Incyte Corporation, in Palo Alto, California, and from 1999 to 2001 as associate director, corporate business development. From 1997 to 1999, he served as manager, pharmacogenomics business development, and was a major contributor to the successful launch of Incyte's pharmacogenomics program. From 1996 to 1997, he served as technical advisor to the intellectual property group at Incyte, contributing to the creation of the largest portfolio of gene patents in the world. Mr. Bondar holds a B.S. degree in biotechnology management from Menlo College and an M.B.A. in corporate finance and health sector management from Duke University's Fuqua School of Business.

- **Richard J. Berman: Board of Directors** since in January 2005, and acts as "lead director." In that capacity, he is responsible for meeting regularly with the chairman of the board and chief executive officer to review monthly financials, agenda and minutes of committee meetings and pertinent board issues, presiding, if requested by the board, as chairman of any of the committees of the board and presiding at any meetings of the independent and non-employee directors. In the last five years, Mr. Berman has served as a professional director and/or officer of about a dozen public and private companies. He is currently CEO of Nexmed, a small public biotech company; chairman of National Investment Managers, a public company in pension administration and investment management; chairman of Candidate Resources, a private company delivering HR services over the web, and chairman of Fortress Technology Systems (homeland security). Mr. Berman is a director of eight public companies: Dyadic International, Inc., Broadcaster, Inc., Internet Commerce Corporation, MediaBay, Inc., NexMed, Inc., National Investment Managers, Advaxis, Inc., and NeoStem, Inc. From 1998 - 2000, he was employed by Internet commerce corporation as chairman and chief executive officer. Previously, Mr. Berman worked at Goldman Sachs; was senior vice president of Bankers Trust Company, where he started the M&A and leveraged buyout departments; created the largest battery company in the world by merging Prestolite, General Battery and Exide to form Exide (NYSE); helped create what is now Soho (NYC) by developing five buildings; and advised on over $4 billion of M&A transactions. He is a past director of the Stern School of Business of NYU where he obtained his B.S. and M.B.A. He also has US and foreign law degrees from Boston College and The Hague Academy of International Law, respectively.

- **Robert B. Shapiro: Board of Directors** since March 2005. During the past six years Mr. Shapiro has served as a member of the board of directors of the New York Stock Exchange, Citigroup, Inc. and Rockwell International, as chairman of Pharmacia Corporation's board of directors and, prior to its merger with Phamacia & Upjohn, as chairman and chief executive officer of Monsanto Company (1995 through 2001). Prior to becoming the chairman and chief executive officer of Monsanto, Mr. Shapiro served in various executive capacities with Monsanto from 1985, and with G.D. Searle & Company, a pharmaceutical and healthcare company, first as its general counsel (1979 through 1982), and then as president of its newly formed NutraSweet Group (1982 to 1985). Mr. Shapiro is a 1959 graduate of Harvard College and a 1962 graduate of Columbia University School of Law.

- **Stephen J. Warner: Board of Directors** since October 2004, and a director of Dyadic International (USA), Inc., wholly-owned subsidiary ("Dyadic-Florida"), since August 15, 2004. Mr. Warner serves as chairman of Maxim TEP, Inc. a private energy company based in Houston, Texas, and Search Energy Solutions, Inc., a private company offering energy savings services for large air conditioning systems based in Palm Beach Gardens, Florida. He also serves as a director of UCT Coatings Inc., a private, metal finishing technology company in Stuart, Florida, and AOI Medical, Inc., a private medical device company in Orlando, Florida. Mr. Warner has over 25 years of venture capital experience. In 1981, Mr. Warner founded Merrill Lynch Venture Capital Inc., a wholly-owned subsidiary of Merrill Lynch & Co. Inc. in New York and served as its president and chief executive officer from 1981 to 1990. Under his leadership, Merrill Lynch Venture Capital managed over $250 million and made over 50 venture capital investments. In 1999, Mr. Warner co-founded, and became chairman and CEO, of Crossbow Ventures Inc., a private equity fund that invests in early and expansion stage technology companies primarily located in Florida and the Southeast, with over 20 venture capital investments in Florida. Mr. Warner earned a B.S. degree from the Massachusetts Institute of Technology and an MBA from the Wharton School of Business, University of Pennsylvania.

- **Harry Z. Rosengart: Board of Directors** since April 2005. During the past ten years, Mr. Rosengart has served (and currently serves) as the president and chief executive officer of HK & Associates, an investment and consulting firm which provides advice to small and medium-sized life sciences companies. Mr. Rosengart is a founder of several privately held companies, including: LigoChem, Inc., a DNA/RNA and macromolecule bioseparations company founded in 1995, of which he is a former president and chief executive officer and a current member of its board of directors; SunPharm Corporation, a polyamine based anti-cancer drug development-stage company founded in 1991, of which he is a former chief operating officer, chief financial officer and member of its board of directors; and Syncom Pharmaceuticals, Inc, a contract sales force organization founded in 1991, of which he has had a variety of interim positions and served on its board of directors. Between 1981 and 1990, Mr. Rosengart spent almost 10 years as a banker and investment banker with the Chase Manhattan Bank, NA focused on the pharmaceutical and chemical industries. Prior to joining Chase Manhattan Bank, Mr. Rosengart spent over 10 years with several pharmaceutical and multinational chemical companies in various managerial positions. Mr. Rosengart holds a B.S. in Chemical Engineering and an MBA from Rutgers University.

- **Ratnesh (Ray) Chandra: Senior Vice President, Marketing-Biotechnology Systems**, responsible for business development for the company's biopharma business activities, since April 2005. Mr. Chandra served as vice president, marketing - biopharma from 2000 through March 2005. Mr. Chandra joined Dyadic-Florida from Genencor International in 2000. He had served at Genencor as the director, new business development since 1993. From 1987 to 1993, he was with Merck & Co. holding the positions of director, business/market intelligence and director, business systems in their human health marketing division. From 1976 to 1987, he was with Schering-Plough Corp. in the positions of director economic analysis, manager capital planning and senior operations analyst. Mr. Chandra has an M.B.A. from New York University and an M.S. in engineering from Columbia University.

- **Daniel Michalopoulos, Ph.D.: Vice President, Marketing, Enzymes** since January 2007. Prior to that he served as our vice president, pulp & paper from February 2005 through December 2006 and is focused on growing our pulp and paper effort. Prior to joining us, he served as senior program manager for Hercules' pulp and paper division from 1998 to 2005 where he managed a staff of 40 people with an annual budget of $8 million. Prior to that, he served as research director at BetzDearborn pulp and paper division and held other research and management positions at Betz PaperChem. Dr. Michalopoulos conducted his post-doctoral work at Rice University, received his Ph.D. in Chemistry from Colorado State University and his B.S. in Chemistry from Lowell Technological Institute.

- **Charles W. Kling IV: Vice President, Sales, Enzymes** since January 2007. Prior to that he served as our vice president of sales and marketing - enzymes from July 2005 through December 2006. Prior to joining us, Mr. Kling served as group manager of Hercules, Inc.'s pulp & paper division, with full P&L responsibility and management of staff of over 60 people, from 1998 to 2005. Prior to Hercules, from 1990 to 1998, Mr. Kling served as global director, technical marketing for BetzDearborn Inc.'s pulp & paper division. From 1986 to 1990, he was division manager, S.D. Warren division of Scott Paper. Prior to that, he served as production manager for Buckeye Cellulose, a division of Proctor and Gamble, Inc. Mr. Kling received his B.S. degree in civil engineering from University of Alabama.

APPENDIX D: DYADIC CODE OF CONDUCT

DYADIC INTERNATIONAL, INC.
CODE OF BUSINESS CONDUCT AND ETHICS

INTRODUCTION

Our Company's reputation for honesty and integrity is the sum of the personal reputations of our directors, officers and employees. To protect this reputation and to promote compliance with laws, rules and regulations, this Code of Business Conduct and Ethics has been adopted by the Audit Committee of our Board of Directors. This Code of Conduct is only one aspect of our commitment. You must also be familiar with and comply with all other policies contained in our employee policy manual.

This Code sets out the basic standards of ethics and conduct to which all of our directors, officers and employees are held. These standards are designed to deter wrongdoing and to promote honest and ethical conduct, but will not cover all situations. If a law conflicts with a policy in this Code, you must comply with the law; however, if a local custom or policy conflicts with this Code, you must comply with the Code.

If you have any doubts whatsoever as to the propriety of a particular situation, you should submit it in writing to our Company's Chief Executive Officer, who will review the situation and take appropriate action in keeping with this Code, our other corporate policies and the applicable law. If your concern relates to that individual, you should submit your concern, in writing, to the Chairman of the Audit Committee. The name and mailing address of each of those individuals is included at the end of this Code.

Those who violate the standards set out in this Code will be subject to disciplinary action.

1. Scope
If you are a director, officer or employee of the Company or any of its subsidiaries, you are subject to this Code.

2. Honest and Ethical Conduct
We, as a Company, require honest **and** ethical conduct from everyone subject to this Code. Each of you has a responsibility to all other directors, officers and employees of our Company, and to our Company itself, to act in good faith, responsibly, with due care, competence and diligence, without misrepresenting material facts or allowing your independent judgment to be subordinated and otherwise to conduct yourself in a manner that meets with our ethical and legal standards.

3. Compliance with Laws, Rules and Regulations
You are required to comply with all applicable governmental laws, rules and regulations, both in letter and in spirit. Although you are not expected to know the details of all the applicable laws, rules and regulations, we expect you to seek advice from our Company's Chief Executive Officer if you have any questions about whether the requirement applies to the situation or what conduct may be required to comply with any law, rule or regulation.

4. Conflicts of Interest
You must handle in an ethical manner any actual or apparent conflict of interest between your personal and business relationships. Conflicts of interest are prohibited as a matter of policy. A "conflict of inter-

est" exists when a person's private interest interferes in any way with the interests of our Company. For example, a conflict situation arises if you take actions or have interests that interfere with your ability to perform your work for our Company objectively and effectively. Conflicts of interest also may arise if you, or a member of your family, receive an improper personal benefit as a result of your position with our Company.

If you become aware of any material transaction or relationship that reasonably could be expected to give rise to a conflict of interest, you should report it promptly to our Company's Chief Executive Officer.

Conflicts of interest are prohibited as a matter of Company policy, except under guidelines approved by the Board of Directors. The following standards apply to certain common situations where potential conflicts of interest may arise:

A. Gifts and Entertainment
Personal gifts and entertainment offered by persons doing business with our Company may be accepted when offered in the ordinary and normal course of the business relationship.

However, the frequency and cost of any such gifts or entertainment may not be so excessive that your ability to exercise independent judgment on behalf of our Company is or may appear to be compromised.

B. Financial Interests In Other Organizations
The determination whether any outside investment, financial arrangement or other interest in another organization is improper depends on the facts and circumstances of each case.

Your ownership of an interest in another organization may be inappropriate if the other organization has a material business relationship with, or is a direct competitor of, our Company and your financial interest is of such a size that your ability to exercise independent judgment on behalf of our Company is or may appear to be compromised. As a general rule, a passive investment would not likely be considered improper if it: (1) is in publicly traded shares; (2) represents less than 1 % of the outstanding equity of the organization in question; and (3) represents less than 5% of your net worth. Other interests also may not be improper, depending on the circumstances.

C. Outside Business Activities
The determination of whether any outside position an employee may hold is improper will depend on the facts and circumstances of each case. Your involvement in trade associations, professional societies, and charitable and similar organizations will not normally be viewed as improper. However, if those activities are likely to take substantial time from or otherwise conflict with your responsibilities to our Company, you should obtain prior approval from your supervisor. Other outside associations or activities in which you may be involved are likely to be viewed as improper only if they would interfere with your ability to devote proper time and attention to your responsibilities to our Company or if your involvement is with another Company with which our Company does business or competes. For a director, employment or affiliation with a Company with which our Company does business or competes must be fully disclosed to our Company's Board of Directors and must satisfy any other standards established by applicable law, rule (including any rule of any applicable stock exchange) or regulation and any other corporate governance guidelines that our Company may establish.

D. Indirect Violations

You should not indirectly, through a spouse, family member, affiliate, friend, partner, or associate, have any interest or engage in any activity that would violate this Code if you directly had the interest or engaged in the activity. Any such relationship should be fully disclosed to our Company's Chief Executive Officer (or the Board of Directors if you are a director of our Company), who will make a determination whether the relationship is inappropriate, based upon the standards set forth in this Code.

5. Corporate Opportunities

You are prohibited from taking for yourself, personally, opportunities that are discovered through the use of corporate property, information or position, unless the Board of Directors has declined to pursue the opportunity. You may not use corporate property, information, or position for personal gain, or to compete with our Company directly. You owe a duty to our Company to advance its legitimate interests whenever the opportunity to do so arises.

6. Fair Dealing

You should endeavor to deal fairly with our Company's suppliers, competitors and employees and with other persons with whom our Company does business. You should not take unfair advantage of anyone through manipulation, concealment, abuse of privileged information, misrepresentation of material facts, or any other unfair-dealing practice.

7. Public Disclosures

It is our Company's policy to provide full, fair, accurate, timely, and understandable disclosure in all reports and documents that we file with, or submit to, the Securities and Exchange Commission and in all other public communications made by our Company. Moreover, federal securities laws impose strict prohibitions and limitations on the company regarding the selective disclosure of material nonpublic information concerning the Company. Accordingly, the policies outlined below are intended to prevent unauthorized disclosure of nonpublic information.

- The Company's Chief Executive Officer or his designee is the designated spokesman for the Company for the dissemination of Company information to the public. No other person is authorized to disclose Company information or speak on such matters with the public. The public includes, but is not limited to, stockholders, analysts and members of the media. Any inquiries for information must be referred to the Company's Chief Executive Officer.
- Unauthorized disclosure of Company information, whether confidential or not, is forbidden. Authorization for release must be obtained prior to release from the Company's Chief Executive Officer or his designee.
- Under no circumstances is a director, officer or employee (other than the Chief Executive Officer) to respond to rumors regarding the Company that may appear on the Internet or any other medium.

8. Confidentiality

You should maintain the confidentiality of all confidential information entrusted to you by our Company or by persons with whom our Company does business, except when disclosure is authorized or legally mandated. Confidential information means information maintained by the Company regarding its operational, financial, legal, and administrative activities that has not been released by the Company to the public through press releases, annual reports or periodic filings with the Securities and Exchange Commission. The above categories are very broad as to the information encompassed and would include, but not be limited to, Company procedures, personnel or customers. The presumption should thus be in favor of particular information being confidential. Generally, confidential information includes all non-public information that may be of use to competitors, or harmful to the Company or its customers, if disclosed. Your obligation to preserve confidential Company information survives your separation from the Company.

9. Insider Trading

If you have access to material, non-public information concerning our Company, you are not permitted to use or share that information for stock trading purposes, or for any other purpose except the conduct of our Company's business. All non-public information about our Company should be considered confidential information. Insider trading, which is the use of material, nonpublic information for personal financial benefit or to "tip" others who might make an investment decision on the basis of this information, is not only unethical but also illegal. The prohibition on insider trading applies not only to our Company's securities, but also to securities of other companies if you learn of material non-public information about these companies in the course of your duties to the Company. Violations of this prohibition against "insider trading" may subject you to criminal or civil liability, in addition to disciplinary action by our Company.

If the Board of Directors adopts an insider trading policy for the Company to promote compliance with insider reporting obligations and to enhance compliance with insider trading prohibitions, which may include, among other things, certain "black out periods," you will be required to comply with the applicable provisions of this policy.

10. Protection and Proper Use of Company Assets

You should protect our Company's assets and promote their efficient use. Theft, carelessness, and waste have a direct impact on our Company's profitability. All corporate assets should be used for legitimate business purposes. The obligation of employees to protect the Company's assets includes its proprietary information. Proprietary information includes intellectual property such as trade secrets patents, trademarks, and copyrights, as well as business, marketing and service plans, engineering and manufacturing ideas, designs, databases, records, salary information and any unpublished financial data and reports. Unauthorized use or distribution of this information would violate Company policy. It could also be illegal and result in civil or even criminal penalties.

11. Interpretations and Waivers of the Code of Business Conduct and Ethics

If you are uncertain whether a particular activity or relationship is improper under this Code or requires a waiver of this Code, you should disclose it to our Company's Chief Executive Officer (or the Board of Directors if you are a director), who will make a determination first, whether a waiver of this Code is required and second, if required, whether a waiver will be granted. You may be required to agree to conditions before a waiver or a continuing waiver is granted. However, any waiver of this Code for an executive officer or director may be made only by the Company's Board of Directors and will be promptly disclosed to the extent required by applicable law, rule (including any rule of any applicable stock exchange) or regulation.

12. Reporting any Illegal or Unethical Behavior

Our Company desires to promote ethical behavior. Employees are encouraged to talk to supervisors, managers or other appropriate personnel when in doubt about the best course of action in a particular situation. Additionally, employees should promptly report violations of laws, rules, regulations or this Code to our Company's Chief Executive Officer. Any report or allegation of a violation of applicable laws, rules, regulations or this Code need not be signed and may be sent anonymously. All reports of violations of this Code, including reports sent anonymously, will be promptly investigated and, if found to be accurate, acted upon in a timely manner. If any wrongdoing relates to accounting or financial reporting matters, or relates to persons involved in the development or implementation of our Company's system of internal controls, a copy of the report will be promptly provided to the chairman of the Audit Committee of the Board of Directors, which may participate in the investigation and resolution of the matter. It is the policy of our Company not to allow actual or threatened retaliation, harassment or discrimination due to reports of misconduct by others made in good faith by employees. Employees are expected to cooperate in internal investigations of misconduct. Please see our Whistleblower Policy for details on reporting illegal or unethical conduct and the protections our Company provides.

13. Compliance Standards and Procedures

This Code is intended as a statement of basic principles and standards and does not include specific rules that apply to every situation. Its contents have to be viewed within the framework of our Company's other policies, practices, instructions and the requirements of the law. This Code is in addition to other policies, practices or instructions of our Company that must be observed. Moreover, the absence of a specific corporate policy, practice or instruction covering a particular situation does not relieve you of the responsibility for exercising the highest ethical standards applicable to the circumstances.

In some situations, it is difficult to know right from wrong. Because this Code does not anticipate every situation that will arise, it is important that each of you approach a new question or problem in a deliberate fashion:

- a. Determine if you know all the facts.
- b. Identify exactly what it is that concerns you.
- c. Discuss the problem with a supervisor or, if you are a director, the Company's General Counsel.
- d. Seek help from other resources such as other management personnel.
- e. Seek guidance before taking any action that you believe may be unethical or dishonest.

You will be governed by the following compliance standards:
- *You* are personally responsible for your own conduct and for complying with all provisions of this Code and for properly reporting known or suspected violations;
- If you are a supervisor, manager, director or officer, you must use your best efforts to ensure that employees understand and comply with this Code;
- No one has the authority or right to order, request or even influence you to violate this Code or the law; a request or order from another person will not be an excuse for your violation of this Code;
- Any attempt by you to induce another director, officer or employee of our Company to violate this Code, whether successful or not, is itself a violation of this Code and may be a violation of law;
- Any retaliation or threat of retaliation against any director, officer or employee of our Company for refusing to violate this Code, or for reporting in good faith the violation or suspected violation of this Code, is itself a violation of this Code and our Whistleblower Policy and may be a violation of law; and
- Our Company expects that every reported violation of this Code will be investigated.

Violation of any of the standards contained in this Code, or in any other policy, practice or instruction of our Company, can result in disciplinary actions, including dismissal and civil or criminal action against the violator. This Code should not be construed as a contract of employment and does not change any person's status as an at-will employee.

This Code is for the benefit of our Company, and no other person is entitled to enforce this Code. This Code does not, and should not be construed to, create any private cause of action or remedy in any other person for a violation of the Code.

Chief Executive Officer
Mark A. Emalfarb

Adopted by Resolution of the Board of Directors
January 12, 2005

APPENDIX E: DYADIC WHISTLEBLOWER POLICY

DYADIC INTERNATIONAL, INC.
WHISTLE BLOWER POLICY

Introduction

As you are probably aware, a great deal of emphasis is placed upon the conduct of employees of public companies such as Dyadic International, Inc. One aspect of this emphasis relates to numerous federal laws and regulations that apply to our company and each of our employees. The Audit Committee of our Board of Directors has adopted the following policies and procedures to notify you of (1) certain specific actions that are explicitly prohibited and (2) the procedures that you and others may follow if there is reason to believe that any laws are being violated.

Prohibited Actions

Commission of any of the following acts will be considered just cause for immediate dismissal and may subject you to criminal liability:

1. Destroying, altering, mutilating, concealing, covering up, falsifying, or making a false entry in any records that may be connected to a matter within the jurisdiction of a federal agency or bankruptcy proceeding, in violation of federal or state law or regulations.
2. Altering, destroying or concealing a document, or attempting to do so, with the intent to impair the document's availability for use in an official proceeding or otherwise obstructing, influencing or impeding any official proceeding, in violation of federal or state law or regulations.
3. Fraudulently influencing, coercing, manipulating, or misleading any independent public accountant engaged in the performance of an audit of the financial statements of the Company for the purpose of rendering such financial statements materially misleading, in violation of federal or state law or regulations.
4. Discharging, demoting, suspending, threatening, harassing or discriminating in any manner against any employee, in violation of federal or state law or regulations, because of any lawful act by the employee in providing information to or assisting in any investigation by a supervisory employee, Congress or any federal agency; filing or assisting in any action alleging a violation of federal or state law or regulations; or knowingly taking any action harmful to any person for providing truthful information to a law enforcement officer relating to the possible commission of a federal offense.

Reporting of Concerns or Complaints

Taking action to prevent problems is part of the Company's culture. If you observe possible unethical or illegal conduct, you are encouraged to report your concerns. Employees and others involved with the Company are urged to come forward with any such information, without regard to the identity of position of the suspected offender.

Employees and others may communicate suspected violations of law, policy, or other wrongdoing, as well as any concerns regarding questionable accounting or auditing matters (including deficiencies in internal controls) by contacting the Company's Chief Executive Officer. If, for any reason, you would prefer to contact someone else, you may call the Chairman of the Audit Committee of the Board of Directors of the Company. You may also call our toll-free telephone number that we have set up for this purpose. That telephone number is 1-800-713-8656. In order to be better able to respond to any information, we would prefer that you identify yourself and give us your telephone number and other contact information when you make the report. You can be assured that any information will be treated with utmost confidence, as detailed below. However, if you wish to remain anonymous, it is not necessary that you give your name and position in any notification and caller ID will be deactivated on the telephone lines above.

Confidentiality

The Company will treat all communications under this Policy in a confidential manner, except to the extent necessary (1) to conduct a complete and fair investigation, or (2) for review of Company operations by the Company's Board of Directors, its Audit Committee and the Company's independent public accountants.

Retaliation

Any individual who in good faith reports a possible violation of the Company's Code of Business Conduct and Ethics, or of law, or reports any concerns regarding questionable accounting or auditing matters, even if the report is mistaken, or who assists in the investigation of a reported violation, will be protected by the Company. Retaliation in any form against these individuals will not be tolerated. Any act of retaliation should be reported immediately and will be disciplined appropriately.

Questions

If you have any questions regarding this information, please feel free to contact us as indicated above.

Adopted by Resolution of the Audit Committee
January 12, 2005

REFERENCES

Business Week. 2010. Executive Profile and Bio: Mark A. Emalfarb. January 28th http://investing.businessweek.com/research/stocks/people/person.asp?personId=9495654 &capId=4272801&previousCapId=93958&previousTitle=MAXYGEN%20INC

Daft, R. L. 2010. *Organization Theory and Design* (10th edition). Thomson: South-Western College Publishing.

Delaware Court Case. 2009. *The 15th Judicial Circuit Court, April, 14th. Case No.: 502009CA010680XXXXMB*, pp. 1-71.

Florida Trend. 2009. *A Matter of Chemistry.* http://www.dyadic.com/pdf/FLTRENDSEPT2009COMEBACK.pdf; pp. 55-57.

Obermueller, A. 2009(a). *Government-Driven Investing: Building Wealth from   Government Actions.* http://www.dyadic.com/pdf/Dyadicpiece.pdf

2009(b). *Government-Driven Investing How a University of Iowa*
Student and a Russian Scientist Created a Fungus that Could Replace Oil Forever. http://www.dyadic.nl/Publications/How%20a%20University%20of%20Iowa%20Student%20Government-Driven%20Investing.pdf

Phone interview with Mark Emalfarb, president and CEO, November 12, 2009.

Sandra ,W.A., Hinz, S.W, Pouvreau, L., Joosten, R. Bartels, J., Jonathan, M.C, Wery, J. Schols, H.A. 2009. Hemicellulase production in Chrysosporium lucknowense C-1. *Journal of Cereal Science.* 50(3): 318-323.

South Florida Business Journal. 2009. *Dyadic CEO makes a comeback. Company sues prominent legal and accounting firms.* http://www.dyadic.com/pdf/SouthFloridaBusinessJournalMay1_2009.pdf

Westcott, P. 2007. U.S. *Ethanol Expansion Driving Changes Throughout the Agricultural Sector.* http://www.ers.usda.gov/AmberWaves/September07/Features/Ethanol.htm

The Prince Edward Island Bioscience Cluster: Creating a Knowledge-Based Economy

STEVEN CASPER[1], JUERGEN KRAUSE[2], AND ADELEE MACNEVIN[2]

[1]*Keck Graduate Institute of Applied Life Sciences;* [2]*University of Prince Edward Island*

- **Summary and key issue/decision:** Prince Edward Island is Canada's smallest province, with a population of under 140,000. Its economy, historically focused on agriculture, fisheries, and tourism, lags behind the rest of Canada. Starting in the late 1990s the PEI Provincial Government launched an ambitious plan to develop an internationally competitive biotechnology cluster. Choosing to focus on the bioactives and natural products segment of the industry, substantial investments were made in upgrading scientific capacities on the island and a variety of commercially oriented policies aimed at creating an infrastructure for start-up companies and attracting investments from existing bioscience companies. The case examines the processes by which biotechnology clusters emerge and become sustainable, and explores whether governments can successfully trigger cycles of industrial upgrading within technology clusters. The case ends with a decision of whether the PEI government should invest some CAN$30 million in the BioCommons Research Park, a technology park.

- **Companies/institutions:** Prince Edward Island Provincial Government; PEI Bioalliance; Canadian National Research Council Institute for Health and Nutrition

- **Technology:** Bioactives and natural products

- **Geography:** Canada

- **Keywords:** biotechnology clusters; technology policy; industrial upgrading; natural products and bioactives

---

*   *This case was prepared as a basis for class discussion rather than to illustrate either effective or ineffective handling of an administrative situation.*

INTRODUCTION

As Deputy Minister Michael Mayne stared out the window from his government office in Charlottetown, several young children played in a leafy park across the street. Mayne wondered what the future would hold for these children. Historically, most residents of Prince Edward Island, Canada's smallest province, would wind up working in agriculture, fishing, tourism, or one of the other predominately low-wage industries that dominated the local economy. With a small population, most of PEI's young people that received university training would eventually leave PEI to seek higher paying jobs in other parts of Canada. As Deputy Minister for Innovation and Advanced Learning, Mayne was the point person in PEI's ambitious attempt to fundamentally change the structure of its economy. If the government's plans succeeded, the young children playing in the park would grow up to find any number of high quality jobs available to them in biotechnology and other knowledge-economy industries.

Mayne, only in his late 30s, had experienced a dizzying career ascent, moving in less than five years from a normal academic job as an associate professor, to becoming Director of a major national research institute, to becoming a Deputy Minister in the PEI Provincial Government. A PEI native, Mayne left the University of Prince Edward Island (UPEI) after receiving an undergraduate degree in biology to pursue graduate studies in Toronto. He then worked for several years as a postdoc at the US National Institute of Health before returning to Canada to pursue a successful career as a neuroscience professor at the University of Manitoba, one of Canada's universities oriented towards science and technology. In 2003 he was recruited by the National Research Council of Canada as Lead Scientist for the newly formed Institute for Nutrisciences and Health (NRC-INH), a key institute created to spearhead PEI's push into biotechnology, and was later appointed Director of Research for the Institute.

An energetic individual, Mayne quickly turned the NRC-INH into a success through creating an innovative design that integrated the activities of academic and commercially oriented researchers. The institute attracted several million dollars in grant funding to pursue a mix of academic and commercially oriented programs, and enticed several promising companies from outside PEI to move into the institute's incubator. In late 2007, after a change in provincial government, Mayne was seconded into the Provincial Government. As Deputy Minister for Innovation and Advanced Learning, he would bear responsibility for developing policies that would eventually channel hundreds of millions of dollars into economic development projects.

In the summer of 2009 Mayne faced an important decision: should the PEI Provincial Government commit upwards of CAD$30 million to construction of the BioCommons Research Park, a 65 acre technology park oriented towards biotechnology? Creating a technology park was a core element in "Island Prosperity: An Agenda for Change", a widely supported government development plan focused largely around biotechnology that was initiated by Mayne's office after his appointment in 2007. While Mayne had the political support to commit funding to the park, there were concerns that building a major technology park was premature, given the nascent origins of the island's biotechnology industry and its inability to meet some development goals. Should the government build the park now, or use the CAD$30 million to fund other bioscience projects, perhaps oriented more towards the building of basic scientific infrastructure?

PRINCE EDWARD ISLAND: ECONOMIC STRUCTURE AND DEMOGRAPHIC TRENDS

Prince Edward Island is Canada's smallest province, with about 140,000 residents. It is located in Atlantic Canada in the Gulf of St. Lawrence, East of New Brunswick and North of Nova Scotia, and about 300 miles north of Maine (see Figure 1). The island is famous in Canada for being the "birthplace" of the Canadian Confederation, in 1864. Internationally, PEI is perhaps best known as the home of Anne of Green Gables, the popular fictional character of Lucy Maud Montgomery.

PEI's economy has long been shaped by its natural resource endowments. The island is marked by a distinctive iron enriched red soil that is ideal for potato farming, and also has a productive dairy industry. In the fisheries sector, the island has strong lobster and shellfish industries. In 1998 agriculture, fishery, and forestry products accounted for 80% of PEI's exports, which totaled CAN$466 million (PEI Provincial Treasury, 2009: 71).

While suffering from harsh winters, PEI has a mild summer that, coupled with miles of beaches and a lush, attractive countryside, create the foundation for a large tourist industry. While many come to visit the beaches, golf courses, and farms located within "Anne's Land" in the central Cavendish region, the island's historic Victorian era capital city Charlottetown has also become a magnet for visitors. Tourism revenue grew substantially during the 2000s following the opening, in 1997, of the 8-mile long Confederation Bridge linking New Brunswick with PEI, and in addition, the increasing number of cruise ships stopping in Charlottetown. In 2008 over a million tourists visited the island, generating CAN$350 million in revenues (The Guardian, 2009).

While PEI's traditional strengths in agriculture, fisheries, and tourism have generated economic in-flows from exports and tourist spending for the Province, this economic structure has lead it to lag in economic growth compared to other Canadian Provinces. A reliance on primary industries and tourism, both of which hire predominantly lower-skilled and seasonal labor, has driven labor productivity in PEI to be the lowest in Canada, more than 40% below the national average (Figure 2). Low wages are the result. In 2006 average weekly wages in PEI were CAN$625, compared to the Canadian norm of CAN$770, and an average in high-growth Alberta of close to CAN$850 (Mayne, 2008: 37).

Figure 1: Map of Prince Edward Island
Source: Wikipedia, 2008

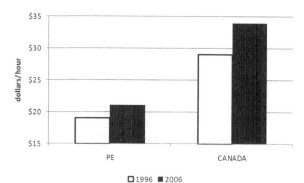

Figure 2: Productivity in real GDP per hour worked
Source: Island Prosperity- A Focus for Change (2008)

Widespread seasonal employment within the agriculture, fisheries, and tourism industries has also driven down income. Many employees within seasonal industries work for only a few months a year and then live off government unemployment benefits during the long winter, at about 50% wages. During the 1990s, for example, a study found that 17.6% of earnings reported by PEI residents on income taxes were from unemployment benefits, compared to a Canadian national average of 5% (MacDonald, 1998). Seasonal employment drives down earnings within the province. Median family income for non-elderly families in PEI, at CAN$47,000 in 2005, were a third lower than the Canadian national average of CAN$64,000 (Mayne, 2008: 38). Unemployment in PEI, which has hovered in the 10-15% range during the 2000s, has long been among the highest in Canada. An important cause of unemployment has been shifts in the structure of primary industries, and particularly agriculture. Over the past century concentration within agriculture has increased due to the purchase of many family farms by corporations seeking economies of scale (see Figure 3).

Demographic trends are also worrisome. Of particular concern is a trend of young, higher-skilled islanders moving to other parts of Canada. Historically, PEI's educational system lagged behind Canada in part due to the relatively low demand for university graduates within the local economy. PEI does have a strong university, the University of Prince Edward Island (UPEI), which is ranked in the top 10 in Canada. About 750 students graduate every year from UPEI. However, the island's industry structure, focused primarily on lower skilled jobs, limits opportunities for university graduates. Local employers must also compete with much higher wages in other parts of Canada, and in particular oil and natural resource rich Alberta (see Appendix A). While precise statistics on yearly out-migration by new university graduates do not exist, Michael Mayne believes this figure could be as high a 70%.

Driven by a strong personal identity as "islanders", many migrants long to return to PEI, and eventually do, as retirees. This trend, however, creates fiscal pressure on the Provincial Government, which

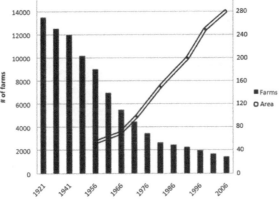

Figure 3: Trends in farm numbers and size 1921-2006
Source: Island Prosperity- A Focus for Change (2008): 24

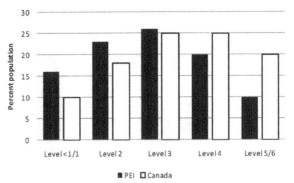

Figure 4: Math achievement, Program for International Student Assessment (PISA).
Source: Island Prosperity- A Focus for Change (2008): 32

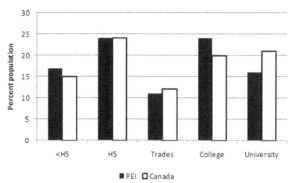

Figure 5: Educational attainment ages 25-64, PEI and Canada, 2006. Within the Canadian context, "colleges" offer a range of certificate programs and 2 to 4 year diplomas in vocation oriented subjects.
Source: Island Prosperity- A Focus for Change (2008)

pays for many islanders' schooling and social costs as retirees, but fails to collect tax revenues during a person's working years. The out-migration trend, coupled with the aging baby-boomer population and lower birth rates, has created a rapidly graying population. In 2006, one in three workers were baby boomers, and, as a result, between 2011 and 2015, the annual growth rates in the labour force will fall by at least 50% (Mayne, 2008: 29).

THE DECISION TO ENTER THE KNOWLEDGE ECONOMY

PEI's push into the knowledge-based industries started with a systematic "Knowledge Assessment" exercise sponsored by the Provincial Government in 1999, followed by a series of road mapping studies and meetings among PEI business, political, and educational elites during the 2000-2003 period. A consensus formed among the island's leaders that the government of PEI needed to aggressively introduce policies to push the province into the knowledge economy. To quote Mayne, "A decision was made to move from public policies generating a high quantity of jobs, most of which were low-skilled, to a policy of creating higher quality jobs—drawing on the skills of PEI citizens." The creation of these high quality jobs, however, would require creative government policies aimed at building capacity and infrastructure to support the new industries, train islanders in the skills to be productively employed within the knowledge economy, and create incentives for existing companies to expand and new companies to set up shop on PEI.

The PEI Provincial Government decided to focus on two industries, aerospace and biotechnology. The aerospace cluster would be developed in and around an old Canadian Air Force base located near Summerside, while the biotechnology industry would be located in and around Charlottetown.

PEI's first capacity building investments were in aerospace. Beginning in 1999 about CAN$5.5 million was invested to convert the old air force base near Summerside into the Slemon Industrial park. An additional CAN$1 million was invested in educational programs aimed at training skilled aircraft mechanics at Holland College, a local vocational school. Over the past decade, several companies involved in the repair and refurbishment of aircrafts, including large commercial jets, have moved to PEI. Over 1000 individuals are currently employed within the Slemon complex, including over 350 at Vector Aerospace, a company specialized in the repair and refurbishing of commercial aircraft engines. Aerospace is PEI's fourth largest industry and second largest exporter, accounting for 25% of PEI's international exports.

While the employment figures are impressive, the PEI Provincial Government has granted all companies located within the complex a 10-year tax holiday, limiting the fiscal impact of this new industry on the Provincial Government's finances. There are debates on whether tax holidays should be extended to prevent companies from relocating once the initial tax holidays expire. Absent this large incentive there is a concern that Vector and other local aerospace companies could easily move to near-by Quebec, another province with a substantial aerospace industry that offers similar tax incentives.

Biotechnology was targeted as the second major growth industry. The selection of biotechnology was strongly influenced by the success of one long-standing local company, Diagnostic Chemicals Limited (DCL), which was started in 1970 by a UPEI chemistry professor and then dean of science, Regis Duffy. DCL became a leading supplier of chemical intermediates within the diagnostics and pharmaceutical industries, employing over 200 individuals by the late 1990s and earning upwards of CAN$35 million in yearly revenues. In the early 2000s Duffy was a member of the UPEI board of governors and highly influential within the small community of PEI political and economic elites.

In addition to DCL, PEI did have a few other companies active in bioscience fields during the early 2000s. The island was home to AquaHealth, a small biotechnology company focused on the development of fish vaccines. While fish vaccines sell for pennies on the dollar, this market has boomed in recent years due to the growth of fish farms, the largest of which can contain millions of fish, and the susceptibility of these farms to epidemics. Other bioscience oriented firms on PEI included Agritech, a contract research company specialized in running sophisticated trials of new seed varieties for international agri-food companies, and Phytocultures, a company that had pioneered the use of plant culture technology within the Atlantic Canada potato industry.

UPEI has both undergraduate and graduate programs in the biological sciences and chemistry, and graduates around 220 science degree students every year. By comparison, the two largest universities in Atlantic Canada, Dalhousie University in Nova Scotia, and Memorial University in Newfoundland graduate 546 and 748 science degree students every year respectively (Table 1). UPEI houses a strong School of Business, which graduated an average of 108 students per year from 2006 to 2009. UPEI is also home to the Atlantic Veterinary College, one of only five veterinary schools in Canada and an acknowledged leader in many animal health related fields, especially the shell and fin-fish areas and dairy. In addition to training veterinarians, most AVC professors operated grant-funded research labs and train PhD students. While PEI does not have a medical school (the closest is in Halifax, Nova Scotia), as a site for preclinical trials involving animals, the AVC could potentially play an important role within the development of local bioscience value chains surrounding human therapeutics.

The Food Technology Center (FTC), a government run contract research organization located in Charlottetown on the UPEI campus, has supported local companies within the agriculture and food industries. The FTC employs several full-time research scientists that consult with food companies on a variety of nutriscience and functional foods areas surrounding ingredient testing, the validation of health claims, and manufacturing methods. The facility also has the capacity to set up pilot-production plants for food companies.

The bioscience industry has a more complex industrial structure, requiring a more sophisticated policy approach than had been employed in aerospace. As with the aerospace industry, the PEI government would create a variety of tax and related incentives aimed at inducing established bioscience companies to move to PEI. However, to make PEI an attractive site for investment, the government would have to make serious upgrades to the scientific infrastructure in and around UPEI and develop a labor market of scientists and skilled technicians available for employment within bioscience companies.

Table 1: 2007-2009 Average graduates from Atlantic Canadian universities

Annual average graduated	UPEI & AVC	Dalhousie University (Halifax, NS)	Memorial University (St.John's, NF)
All Bachelor's Degrees	624	2469	2449
Bachelor of Science Degrees	152	624	665
Science Post Grads	70	145*	83
All Science Degrees	222	769	748

*Includes Dalhousie's Faculty of Medicine

Source: Dalhousie Data Requests, 2009; Gong, 2009; Fowler and Leake, 2008

Moreover, biotechnology clusters are closely associated with the existence of research-focused entrepreneurial start-up companies. While the success of DCL and the local prominence of Duffy could create an important demonstration effect, the region had yet to develop a tradition of technology-oriented entrepreneurship. Moreover, during the early 2000 period the science infrastructure on PEI was relatively weak, the university had only begun to promote commercialization processes, and there were no local venture capital companies on PEI.

PUBLIC POLICY TOWARDS BIOTECH: THE "PEI MODEL"

To promote biotechnology, the Provincial Government announced in 2003 that it would invest $100 million over several years to create a bioscience industry in PEI. Much of this funding would initially be spent on capacity building programs in and around UPEI. However, in partnership with the programs developed through the federal government, the PEI Provincial Government also developed a variety of initiatives that would allow direct investment in companies. Accompanying these investments were ambitious policy goals. By 2010, the province aimed to employ over 1000 people in private sector bioscience firms and achieve at least $200 million in revenue and $40 million in private sector R&D spending (ACOA, 2005).

In developing a strategy, PEI could draw on two intrinsic advantages that are similar to successful bioscience clusters in Israel, Singapore, and Denmark (Porter, 1998; Ahn & Meeks, 2007; Fair, 2007). First, there are benefits of being small. Charlottetown is a small city, with a population of about 30,000. Elites from political, business, and educational circles know one another and, once a strong consensus was formed around the need to invest in the biosciences, could develop an orchestrated approach to reaching consensus around economic development goals, developing policies, and implementing them. In 2005 an industry association called the PEI BioAlliance was formed. While a private organization, the BioAlliance worked closely with government in developing and implementing policy. The BioAlliance also worked with the government on prospecting, the recruitment of existing companies onto PEI.

Second, PEI has an abundance of government-controlled resources that can be channeled into the local economy. In addition to economic development funds controlled by Provincial Government, the Canadian Federal Government has set up the Atlantic Canada Opportunities Agency (ACOA), which delivers numerous programs oriented towards economic development, and has several local offices in Charlottetown and in the region. Funding through federal programs could be used to augment Provincial funding, allowing more money to be spent on both capacity building and company investments.

A variety of policies were introduced and refined over the years. During the 2005-2009 years a distinct "model" for the PEI cluster had emerged and had several elements:

NATURAL PRODUCTS AND BIOACTIVES AS A FOCUS FOR CAPACITY BUILDING

In order to foster synergies across companies and allow more targeted investments in scientific capacity, PEI decided to channel most of its capacity building investments in biotechnology into the broad area of natural products and bioactives. Pharmaceutical companies had long engaged in bioprospecting activities – hoping to locate bioactive compounds within naturally occurring substances gathered from different areas of the world. Through the bioactives focus, scientific capacity in a variety of advanced screening and assay technologies could be developed and used to screen naturally occurring compounds in and around PEI for biological activity. The focus on bioactives also helped garner local political support for biotechnology investments, as most islanders closely associate PEI with its natural environment, and particularly its rich marine resources. This environment could now possibly be tapped by local biotechnology companies to create new wealth for the island.

Another important advantage of the natural products and bioactives focus, however, is the ability to link technical resources in this area to other industries outside of pharmaceuticals. Natural products screening technologies could also be used to work with food and nutraceutical companies to identify and validate health claims, or possibly develop new products. Animal health companies could also participate in the cluster, both through the identification of bioactives that could possibly serve as veterinary therapies and diagnostics, and through developing and validating health claims associated with additives and foods. As PEI has a presence in several food, dairy, livestock, and marine related industries, companies in these industries could potentially tap into biotechnology capabilities on the island to upgrade their capabilities and product offerings.

The centerpiece of PEI's capacity building initiatives was the opening in 2007 of the National Research Council Institute for Nutrisciences and Health (NRC-INH) on the UPEI campus. The Canadian NRC is a federal agency that runs and operates 27 specialized research institutes, each with an applied research focus. About CAD$40 million was initially invested in the center, including CAD$13.5 million for constructing the facility, CAD$20million for operations, and CAD$6.5 million to purchase core equipment. The three-story building includes 50,000 square feet of working space and could house about 20 research scientists and teams, over 100 people in total.

As the founding Director and Project Leader of the NRC-INH, Mayne was responsible for overseeing the research orientation and general design of the institute. The initial research focus of the institute was screening bioactives against neurological disorders, obesity, and infections. In addition to recruiting research teams to operate labs in each of these areas, the NRC-INH developed several core or service oriented facilities oriented towards the screening of natural products for bioactivity. Initial core labs were oriented around analytical chemistry and molecular biology, and include a wide range of research tools, including mass spectrometry, NMR, cell sorting flow cytometry, and a variety of assays using cells and, in conjunction with laboratories at the AVC, animal models.

The facility was designed to house companies: a technology incubator is contained in the facility with space for six companies, which are allowed to make use of the core labs and are given laboratory and office space in close proximity to the academic labs. Scientists and companies within the center could also make use of the NRC's strong capabilities in commercialization. These would include a technology licensing office within the institute and access to NRC-controlled funding for commercialization related projects through the Industrial Research Assistance Program (IRAP).

Major investments have also been made to the Atlantic Veterinary College. About CAD$32 million have been invested in the facility in the mid to late 2000s. This included the creation of a new research building, the Atlantic Center for Comparative Medical Research. New facilities have been also added to the Food Technology Center, which received CAD$7.5 million in funding from the "Atlantic Natural Products Development Initiative" to build a 5,000-square-foot-natural-products facility. The center has developed expertise in extracting bioactive compounds from a variety of food and other natural products.

In addition to facility upgrades, an effort has been made to attract prominent professors to UPEI under the guise of a national program to designate and fund "Canada Research Chairs." The university currently has six Canada Research Professors, four of which are in the biomedical sciences. One example was the appointment in 2006 of Russ Kerr to a Canada Research Chair at UPEI. Kerr is a prominent scientist involved in the identification, isolation and synthesis of bioactive compounds found in marine environments. His chemistry lab has more than 20 team members, and has identified compounds located within coral reefs that have anti-inflammatory properties.

To further promote commercialization, in 2006 UPEI created Three Oakes Innovations Inc. (TOI), a tech transfer organization located on campus. TOI's mandate is to promote the commercialization of research at UPEI, and they work with researchers to identify, evaluate, protect, and commercialize the intellectual property that results from research. In addition to licensing technology to established firms, TOI had successfully worked with UPEI faculty to launch several start-ups.

GOVERNMENT FUNDING AND INCENTIVE PROGRAMS FOR COMPANIES AND ENTREPRENEURS

Lacking a venture capital sector, bioscience companies in PEI have turned to a variety of government sources for funding. Most bioscience companies active on PEI have received a mix of federal and provincial funding. The most important funding agency on PEI is the federal Atlantic Canada Opportunities Agency (ACOA), which through its Atlantic Innovation Fund (AIF) has invested over CAD$500 million in a variety of non-profit and for-profit research and development projects across Atlantic Canada during the 2001-2009 period. Most funding from ACOA is competitive and based on a peer-review process. However, PEI has received a large share of bioscience investments: between 2001 and 2008, a total of CAD$130 million funded 24 projects on PEI, including CAD$65 million that was invested directly into companies. PEI has succeeded in part due to the coordinated effort of its policies towards biotechnology. Most AIF investments require matching funds from public or private sources, which were often provided or orchestrated through the PEI Provincial Government. The PEI Provincial Government has also at times hired local consultancies to provide support for local research labs on companies in developing commercialization plans as part of AIF proposals.

Most AIF grants, when matching funds are included, total between CAD$1 million and $9 million, with many grants to entrepreneurial companies totaling around CAD$5 million and lasting between 3 and 5 years. AIF funding is generally provided towards projects with clear goals, such as developing a particular R&D capacity or completing a well-defined research project. An example of a recently funded project is the CAD$2.9 million dollar grant to AquaBounty, a PEI-based subsidiary company, to conduct research on PEI towards creating a transgenic salmon species that would grow to maturity faster; including matching funds and private capital raised by AquaBounty, the total budget for this project is CAD$4.6 million. The AIF also funds commercialization-oriented projects within academic labs. For example, in 2007 Russ Kerr received a CAD$3 million grant from the AIF to fund his lab's research efforts to create synthetic versions of marine compounds, funds matched by an additional CAD$4.3 million that were raised through local PEI funding and additional grants, primarily from federal sources (ACOA, 2007a).

The PEI Provincial Government has worked closely with ACOA to fund start-ups. Each of the companies located in the NRC-INH incubator has received funding from the AIF. While the AIF only funds R&D projects with defined milestones and commercialization plans, many local entrepreneurial companies have received smaller grants that can be used for operational expenses from the PEI Provincial Government. This has allowed start-up bioscience companies to be launched on PEI with essentially no private funding. An example is Nautilus, a company founded by Professor Russ Kerr to commercialize marine environment discoveries made in his lab. Another prominent example is the Atlantic Center for BioProducts Valuation (ACBV), a non-profit company organized along the lines of a start-up that is located within the NRC-INH incubator and managed by AVC professor Tarek Saleh. The ACBV has developed sophisticated mouse models in the areas of neurodegeneration, inflammatory diseases, and metabolic disorders in which small molecules can be tested. The ACBV is considered by scientific elites within Charlottetown as a crucial capability, as the mouse models can be used to perform efficacy related preclinical trials on many compounds discovered by PEI-based labs. The ACBV aims to raise over CAD$9.7 million in funding over its first five years, with CAD$3 million funded by the AIF and the rest through a variety of Canadian funding agencies such as the NRC-IRAP program, and the PEI Provincial Government (ACOA, 2007b).

AIF rules regarding financing have strongly impacted the strategies of commercially oriented actors on PEI. Funding channeled directly to companies is considered loans, generally repayable at a time when the company becomes profitable. AIF funds given to universities and other non-profit institutes are considered grants. As a result, organizations that in most respects are start-ups have an incentive to initially incorporate as non-profits in order to avoid having to pay back the AIF portion of their public funding immediately. AIF funding rules also encourage local companies to develop collaborations with academic research labs when seeking funding. In most cases intellectual property from such collaborations will flow to the company, while any funding that flowed to non-profit research organizations is treated as a grant.

A TARGETED PROGRAM TO ATTRACT COMPANIES TO THE ISLAND

PEI has developed a coordinated effort to attract bio-business to the island. A variety of tax and investment incentives are in place to attract investors. As in the case of aerospace, established companies employing at least 10 people and with over CAD$750,000 salaries and wages qualify for a 10 year tax holiday when moving to PEI. The Provincial Government also offered rebates on research and development expenditures for most start-ups and some established companies moving to PEI. In an effort to attract talented personnel, PEI has also set up a program in which many scientists and managers moving for the first time to PEI could apply for a multi-year income tax holiday (Jollymore & Johnston, 2007). Combined, these corporate and personal tax incentives were regarded in PEI as the strongest in Canada.

The PEI bioscience model is most strongly defined by a shared vision and commitment to building biotechnology across the island's elites. This shared vision, combined with the small-size of the local community, has allowed the PEI Government to effectively court prospective bioscience companies interested in investing on PEI.

An example of a company recruited to PEI as a result of strong government-industry coordination is Nature's Crops International (NCI). NCI is a North Carolina based agricultural company that produces a variety of specialty oils used in pharmaceuticals and cosmetics. NCI's oils are derived from a variety of niche-market seed crops. The company generally contracts out production of crops to farmers, who are paid a pre-arranged price for their crops. Due to preferable growing condition for some of its crops, the company began in 2007 to search for an Atlantic Canada base of operations. NCI initially settled on creating a subsidiary in New Brunswick, but after two years of stalled negotiations, decided to locate instead to PEI.

PEI's success in recruiting NCI illustrates the advantages of having coordinated policies linking government and industry elites. In late 2008 representatives from NCI met members of the PEI BioAlliance at a networking event. The BioAlliance members arranged for NCI's senior management to visit PEI and meet key business, government, and educational leaders on the island during a breakfast meeting and tour key bioscience facilities on the island. Within three months NCI, in response to tax and investment incentives and an agreement on regulatory rules surrounding the crops, committed to using the province as a base of operations. In addition to working with local farmers to produce its specialty crops, NCI committed to build a biorefinery on PEI to process the crops. As of Summer 2009, NCI was also in negotiations with the Provincial Government and NRC-INH to launch a research collaboration focused on developing new oil-based pharmaceutical products on PEI, a project potentially fundable through the AIF and other agencies.

Several other companies have moved to PEI, often through acquisitions. Novartis, for example, acquired the fish vaccine R&D company AquaHealth in 2005. Since the acquisition, Novartis has decided to move most of its research and development activities in the area of aquaculture to PEI. This now includes the construction of a manufacturing facility to produce its fish vaccines, which employs over 70 people, in addition to renewed R&D investments in this area. In 2007, Genzyme Diagnostics acquired the diagnostics division of DCL in a deal worth CAD$56 million. Genzyme Diagnostics will maintain and possibly expand its newly acquired PEI facilities to manufacture a variety of intermediate chemicals sold to equipment manufacturers for use in consumable diagnostic kits. The pharmaceutical division of DCL, BioVectra, was spun off as an independent company and continues to develop and manufacture chemical intermediaries for the pharmaceutical industry (see Appendix B).

PEI'S BIOSCIENCE INDUSTRY LANDSCAPE IN 2009

As of August 2009, eighteen bioscience companies with R&D capabilities were located on PEI. These include eight development stage biotechnology companies, many of which are housed in the incubator facility at the NRC-INH. PEI is also home to four "mature" bioscience companies, all with manufacturing facilities, including the subsidiaries of Genzyme Diagnostics and Novartis Animal Health. In addition to these core bioscience companies, the island is home to four R&D oriented companies in supporting and relating industries to bioscience, and four non-profit organizations founded by professors from the Atlantic Veterinary College, each of which had received funding from ACOA and was engaged in activities that could possibly be commercialized into for-profit ventures (see Table 2 and Figure 6 for a broader overview of the Canadian biotechnology industry).

Combined, 455 individuals are employed in R&D focused bioscience companies on PEI. The companies are active in a variety of product segments, including human health (9 firms), nutraceuticals (6), animal health (6), aquaculture (6), and agriculture (6). Firms are also involved in a variety of value chain activities, including research and development (13), manufacturing (8), and contract research (6). Fourteen of the firms had revenue, totaling about CAN$75 million total, though most of these firms (11/14) have revenues of less than CAN$1 million. Fourteen of the companies had successfully developed patents and sixteen companies, when surveyed, claimed to have a "key technology."

Many of the eight development stage biotechnology companies moved to PEI after being founded elsewhere. Reasons for the move vary, but include access to the incubator and facilities at the NRC-INH, the availability of government funding, and access to a growing, and relatively low-wage, skilled labor market pool on PEI. Examples of early stage biotechnology firms that have moved to PEI include

Table 2: Biotechnology company concentration by region

Region	Percentage
Alberta	8%
British Columbia	30%
Manitoba	2%
New Brunswick	2%
Newfoundland	1%
Nova Scotia	4%
Ontario	26%
Prince Edward Island	3%
Quebec	20%
Saskatchewan	4%

Source: BIOTECanada, 2010

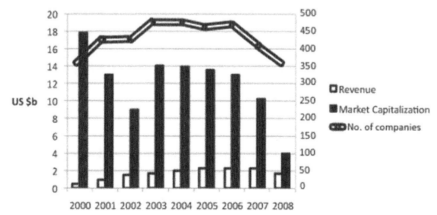

Figure 6: Canadian biotech industry indicators 2000-2008
Source: Beyond Borders: Global Biotechnology Report (2009)

Neurodyn, a neuroscience company originally founded at an NRC institute in Ottawa that moved to PEI primarily to collaborate with scientists at the NRC-INH. Other previously established companies that have moved into the NRC-INH incubator include Phyterra, a natural products company focused on identifying therapeutic benefits from micro-algae and Chemaphor, a chemistry oriented company focused on exploring how beta carotenes and other naturally occurring compounds can be used for animal and human health. An advantage of recruiting previously established start-ups is that the firms have likely secured venture or other forms of private capital, and have established company boards and scientific advisors. However, each of these companies has received R&D related funding from the AIF and the Provincial Government, often to participate in collaborative projects with NRC-INH scientists.

The upgrading of scientific capabilities in and around Charlottetown has created new markets, as well as upgrading possibilities, for companies on the island in related and supporting industries. For example, a local specialty metal fabrication company, DME, had previously developed a competency in designing and manufacturing vessels used by micro-brewers in the beer industry, but with the growth of biotechnology in the region has found local companies interested in purchasing somewhat similar vessels customized for pilot-scale bioprocessing activities. Based on success with local PEI clients, DME has now turned the fabrication of bioprocessing vessels into a successful export business, and considers biotechnology as a core segment for future growth.

An interesting example of upgrading surrounds Phytocultures, the potato plant culture company mentioned on page 5. While profitable for over two decades, the owner of the business, Don Northcott, had repeatedly failed in several attempts to expand the business by entering into plant culture ventures for other agricultural products, primarily due to the lack of local demand, and has never been able to grow the business to revenues above a few hundred thousand dollars. However, Phytocultures is currently collaborating with a scientist at the NRC-INH to validate potential health claims in new varieties of potatoes for which the company has developed patents. If successful, these potatoes could be marketed at a premium. Northcott has also partnered with several local investors to establish a distribution company, called Real Potatoes, which will market and distribute the new products.

Another local company formed to exploit the region's emerging capabilities in natural products technology is Abbey Foods, founded by John Rowe, a local serial entrepreneur. Rowe, while working in the British Columbia, had the idea to develop a method to manufacture an all natural form of honey that is not sticky and could be packaged and marketed in one ounce cubes, at a substantial market premium. Rowe was able to contract out research to the Food Technology Center (FTC) to develop a scalable manufacturing process for his technology, and eventually decided to lease space to house a temporary production facility at the FTC. During 2009, Abbey Foods was able to successfully distribute its non-stick honey to supermarkets across Canada, and recently reached an agreement to distribute the product to the US based Whole Foods Supermarket on a regional basis. By the end of 2009, Abbey Foods transitioned into their own manufacturing facility.

While PEI has experienced a marked increase in biotechnology related activity in the fast few years, policy-makers and industry experts on the island frequently debated whether the fledging cluster was sustainable. To help answer this question, a survey of all 18 local companies was conducted in the summer of 2009. This survey revealed several promising trends and also uncovered some problematic aspects of the emerging PEI bioscience cluster.

LABOR MARKET AND SKILLS

PEI biotechnology companies were asked whether they could find or develop the human resources they needed to succeed, which has been shown to be the most critical factor in developing a thriving biotechnology cluster across countries (Fair, 2007). Data was gathered on the skill composition of local companies (Figure 7). Less than one-third of the employees of PEI biotechnology companies are working as scientists, and only 52 of the 455 total employees had PhDs. In part due to the establish-

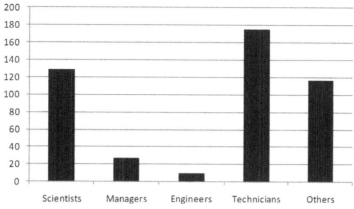

Figure 7: Skill composition of PEI biotechnology firms
Source: Survey data, 2009

ment of medium-size manufacturing facilities at Novartis Animal Health, Genzyme Diagnostics, and Biovectra, a large number of technicians were employed on the island. As most companies preferred to recruit technicians locally, this created a strong demand for the locally trained individuals. To help meet this demand, Holland College, PEI's main vocational training school, has developed and is expanding a program to train and certify bioscience technicians.

Most PEI firms reported that labor market turnover is very low, at less than 5%. They also thought that wages at their company were generally higher than average wages on PEI, but significantly lower than bioscience industry norms, particularly compared to the United States. While this labor market structure creates a competitive advantage for local firms, managers wondered if wages would begin to rise once additional firms entered the local market, creating stronger competition for skilled employees. Other managers worried that low wage levels on PEI made it difficult to recruit skilled scientists to the island. Managers were particularly worried about a shortage of university-trained scientists with experience in molecular methods. One respondent, the manager of an R&D-focused company who had work experience in large US biotechnology clusters, mentioned that he could find student interns in San Diego with scientific skills equivalent to fully employed research associates in PEI.

Managers also frequently noted that PEI workers, particularly technicians and other skilled-workers, have a strong work ethic and exhibit strong loyalty to their firms. The managers of two US companies that ran subsidiary operations on PEI had a strong desire for their operations to succeed due to the strong work ethic at PEI plants and, if possible, would push their parent firms to locate additional capacity to PEI.

INTER-FIRM COLLABORATIONS AND NETWORKING

A common concern held by bioscience managers in PEI is that, due to the remote location of the island, companies located there would find themselves "off the grid" of the industry, unable to keep up with technical and competitive trends. The difficulty of traveling to PEI, especially during the non-summer months, compounded this issue. During most of the year there are no direct flights to the United States from PEI.

Local companies on PEI need to network in order to gain access to resources and knowledge within the wider bioscience community. To address this geographical disadvantage, the PEI Provincial Government has emphasized networking and inter-organizational collaboration. This is accomplished in part through extensive "prospecting" initiatives, in which representatives of the local cluster are sent to biotechnology industry events and to visit companies and scientific research institutes within regional biotechnology clusters around the world.

To help assess the degree to which PEI companies collaborated with other organizations, managers of companies were asked to list all collaborations they had developed, both with non-profit organizations (for example, universities or funding agencies) and with companies, and to list the nature of the collaboration (research, financing, or commercialization activities). Using social network analysis methods, ties linking PEI bioscience companies were mapped (see Figure 8).

The results of this analysis are mixed. On a positive note, most PEI bioscience companies are active collaborators, particularly with local organizations. Ten of the 18 companies, for example, had established collaborations with the NRC-INH, and many companies had also established collaborations with the AVC and the FTC, the two other prominent local research institutions. These results confirm a common perception among UPEI faculty and administrators that the institution is extremely open to collaborations with industry. Another clear finding from the networking analysis is that PEI bioscience companies, particularly those located on the UPEI campus through the NRC-INH incubator and AVC facilities, frequently collaborate with one another. This is less so among the larger companies on the island, such as Novartis Health Sciences and Genzyme Diagnostics.

However, PEI companies face difficulties establishing collaborations outside PEI. Of the 219 collaborations identified, 74% were within Atlantic Canada, and more specifically, 60% were with organizations located on PEI (Figure 9). Only 12% of collaborations were with organizations residing outside of Canada, mostly in the USA. Moreover, 70% of collaborations were with non-profit organizations (see Table 3). In other words, only 56 of 219 collaborations were with for-profit entities. While policy-

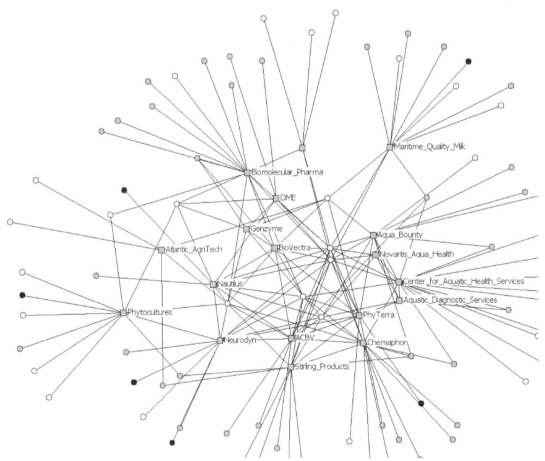

Figure 8: PEI bioscience cluster: Inter-organizational ties
Gray squares with labels: PEI bioscience companies
Clear circles: PEI organizations (both public and private)
Gray circles: Canadian organizations (public and private)
Black circles: Foreign organizations (public and private)
Source: Survey data, 2009

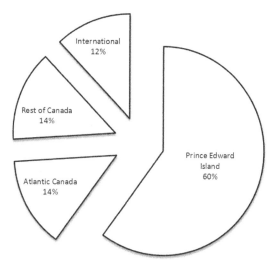

Figure 9: Geographical organization of PEI bioscience organizational ties
Source: Survey data, 2009

Table 3: Composition of PEI bioscience inter-organizational ties

Type	Percent	Number
Public PEI	42%	91
Public Atlantic Canada	15%	33
Public rest of Canada	10%	22
Public international	3%	7
Private PEI	14%	31
Private Atlantic Canada	2%	4
Private test of Canada	5%	11
Private international	9%	20
Total		219

Source: Survey data, 2009

makers and industry executives are pleased with the overall level of inter-organizational collaboration, particularly within PEI, they wondered why local firms had not done more to collaborate with organizations better centered within the mainstream biotechnology marketplace, particularly in centers such as Boston, which is only a few hundred miles away.

CORPORATE GOVERNANCE AND FINANCE

A final area of concern is governance and finance. Results from the survey confirmed what everyone on PEI already knew: the local biotechnology industry was being primarily financed by government (Figure 10). Fifteen of the 18 local companies had received public financing, while only one reported receiving a venture capital investment. While half the firms had achieved at least some retained earnings via sales, the primary source of venture financing on PEI was the government, which raised the question of whether government financing would lead to the creation of sustainable firms, both in terms of business models and corporate governance.

One issue of concern was corporate governance. PEI bioscience companies had a variety of ownership structures. Half the companies on the island had moved to the island, either as subsidiaries of larger companies or in order to develop technology as part of the NRC-INH incubator. These companies tended to have traditional ownership and governance structures for technology companies, and indeed six of the companies were publicly traded. The other 9 companies, however, were founded on PEI and overwhelmingly had concentrated ownership structures; they are still majority owned by one or more founders. Most of these companies had relied primarily on public funding, and none had received traditional venture capital. As a result, none of the PEI owned companies had developed traditional governance structures normally seen in technology start-ups. In particular, these companies had not developed company boards of directors (see Table 4).

Policy-makers on PEI wondered how these atypical governance arrangements impacted company growth trajectories. boards of directors usually are created through the process of receiving venture funding, as VCs demand board representation to monitor their investments. Over time company boards often recruit industry experts in the firm's field of operations. The decision by PEI owned companies not to use boards might strongly impact the ability of these companies to eventually raise private funding. It might also help explain why most inter-organizational ties were regional in orientation; companies lacked access to the industry networks of seasoned company executives that typically populate boards.

In the short-term, most companies on PEI continued to view the government as their primary funder. When surveyed, 13 of 16 companies expected to receive additional funding from government sources, and none expected to receive venture capital (see Figure 11). Given the lack of private venture capital on the island and the general availability of government funding, this might be seen as a rationale response.

Policy-makers on PEI wondered if local companies, if sufficiently incubated, could eventually tap into more established venture capital markets that existed in Toronto or possibly Boston. They also pondered whether a local market would ever develop on PEI. Encouragingly, Regis Duffy, who was now semi-retired and had reaped a financial windfall from the Genzyme Diagnostics acquisition, pledged in July of 2009 to start a CAD$3 million angel fund to invest in promising biotechnology companies. However promising, these funds, when spread across multiple investments, would pale in comparison to the multi-million dollar investments frequently made by government sources such as ACOA.

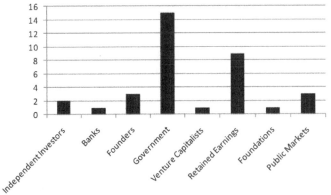

Figure 10: Sources of recent finance for PEI bioscience firms.
Number of respondents = 18; multiple answers possible
Source: Survey data, 2009

Table 4: Use of corporate boards across PEI bioscience companies

Geographic origin of company	Company has a board of directors	Company does not have a board of directors
PEI founded and owned	0	9
PEI founded but acquired	5	0
Company that moved to PEI	2	2

Source: Survey data, 2009

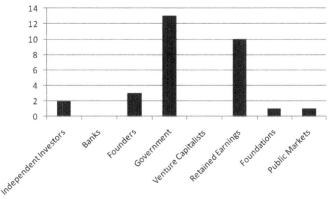

Figure 11: Expected funding sources for PEI bioscience companies
Source: Survey data, 2009

THE BIOCOMMONS RESEARCH PARK DECISION

Should PEI invest in the BioCommons Research Park? Mayne wrestled with this decision in early August 2009. The park, as envisioned, would sit on a 65-acre site on the outskirts of Charlottetown near an existing business park that currently housed BioVectra, Genzyme Diagnostics, and Novartis Animal Health. As designed, the park would contain a larger incubator and commercialization facility into which development stage companies could move. In addition to lab and commercial space to house companies, the BioCommons Research Park would have extensive shared facilities that tenants could tap into for pharmaceutical development activities and manufacturing. It was hoped that an anchor tenant, perhaps a large multinational contract research firm, would be recruited to the complex and would help manage the shared facilities (see Figure 12).

It was also hoped that the BioCommons Research Park would facilitate the creation of a staged development process on PEI, whereby companies located in the NRC-INH incubator could "graduate" to the BioCommons. As originally designed, the NRC-INH incubator was supposed to house companies working on early stage technical validation projects. It is likely that at least two or three of the existing NRC-INH tenants could immediately move to the BioCommons. This would free up space for new start-ups to take up residence at the NRC-INH. According to Paul Neima, Business Team Leader at the NRC-INH, the facility has received a steady stream of expressions of interest from entrepreneurs wanting to explore the possibility of using the incubator facilities, requests that had to be denied due to the lack of space.

However, the BioCommons Research Park also posed risks. Was the PEI cluster ready for a larger facility; did it have a critical mass of companies that could become tenants in the facility? As of 2009, no start-up company on PEI had achieved sustainable financing from private sources—and the situation is becoming more challenging. As an Ernst & Young biotechnology industry report concluded: "venture funding will be limited to a handful of successful private companies. And the pool of those companies is likely to dwindle as they move away from Canada in search of financing" (Ernst & Young, 2008). Would the development of the technology park implicitly commit the government to continue to fund the development of local companies, particularly as they needed progressively more funding to finance commercialization? Others wondered if the money would not be better spent on further efforts to build basic science capacity, possibly through recruiting additional Canada Research Chairs and/or individuals with outstanding international reputations, or through creating more graduate programs at UPEI in biology, chemistry, and other fields impacted by the biosciences.

There was also concern that the technology park ignored the needs of PEI's larger and more successful manufacturing oriented companies, such as BioVectra, Novartis Animal Health, and Genzyme Diagnostics. These three firms were the island's largest bioscience employers. Yet they were farther

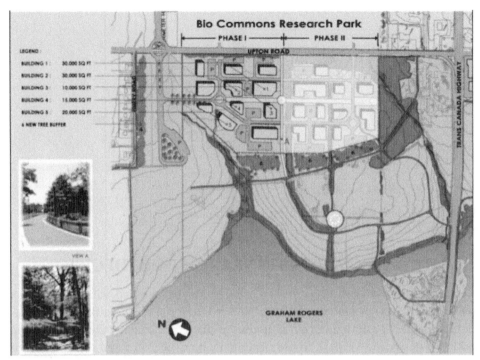

Figure 12: Bio Commons Research Park plan
Source: Personal communication from Michael Mayne, 2010

removed from the natural products and bioactives capacity building emphasis, and might not directly benefit from the new technology park. Would the CAD$30 million, or some part of it, be better spent on capacity building and/or education and training programs aimed at servicing the growing manufacturing sector?

Finally, there was the political risk. Most citizens of PEI were firmly behind the push towards the knowledge economy, and especially biotechnology. While most would acknowledge that issues of sustainability persisted, there was no doubt that tremendous momentum had been achieved—momentum that created important political capital that the new government, which Mayne served, must preserve in order to pursue its agenda. If the BioCommons failed, the political fallout would be severe.

A consensus existed across PEI's elites that the BioCommons Research Park was an important "next step" in the development of the island's bioscience cluster. But was 2009 the right time to make this investment?

APPENDICES

APPENDIX A: GRADUATING ELECTRICIANS HEAD TO ALBERTA.
Source: CBC News (2008)

CBC News: May 5, 2008

Most of this year's graduating class of electricians from P.E.I.'s Holland College have already accepted jobs in Alberta.

'There's no reason to stay on the Island.
—graduate Neil Arnold

While the exodus of trades-people to Alberta from P.E.I. is not new, instructor David Webster, who has been teaching the electrical technology program for almost 20 years, told CBC News he's never seen a year like this. Eight of 13 are on their way, and more could follow.

Typically, said Webster, only one or two graduates leave immediately, but this year a former student dropped by with recruiting brochures. Two were hired early.

"Then the word came back that those two people had got hired, and they were looking for 20 more if there was anybody else available," said Webster.

Graduates can expect to start earning $12 to $13 an hour on P.E.I., but $25 an hour in Alberta. That's incentive enough for 22-year-old Neil Arnold of Miscouche to climb aboard the wagon train heading west.

"It's time to make a move," said Arnold.

"The biggest reason is just the money, like they're just crying for people out West. They're willing to do everything they can for you to get you out there and get you set up. There's no reason to stay on the Island. The pay is not even comparable."

Webster said a change of policy by Alberta companies explains the exodus of the graduating class. Just two or three years ago, he said, Alberta companies would only hire electricians with four years' experience. Now the companies are so desperate they'll take students before the ink is dry on their diplomas.

APPENDIX B: NOVARTIS ANIMAL HEALTH EXPANDING PRINCE EDWARD ISLAND VACCINE FACILITY.
Source: Burkett (2007)

Novartis Animal Health US, Inc.
3200 Northline Avenue Suite 300
Greensboro NC 27408

NOVARTIS
ANIMAL HEALTH

MEDIA RELEASE

CONTACT: Joseph Burkett
(336) 387-1006
joseph.burkett@novartis.com

Novartis Animal Health Expanding Prince Edward Island Vaccine Facility

CHARLOTTETOWN, Prince Edward Island (January 11, 2007) – Novartis Animal Health is expanding its vaccine production facility located in the West Royalty Business Park near Charlottetown, the company's primary global location for producing disease prevention vaccine products used in the aquaculture industry.

The expansion represents a consolidation of aquaculture vaccine production facilities for the company. Manufacturing processes currently taking place at a plant in Montreal are being moved to the expanded facility in PEI, improving the company's potential to meet production needs based on future market demand for aquaculture vaccines. Up to seven new jobs will be created at the PEI facility.

The driving factors behind the decision to invest in a new vaccine manufacturing facility were the 2005 registration of Apex IHN, which is used in farmed salmon, and the expansion prospects for the aquaculture vaccine business overall.

As well, the recent announcement by the Provincial Government of a comprehensive tax incentive program aimed at bioscience and research companies made Prince Edward Island an attractive location for Novartis.

According to Grant Roberts, president and CEO of Novartis Animal Health Canada, the company hopes to achieve significant growth in its global aquaculture business by 2010.

"The aquaculture vaccine business is one of the most dynamic segments of our industry, as well as being one where we can extend our leadership position and grow as the world increases its appetite for farm raised fish products," Roberts says. "With the projected growth in fish consumption continuing to rise through the year 2020, we see many untapped opportunities for disease-preventing vaccines – not only in our core salmon farming markets, but also in warm water species and marine fish."

Last year, Novartis Animal Health named the Prince Edward Island research facility as one of six R&D Centers for worldwide animal health business, which also includes medicines

REFERENCES

ACOA. 2005. *Strategy Unveiled to Further Develop Prince Edward Island's Bioscience.* http://mediaroom.acoa-apeca.gc.ca/e/media/press/press.shtml?3448. Charlottetown, PE: Atlantic Canada Opportunities Agency.

ACOA. 2007a. *Canada's New Government Invests up to $67 Million in R&D Projects in Atlantic Canada, $15 Million in PEI.* http://mediaroom.acoa-apeca.gc.ca/e/media/press/press.shtml?3734. Summerside, PE: Atlantic Canada Opportunities Agency.

ACOA. 2007b. *New Prince Edward Island Research Centre to Play a Critical Role in Advancing Heath Care.* http://mediaroom.acoa-apeca.gc.ca/e/media/press/press.shtml?3742. Charlottetown, Prince Edward Island: Atlantic Canada Opportunities Agency.

Ahn, M & Meeks, M. Building a conducive environment for life science-based entrepreneurship and industry clusters. 2007. *Journal of Commercial Biotechnology* 14, 20-30. doi:10.1057/palgrave.jcb.3050076

BIOTECanada, 2010. *Facts + Figures: Biotechnology company concentration by region.* Statistics for Canada, Canadian Lifesciences Database. http://www.biotech.ca/en/what-biotech-is/facts-figures.aspx, January.

Burkett, J. 2007. *Novartis Animal Health Expanding Prince Edward Island Vaccine Facility.* http://www.peibio-alliance.com/?q=novartis-animal-health-expanding-prince-edward-island-vaccine-facility, January 12.

CBC News. 2008. *Graduating electricians head to Alberta.* http://www.cbc.ca/canada/prince-edward-island/story/2008/05/05/electricians-alberta.html, May 5.

Dalhousie Data Requests. 2009. "Graduates Statistics Request." E-mail to Adelee MacNevin. 4 Dec.

Ernst & Young, 2008. *Beyond borders: Global biotechnology report* 2008.

Ernst & Young. 2009. *Beyond Borders: Global Biotechnology Report 2009.* http://www.ey.com/Publication/vwLUAssets/Beyond_borders_2009/$FILE/Beyond_borders_2009.pdf

Fair, B. (2007) Building a Bioscience Cluster, *Genetic Engineering News*, Feb, (Vol. 27, No. 4), 2.

Fowler, Anna, and Ruth Leake. 2009. "Fact Book 2008." *Centre for Institutional Analysis and Planning.* Memorial University. http://www.mun.ca/ciap/Fact_Book_2008.pdf.

Gong.Yuqin. 2009. "Graduates Statistics Request." E-mail to Adelee MacNevin. 17 Dec.

The Guardian. 2009. *TRC releases full-year results of visitor exit survey.* http://www.trc.upei.ca/files/January_2009.pdf, January 3.

Jollymore, M, & Johnston, N. (2007, January 2). A New kind of island holiday. *Research and Discovery*, 2, Retrieved from http://www.progressresearchanddiscovery.com/issue4/20_21_PEI_BIZ.pdf

MacDonald, W. Nov 1998. *Employment and Unemployment in PEI: Trends and Patterns.* http://www.upei.ca/iis/art_wm_1. Charlottetown, PE: University of Prince Edward Island Institute of Island Studies

Mayne, M. 9 Apr 2008. *Island Prosperity- A Focus for Change.* http://islandprosperity.com/Island_Prosperity.pdf. PEI Provincial Government.

Porter, M. (1998) Clusters and the New Economics of Competition, *Harvard Business Review*, 78-90, Boston.

Prince Edward Island Provincial Treasury. 2009. *35th Annual Statistical Review.* http://www.gov.pe.ca/photos/original/pt_annualreview.pdf, June.

Wikipedia. 2005. *Prince Edward Island-map.* http://commons.wikimedia.org/wiki/File:Prince_Edward_Island-map.png, March 24.

Wikipedia. 2008. *Prince Edward Island contour.* http://commons.wikimedia.org/wiki/File:Prince_Edward_Island_contour.png, January 10

The Founding and Growth of On-Q-ity: Developing Advances in Personalized Medicine

DANIEL DORNBUSCH[1] AND MARK J. AHN[2]

[1]*Novartis International AG;* [2]*Atkinson Graduate School of Management, Willamette University*

- **Summary and key issue/decision:** This case focuses on On-Q-ity, a personalized medicine diagnostics company uniquely positioned to capitalize on current and future diagnostics market opportunities. The company faces critical strategic, therapeutic, and investment decisions about how to develop high-risk, high-reward products. . On-Q-ity will need to balance lowering development risks with increasing product opportunities and competitive positioning.

- **Companies/institutions**: On-Q-ity, US Food and Drug Administration (FDA) Division of Laboratory Services

- **Technologies:** circulating tumor cells (CTCs) and DNA repair pathways, important elements of detecting and diagnosing cancer.

- **Stage of development at time of issue/decision:** On-Q-ity was formed as a merger of two companies and raised a significant sum from prominent venture capital investors.

- **Indication/therapeutic area:** Oncology

- **Geography:** Worldwide

---

* *This case was prepared as a basis for class discussion rather than to illustrate either effective or ineffective handling of an administrative situation.*

INTRODUCTION

Mara Aspinall, president and CEO of On-Q-ity, sat alone for the first time in the company's new offices and reflected on the past few months: merging of two companies, advancing of the company's core technologies, hiring of new personnel, and bringing on new investors to recapitalize the company. Yet her job had only begun.

The company's leading position in two separate but related diagnostic technologies and approaches uniquely positioned it to define a new generation of predictive as well as companion diagnostics. However, explaining the difference between a predictive and prognostic diagnostic, ensuring wide clinical adoption and working with a pharmaceutical firm to embrace a companion diagnostic, let alone explain the intricacies of rare cell capture and DNA repair pathway predictive algorithms remained a challenge on multiple fronts. Some potential customers remained unconvinced of the product utility. Some investors remained skeptical of the timelines. Some scientists were unconvinced of the technology potential. But the winds of the cancer medicine development storm were blowing at the backs of On-Q-ity; the most recent American Society of Clinical Oncology (ASCO) meeting declared that now is the time of "Personalizing Cancer Care" (Figure 1) and new discoveries in cancer genetics were being made on almost a weekly basis. At that ASCO meeting, the largest annual gathering of oncologists, DNA repair was called the "new big science" in understanding cancer therapy response and circulating tumor cells was the "excitement of the meeting" with talk about the potential of this technology.

As the business of personalized medicine formed and grew, Mara had held a front-row seat. She grew Genzyme Corporations' Genetics Business Unit to more than $300 million in revenues and 1,900 people from $50 million only 2 years earlier. Most significantly, she moved the division into cancer diagnostics, completing several acquisitions, including the purchase of IMPATH Corporation, at the time, the largest independent cancers lab in the country. She sat on the board of the Dana Farber Cancer Institute as the first personalized therapeutics and diagnostics emerged: human epidermal growth factor receptor 2 (HER2) testing and Herceptin® (trastuzumab, Genentech), bcr-abl testing and Gleevec® (imatinib mesylate, Novartis), CCR5 diagnosis and Selzentry® (maraviroc, Pfizer) to name a few. In order to secure Genzyme's place in this revolution, she successfully secured exclusive diagnostic rights to another gene mutation, epidermal growth factor receptor (EGFR). The identification of this mutation is critical in the accurate diagnosis of those lung cancer patients who would be most likely to

Figure 1: Outside of the 2009 ASCO annual conference
Source: D. Dornbusch

respond to a new class of tyrosine kinase inhibitor (TKI) drugs.

"When you ask people about their great fear in life, their number one fear is cancer," says. Aspinall. "Science has moved especially quickly in the last five years with the completion of the Human Genome Project and the explosion of new diagnostic capabilities. In our medical system, however, new science has to clear a high hurdle before it has an impact on practice. This caution is sensible in some arenas, but personalized medicine results in safer and more effective therapy choices, not higher risk. Moving forward, we need to embrace the more precise information that personalized medicine brings. This is particularly important in fast-moving diseases, where the window for a positive outcome is small (HBS, 2008)."

Facing the company now was the demonstration and commercialization of two, groundbreaking diagnostic strategies that if successful, might redefine how cancer is identified, diagnosed, and treated.

THE DILEMMA

With a fresh cash infusion from such leading venture capital investors as MDV, Atlas Venture, Bessemer Venture Partners, Physic Ventures, and Northgate Capital, On-Q-ity had a fresh opportunity to advance the company, its technologies, and the commercialization of its products. However the $26 million series A financing round would only take the company so far. Should Mara focus the company on a single technology and single product to gain revenue as fast as possible? Would such a strategy cause the other technologies languish? If so, which technology should be the primary focus?

Or, should the company develop its multiple technologies in parallel? Would this strategy create the most compelling and therefore useful product or service offering? Could the company increase the enterprise value most substantially in this way?

BACKGROUND OF MOLECULAR DIAGNOSTICS MARKET

Molecular diagnostics represent a range of platform technologies which provide valuable tools for researchers and clinicians, promise improved therapeutic outcomes, and have the potential to reduce healthcare costs. Molecular diagnostics describes test methods that identify a disease or a predisposition for a disease through analysing nucleic acids (DNA and/or RNA) of an organism. Its genesis lies in the characterization of certain diseases as *molecular diseases* with the discoveries in the 1940s and 1950s of nucleic acid structures and inheritance patterns (Pauling, Itano, Singer, & Wells, 1949). Scientific breakthroughs included recombinant DNA cloning and sequencing which helped spark the modern era of molecular diagnostics (Patrinos & Ansorge, 2005).

DNA sequencing, in particular, facilitated the generation of DNA probes which enabled analysis of specific regions of the genome, allowing mutant alleles to be tracked and analyzed. In the late 1970s, the first prenatal diagnosis of α-thalassemia (an inherited blood disease where a genetic defect affects haemoglobin synthesis) was followed by an analysis that pinpointed sickle cell alleles (an inherited blood disorder) in patients of African descent. It provided a breakthrough foundation for similar diagnostic approaches for other genetic diseases including phenylketonurea and cystic fibrosis (Patrinos & Ansorge, 2005).

However, despite intense efforts from multiple laboratories around the world, diagnosis of inherited diseases at the DNA level was not implemented for routine patient analysis due to the complexity, costs and time requirements of the available technology for decades after its discovery. The invention and rapid optimisation of Polymerase Chain Reaction (PCR) in the 1980s permanently changed these restrictions (Mullis & Faloona, 1987; Saiki *et al.*, 1985). PCR generates large numbers of copies of target sequences, speeding what formerly took months to a matter of hours or minutes. The technology also significantly increased the safety of molecular testing by decreasing the need for radioactive materials. These factors enabled clinical laboratories to screen populations for known mutations and in more recent years, identify new mutations. By 2000, the sensitivity and specificity of molecular testing had improved false negative and false positive results from about 30 to five percent or less (Patrinos & Ansorge, 2005).

The industrialisation of PCR has led to over 50 different mutation detection platforms, creating a challenge for labs to coordinate efficient, integrated systems to support their particular laboratory settings (Patrinos & Ansorge, 2005). At the same time, the elucidation of genome sequences resulted in an increased number of diagnostically and clinically relevant targets which, coupled with increased understanding of inherited diseases, led to growing volumes of molecular diagnostic tests. Currently, vast numbers of tissue and blood samples are routinely analyzed annually around the world with the number of new genetic tests steadily increasing. According to genetests.org, in early 2010, there were tests available for 1982 different genetic diseases (1727 in clinical testing with 255 in research testing). This has grown more than 15% per year on average since 2000.

Almost all hospitals now have molecular diagnostic capabilities in their hospital labs or a department that focuses exclusively on the testing (Patrinos & Ansorge, 2005). For example, newborns across the US are routinely screened for phenylketonurea and other treatable genetic diseases. Parents now have the option of being informed about their carrier status for many recessive diseases prior to deciding to start families. Diagnostics have entered the consumer world as well where, for a fee, anyone can send in their saliva, get their DNA extracted and analyzed and receive their risk profile for various diseases from Parkinson's disease to Lupus and diabetes. Further, like the cost of microchips, the last two decades witnessed a dramatic collapse in the cost of DNA testing, from multiple dollars per SNP (Single Nucleotide Polymorphism) to pennies SNP (Upstream Biosciences, 2008).[1] Next generation sequencing technologies promises a $1000 full genomic analysis within the next decade.

Diagnostics have similarly revolutionized research and clinical development (Sauter and Simon, 2002). Examples include addressing the problem of antibiotic resistant organisms for the identification and determination of treatment susceptibility (Whitley, 2008); and while oncologists have historically predicted the likelihood of cancer recurrence on epidemiological factors such as age, a molecular prognostic test such as OncoType Dx (Genomic Health) quantifies the likelihood of tumor recurrence which assists the key treatment decision of whether a patient should or should not receive chemotherapy. Moreover, molecular diagnostics is rapidly changing pharmacogenomics and personalized medicine making individualised drug therapy possible based on the understanding of the genetic make-up and predisposition of patients as well as the genetic make-up of the tumor itself. The missing link however, has been access to this information early enough and regularly enough to choose therapy most effectively.

THE ON-Q-ITY PROMISE

Today's cancer challenges can be divided into three main questions once a person is positively diagnosed: (1) what drug should I take? (2) is the drug working? and (3) has the cancer recurred? The ability to answer these questions accurately is a major contributor to the treatment outcome. The currently possible answers and potential future answers that On-Q-ity could address are shown in Figure 2. Operationally, multiple options face the company. Should On-Q-ity move forward independently to bring its technologies to clinical products as quickly as possible or move forward aggressively to work with pharmaceutical companies and academic centers to accelerate research in personalized medicine in multiple product areas?

---

1   Single Nucleotide Polymorphisms (SNPs) are DNA sequence variations that occur when a single nucleotide (A,T,C,or G) in the genome sequence is altered. Although more than 99% of human DNA sequences are the same and SNPs are evolutionarily stable, minor variations in DNA can have a significant impact on how patients respond to a disease which make SNPs an invaluable diagnostics tool.

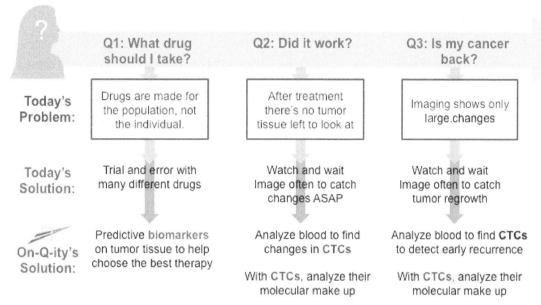

Figure 2: Cancer challenges and solutions
Source: On-Q-Ity (2010)

Clearly, time is a priority and only limited development pathways are feasible to commercialize these potentially life-saving technologies.

COMPANION DIAGNOSTICS

At the crux of companion diagnostics is the insight that diseases are complex, with varied causative factors and varied treatment patterns. Even when specific tumors are grouped together due to phenotype (e.g., breast cancer or non-small cell lung cancer), for example, hey can act and respond to therapy differently due to their genetic composition.

Companion diagnostics are assays that are essential to the diagnosis of specific subtypes of diseases and therefore the delivery of correlated therapeutics. One of the first examples of a companion diagnostic was with Genentech's approval of Herceptin® (trastuzumab). Normal breast cells contain multiple copies of the HER2 gene which produce specific ratios of HER2 proteins. These proteins bind extracellular growth factors and internalize them to signal normal cell growth and division (Pegram & Slamon, 2000).

In approximately 25 percent of women with breast cancer, there is a genetic alteration in the HER2 gene that produces an increased amount of the HER2 growth factor receptor protein on cell surfaces. This overexpression increases the rate of cell division, multiplication, and growth. HER2 positive breast cancer is thus correlated with more aggressive forms of breast cancer, poorer prognosis, and decreased survival compared HER2 negative breast cancer.

In HER2 positive disease, Herceptin binds to the abnormal HER2 receptors and induces antibody-dependant cellular cytotoxicity (ADCC), thus selectively culling the malignant cells. Detection of HER2 protein over-expression is necessary for selection of patients appropriate for Herceptin therapy because these are the only patients for whom drug benefit has been shown. Several FDA-approved commercial assays are available to aid in the selection of patients for Herceptin therapy. These include HercepTest™ and Pathway® HER-2/neu (IHC assays) and PathVysion® and HER2 FISH pharmDx™ (FISH assays) (Genentech, 2010).

GUIDING THERAPY CHOICE: MEASUREMENT OF DNA REPAIR PATHWAYS

Every day, in every person, DNA damage occurs tens of thousands of times. This damage is caused by a variety of both natural and environmental factors including chemicals in metabolism, tobacco use, UV light, or inflammation. While most DNA errors are corrected by cells' DNA repair pathways. If not repaired, the specific DNA damage can give rise to unmitigated cell growth and therefore cancer. Thus individuals with damaged or inherited impairments to the DNA repair pathways are at greater risk for cancer (Bernstein, C., Bernstein, H., Payne & Garewal, 2002).

Because human somatic cells are diploid, mutant genes are most often recessive when they arise in a somatic cell, and are not usually expressed because of the presence of the wild-type copy of the homologous gene. However, when there is a loss of heterozygosity by such mechanisms as mutations in the homologous gene, occurrence of aneuploidy (loss of a chromosome), or by recombination with the homologous chromosome, recessive mutations can be expressed (Bernstein, 2009).

Chemotherapeutics that are specific inhibitors of DNA repair pathways have been demonstrated to be efficacious when used on tumors with corresponding damaged pathways. Radiotherapy has also been shown to be effective in cancers that incorporate DNA repair abnormalities. There is evidence that drugs and radiation that interfere with these pathways could prove useful as single-agent therapies and if so, increase the efficacy and reduce the side effects of cancer treatments (Helleday *et al*, 2008).

A major hindrance to this improvement in therapy is detecting which DNA repair pathway is important and predictive, especially when these mutations are homologous. Several biomarkers in the Homologous Recombination (HR) pathway have been identified as beneficial and even synergistic. *Rad51*, *FANCD2*, *BRCA1*, *ATM*, and *XPF*, to name a few, have been correlated with increased survival outcomes with certain therapies (Seiwert, 2009).

The promise, however, is that a diagnostic could detect the relevant DNA repair pathway mutations and therefore identify the optimal, tailored therapy for a specific patient and their tumor. Scientists at On-Q-ity helped pioneer the field of DNA repair and signature prediction. Specifically, the On-Q-ity process (Figure 3):

1. Measures multiple DNA repair markers to characterize status of repair pathways in the tumor cell.
2. Based on the combined profile of DNA Repair markers, determine if particular chemotherapy will be effective in that specific tumor or if alternate therapy should be used.

They have also generated early stage, clinical success. A clinical trial conducted at the Dana Farber Cancer Institute demonstrated that DNA repair biomarkers can separate which patients may or may not benefit from anthracycline therapy in breast cancer therapy (Figure 4).

CIRCULATING TUMOR CELLS (CTCS): IMPACT ON NON-INVASIVE SCREENING DIAGNOSTICS

Metastasis, the primary cause of mortality in cancer patients, is caused by tumor cells that break off from a primary tumor, gain entry into the bloodstream, travel through the circulation, and deposit at a distant location where it develops into a secondary tumor. Metastases depend on cancer cells developing two features: increased motility and invasiveness.

In recent years, CTCs have been shown to exist in the blood of cancer patients in extremely low concentrations (often less than 10 cells per billion of normal cells) making their isolation and identification extremely difficult. The first generation of CTC diagnostics has demonstrated that detection and

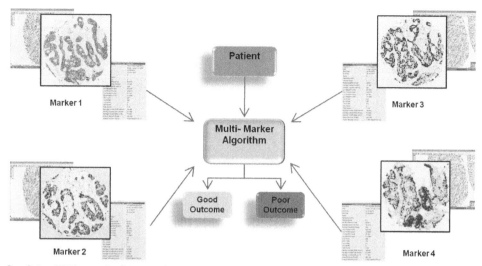

Figure 3: On-Q-ity DNA repair diagnostic process
Source: On-Q-ity (2010)

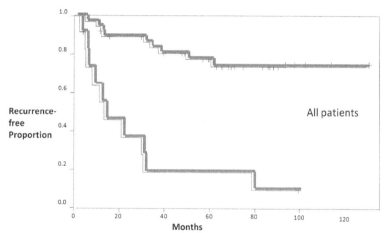

Figure 4: Dana Farber Cancer Institute triple negative study. Patients in study n=152, p=9.05E-07, 76% effective, 24% failure.
Source: Alexander B, et. al. (2010).

measurement of CTCs is an accurate predictor of patient outcomes. Multiple clinical trials showed that the CellSearch System (Veridex) accurately predicted progression-free survival (PFS) and overall survival (OS) in patients with metastatic breast, colorectal, and prostate cancer. Patients with more than 5 CTCs per 7.5 ml of blood had increased disease progression and shorter survival times. While approved by the FDA, ASCO does not recommend the CellSearch System for treatment decisions because the assays have not demonstrated quality of life or long term survival improvements.

In order to improve the diagnostic capabilities of CTC diagnostics, the next generation of technologies will not only count the number of CTCs present in a sample of blood, but will also examine the underlying molecular characteristics of the individual CTCs. The On-Q-ity technology employs a customized chip that isolates the CTCs and allows for subsequent molecular analysis (Figure 5).

THE DECISION FOR INVESTMENT

The On-Q-ity management team frequently discussed and actively debated the optimal development commercialization strategies for the company. At no time were these discussions more intense than

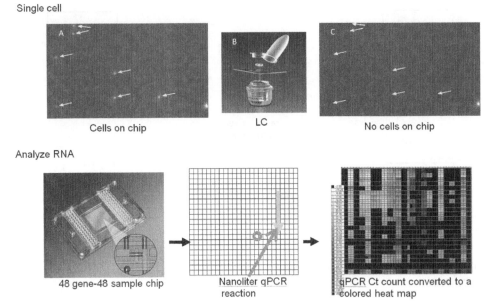

Figure 5: On-Q-ity CTC extraction and single cell analysis
Source: On-Q-ity (2010)

when the team prepared for their road show for their next round of venture funding.

First, should On-Q-ity focus on, develop and launch their ideal product or the one that can make it to market in the shortest period of time? In this case, On-Q-ity could launch a product with a single technology or one that combined both technologies? Should the first product predict drug response on tumor tissue from DNA repair mutations or, should the first product combine both the CTC and the DNA repair technologies for a user to analyze a cancer type and sensitivity from a simple blood test?

Second, which market should the company pursue first? Should On-Q-ity develop their first product for the research market or for the clinical market? The former, an RUO (Research Use Only) product, has a faster and less expensive route to market. No FDA approval is required. In Europe, the company could gain a CE mark, requiring only a self-registration. Interest from the research market on the company's technologies was strong. While it was unclear if the research market's profit margins could be as attractive as the clinical market, it was clear that to serve the market, it would take significant corporate resources and appropriate infrastructure to service the market well.

Alternatively, the company could pursue the coveted clinical market, bringing its products directly to oncologists and patients. This market would require an FDA approval and most likely, additional, follow-on studies demonstrating continued long term efficacy. In addition, a clinical product would likely require early pharmacoeconomic studies for payors to justify appropriate reimbursement.

Answers to these questions are paramount to the potential of the company. The decisions would significantly influence the valuation of the company, the required investment to the next value inflection point and the timeline for which these important technologies would reach patients.

THE NEXT STEP

As Mara arrived at the office in the dark of morning, she mulled these critical questions: which product to pursue, which market to pursue, which partners to work with and which development strategy to follow. Knowing that the decisions the company makes now could influence its success and the potential to bring a life-saving diagnostic to thousands of cancer sufferers.

APPENDICES

APPENDIX A: MOLECULAR DIAGNOSTICS (MDX) MARKET

While human therapeutics has been the largest segment of the pharmaceutical and biotech industry, significant technology spill over effects are translating into advances in intractable areas of need which have created several large, high-growth markets in molecular diagnostics ($3.5 billion), nutraceuticals and functional foods ($50 billion), environment ($30 billion), biosecurity ($7 billion), and biofuels ($6 billion) (Figure 6) (BIO, 2007; Burrill, 2007; Kalorama, 2006).

Molecular diagnostics forms part of the wider *In Vitro* Diagnostics (IVD) market, which is worth over US$32B. North America (44%), the European Union (30%) and Japan (11%) make up 85 percent of the total IVD market, however their share is expected to decrease to 80 percent by 2010, with the rest of the world becoming increasingly important (Kalorama Information, 2006). While traditional tests such as blood cultures (clinical chemistry, immunoassay, haematology and microbiology) comprise the majority of the IVD market, this category is projected to have relatively flat growth. In contrast, molecular diagnostics is one of the fastest growing segments of the IVD market, and will gain significant market share by 2010 due to high value-added tests that command premium pricing due to the quality and critical nature of the information provided to clinicians (Scientia Advisors, 2007b).

The MDx market is projected to increase from US$3.5 to US$7.2 billion in 2012. In particular, significant growth is expected in the oncology and infectious disease segments—driven by the availability of tests to address critical unmet needs, increase turnaround time, accuracy, and reduce healthcare costs. For example, in 2005 molecular test sales for HIV were $515 million and expected to grow to US$900 million by 2010. Combined, molecular diagnostics for oncology and infectious diseases alone are expected to grow from US$1.5 billion in 2006 to US$4.3 billion in 2012. In the future, theranostics (use of molecular diagnostics to guide the selection, dosing and timing of biopharmaceutical treatments) should significantly impact growth as more validated biomarkers become available (Scientia Advisors, 2007a) (see figure 7).

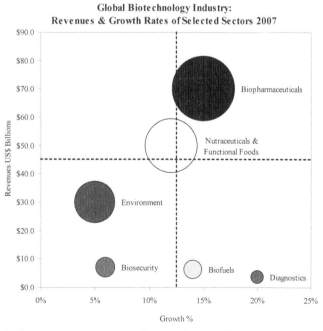

Figure 6: Biotechnology industry sector revenues & growth rates %
Source: Adapted from Burrill, 2007

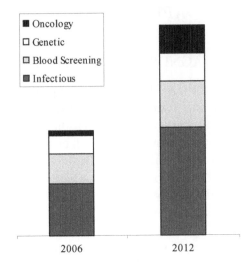

Figure 7: Molecular diagnostics market
Source: Adapted from Scientia Advisors, 2007

REFERENCES

Alexander B, *et al.* 2010. *DNA Repair Biomarkers in Triple Negative Breast Cancer. Poster Presentation.* San Antonio Breast Cancer Symposium.

Bernstein, C; Bernstein, H, Payne CM, Garewal, H. 2002. DNA repair/pro-apoptotic dual-role proteins in five major DNA repair pathways: fail-safe protection against carcinogenesis. *Mutation Research*, 511: 145-178.

Bernstein, C. *DNA Damage and Cancer.* 2009. http://www.scitopics.com/DNA_Damage_and_Cancer.html, 10 Apr 2010.

BIO (Biotechnology Industry Organization) 2006. *BIO 2005-2006: Guide to biotechnology.* www.bio.org, accessed 12 Dec 2009.

Burrill & Co. (2007). *Biotech 2007 Life Sciences: A Global Transformation.* Burrill Life Sciences Media Group.

Genentech. 2010. *Herceptin Package Insert.* www.gene.com, accessed 1 Dec 2009.

Harvard Business School Alumni Bulletin, September 2008. http://www.alumni.hbs.edu/bulletin/2008/september/oneonone.html, accessed 16 Dec 2009.

Helleday, T. *et al.* 2008.. DNA Repair Pathways as Targets for Cancer Therapy. *Nature Review Cancer*, 8(3):193-204.

Kalorama Information. 2006. *The worldwide market for in-vitro diagnostic test, 5th edition.* New York.

Pegram M, Slamon D. 2000. Biological rationale for HER2/neu(c-erbB2) as a target for monoclonal antibody therapy. *Seminars in Oncology*, 27 (suppl9): 13-19.

Scientia Advisors. 2007a. *Established diagnostic and therapeutic players to build on molecular diagnosis franchises.* www.scientiaadv.com, accessed 16 Dec 2009.

Scientia Advisors. 2007b. *Strategic Review of in vitro Diagnostics.* Cambridge, MA.

Scientia Advisors. 2007c. *Strategic Review of Molecular Diagnostics.* Cambridge, MA.

Seiwert, T. *et al.* 2009. *Use of DNA Repair Pathway Analysis to Predict Overall Survival in Head and Neck Cancer Patients Treated with FHX based Chemotherapy.* Abstract 6033, Poster 22. ASCO (Amercian Society of Clinical Oncology).

Airway Tools Company: Changing Medical Device Standards of Care

ANDREW MAXWELL[1], BAHRAM BEHNAM[1], MICHAEL ALVAREZ[2]
[1]University of Toronto; [2]Stanford University

- **Summary and key issue/decision:** Airway Tools Company is a startup medical devices company centered on the "Airway Tool" technology. Its founders, Dr. Adam Law (anesthesiologist), Dr. Michael Gross (orthopedic surgeon), and Dr. George Kovacs (emergency physician) formed the company in 2004 to finance and develop a novel tracheal intubation device that would address flaws they identified with current intubation devices. While the initial development work focused on making a product that was incrementally better than competitors, the developers decided to provide significant added value that would allow them to justify the creation of a new venture, rather than simply licensing the design to existing manufacturers. However, as the inventors moved from the development of a product to a business, they encountered several constraints that have slowed down the introduction of the "Airway Tool" into the market. These constraints are primarily around the current nature of competition in the market place, current usage patterns and several intellectual property challenges. To achieve Airway Tools, Inc.'s long-term objective of becoming a major device manufacturer in the intubation market, they are developing strategies to both add value in the existing Operating and Emergency Room environments, and possibly create a new market among first-response practitioners.

- **Companies/institutions:** Airway Tools Company Inc., Halifax Biomedical Engineering, Panacis Medical

- **Technology:** The "Airway Tool" is a portable intubation device aimed for airway stimulation. Its added features allow for "round the corner" visualization, ease and consistent tube passage through the tube delivery channel, and effortless soft tissue control with midline stability. In addition, its video-based technology with an on-handle screen assists the user in avoiding serious complications commonly caused by intubation procedures.

- **Stage of development at the time of issue/decision:** Airway Tools Company is currently in the midst of pre-production development, finalizing a prototype and validating the "Airway Tool" using manikins and difficult airway simulators and clinical-grade cadavers. A single formal manikin study has been completed and a provisional patent has been filed.

- **Indication/therapeutic area:** Several circumstances would require the use of the "Airway Tool" to support the respiratory function of patients. Common intubation applications include the use of the apparatus to protect the airway of comatose or intoxicated patients, or during general anesthesia where the patient's spontaneous respiration may be decreased or absent. The "Airway

* *This case was prepared as a basis for class discussion rather than to illustrate either effective or ineffective handling of an administrative situation.*

Tool" can also be used to assist in certain respiratory airway procedures, such as bronchoscopy, laser therapy, and stenting of the bronchi. In addition, the device has applications among First Responders, but there are concerns about the effectiveness of the procedure by unfamiliar users that have reduced the use of the product. In addition, the device may allow people with less training to perform intubations and increase the number of positive outcomes.

- **Geography: Canada**, US, UK, Australia, New Zealand, South Africa
- **Keywords:** intubation, laryngoscope, airway stimulation, business models, adoption issues, freedom to operate.

INTRODUCTION

In Spring 2008, the three founders of The Airway Tools Company—Dr. George Kovacs, Dr. Adam Law, and Dr. Michael Gross—met in the "company café", also known as the Toronto-based hospital cafeteria where they all practiced medicine, to discuss their startup company. Dr. Gross began, "I'm excited about the latest Airways ETI prototype and can't wait to try it in the clinic. It will be easy to use for a wide range of healthcare providers including nurses, EMTs, and physicians. On the other hand, while the product development side is progressing on plan, we're running into some challenges in raising additional financing. Venture capitalists are concerned about the product differentiation and IP," Dr. Law mused: "I thought the technology issues would be the difficult bit and financing would be the easy part!" Dr. Gross added: "the good news is we have a product developed by users for users, and we know the Airway Tool offers significant advantages if we can position our product successfully and put in the hands of healthcare professionals. We need the right business plan and the right marketing muscle behind our efforts."

BACKGROUND ON THE AIRWAY TOOLS COMPANY

The Airway Tools Company is a start-up company founded in 2004 by Dr. George Kovacs, Dr. Adam Law, and Dr. Michael Gross, who saw a need for developing a new laryngoscope that would facilitate easier endotracheal intubation (ETI) and reduce the patient's risk of complications associated with this procedure. Although these individuals have a strong medical background, they were concerned that their lack of business experience or knowledge of the industry marketing practices may impede company development.

The Airway Tools Company was founded by Dr. George Kovacs, Dr. Adam Law, and Dr. Michael Gross (see Appendix A for the company's organization chart). Dr. Michael Gross, president and director for Airway, is professor of surgery at Dalhousie University in Halifax, Nova Scotia for the past 14 years. Dr. Gross is also the Medical Director for the QE II Health Sciences Centre Tissue Bank, the largest comprehensive tissue bank in Canada. In addition, he offers consultation to the orthopaedic appliance industry and holds patents on devices in reconstructive orthopaedic surgery. Dr. Gross has served on the boards of directors of two publicly-traded companies. He joined the Linear Gold Corp. board of directors in 2002, the Linear Metals Corporation Board in 2006, and is chairman of the NWest Energy Corporation board.

Co-founder and director, Dr. Adam Law is an attending anesthesiologist at the QEII Health Sciences Centre in Halifax, Nova Scotia. He has subspecialty training in and is section chief of neuroanesthesia, and is Professor of Anesthesiology and Surgery at Dalhousie University. He is also medical director of the Atlantic Health Training and Simulation Centre in Halifax. Dr. Law's teaching and research interests lie primarily in airway management. In addition to having authored or co-authored numerous book chapters on the topic, he teaches the subject nationally and internationally on a regular basis.

The third co-founder, investor, and advisory board member, is Dr. Michael Murphy, FRCPC (EM), FRCPC (ANES). Dr. Murphy is board certified in both emergency medicine and anaesthesiology in both the US and Canada. He completed his emergency medicine residency training at Denver General Hospital in Denver, Colorado, and his anaesthesiology training at Dalhousie University in Halifax, Nova Scotia, Canada. Dr. Murphy was appointed as the first Executive Director of Emergency Medical Services for the province of Nova Scotia, Canada, and was responsible for designing and implementing a high performance, full service, advanced life support EMS system in the province, using advanced training systems. He identified the advantages offered by simulation in training EMTs and Paramedics and implemented a simulation center to undertake such training in the Canadian Province of Nova Scotia. Dr Murphy is an internationally recognized educator in the field of airway management.

AIRWAY'S MARKET, TECHNOLOGY AND PRODUCTS

ENDOTRACHEAL INTUBATION (ETI)

In medicine, intubation refers to the placement of a tube into an external or internal orifice of the body. Although the term encompasses endoscopic procedures, it is often indicative of tracheal intubation. Tracheal intubation refers to the placement of a flexible plastic tube into the trachea to protect the patient's airway allowing for mechanical ventilation. The most common tracheal intubation is orotracheal intubation. In this procedure, an endotracheal tube is passed through the mouth, larynx, and vocal cords, into the trachea with the assistance of a laryngoscope. A bulb is then inflated near the distal tip of the tube to help secure it in place and protect the airway from blood, vomit, and secretions. Alternatively, nasotracheal intubation (NSI) involves passing the tube through the nose, larynx, vocal cords, and into the trachea.

Health practitioners perform tracheal intubations in several different medical situations: when a patient is comatose or intoxicated, often they are unable to protect their airways due to throat muscle relaxation or the lack of protective airway reflexes, such as coughing and swallowing. Similarly, during general anesthesia, the patient's spontaneous respiration may be decreased or absent due to the effect of the anesthetics, opioids, or muscle relaxants. In these cases, tracheal intubation allows for sustainable respiratory support. Tracheal intubation is also necessary during diagnostic manipulations of the airway, such as bronchoscopy, or endoscopic operative procedures of the airways, such as laser therapy or stenting of the bronchi. In addition, intubation has been tried by paramedics to keep patients alive until they can be taken to hospital. Unfortunately, this application has proven to be problematic, with low efficacy and the potential for liability issues. As a result, this application has become increasingly unpopular.

Historically, the most common device used for endotracheal intubation (ETI) is the laryngoscope. The laryngoscope consists of a handle, usually containing batteries, and an interchangeable blade. Currently, there are two styles of laryngoscope blades commercially available: the straight blade, and

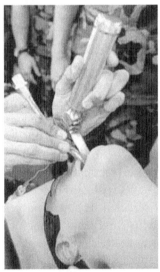

Figure 1: Typical intubation procedure. Endotracheal intubation is a medical procedure in which a tube is placed into the windpipe (trachea), through the mouth or the nose. In most emergency situations it is placed through the mouth. Endotracheal intubation is done to open the airway to give oxygen, medication, or anesthesia, and to help with breathing. It may also be done to remove blockages (foreign bodies) from the airway or to allow the doctor to get a better view of the upper airway.
Source: Lumina

the curved blade. The technique required for proper intubation slightly varies depending on the type of blade being used. In adults, practitioners often prefer to use the curved blade, while the straight blade is often used on infants. There are many other styles of straight and curved blades with accessories, such as prisms for enlarging the field of view or ports for the administration of oxygen. These specialty blades are primarily designed for use by an anesthetist.

Due to the complexity and the difficulty of ETI, there is a steep learning curve that requires years of experience and training. In fact, a significant amount of evidence suggests that reliable intubation is only achieved after the practitioner has completed more than 50 intubations. Incorrect placement of the tube can be detrimental to the patient's health, such as oxygen deprivation, which could be lethal. Other complications as a result of an error during ETI include oxygen desaturation, heart rhythm disturbances, breathing tube dislodgement, and unrecognized tube misplacement. In addition to these risks, unskillful handling of the laryngoscope can lead to problems, such as dental damage or mandibular trauma. As a result, successful intubation requires a skilled practitioner with experience in the science of airway management. Medical practitioners in anesthesiology, emergency medicine, and critical care medicine spend many hours training and practicing ETI in tightly controlled settings, while most emergency staff, including paramedics, receives only a fraction of this training. According to a 2006 study by the University of Pittsburgh, it was found that ETI errors occurred in 22% of paramedic intubations (Wang, 2006).

THE ENDOTRACHEAL INTUBATION (ETI) MARKET

According to the 2005 Saint Mary's Business Development Study, the estimated existing North American intubation device market size for alternative airway management devices is approximately $137M (see Appendix D). A further breakdown revealed that about $51M was generated from the operating room segment, $10M from emergency room sales, and $76M from Level 3 EMT segments. Also, it is important to note that these numbers were estimates for the intubation device only, and did not include replacement blades.

Based on the same study, there are an estimated 24 million intubations performed annually in North America, 22 million of which take place in a hospital setting, and the remaining 2 million in a pre-hospital emergency situations. Operating room anesthesiologists and anesthetists, who are very skilled at using intubation protocols, perform on average an estimated 600 intubations per year; while emergency room physicians and Level 3 EMTs perform 20 and 2 intubations per year, respectively.

There are two key market trends in today's intubation industry: the move towards video application (fiber optics & digital) integration, and increase in the use of disposable blades. The video applications are meant to reduce the difficulty of the procedure, and the use of disposable blades is believed to address concerns regarding the efficacy of sterilization and the possible negative impact from the use of sterilization agents. Currently, there are several products on the market for intubation management (see Appendix C).

TECHNOLOGY: THE AIRWAY TOOL

In an attempt to reduce the difficulty with ETI procedures and prevent some of its associated complications, the founders of the Airway Tools Company identified an unmet need and a potentially successful new venture. As a result of years of experience with ETI procedures and training other practitioners in achieving successful intubation, the founders designed and created a novel laryngoscope known as the Airway Tool. This innovation combines excellent views of the tracheal passage with the ease of use needed by inexperienced users. In comparison, alternative products currently in the market falter in one or both of these categories.

The Airway Tool is designed as a tool for both the expert and non-expert practitioners with effective visualization, user-friendliness, and an ergonomic structure. The visualization of the airway is

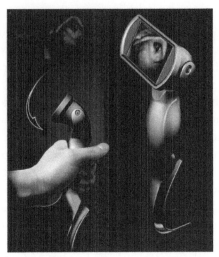

Figure 2: Visualization of Proposed Airway Tool
Source: Lumina

made possible with the use of fiber optics or a camera connected to an external monitor. The additional feature of having video feedback will improve the ease of intubation, and address liability concerns by recording the procedure and monitoring the results. Also, the Airway Tool is a portable device that can be used on patients of different age groups from pediatrics to adults. The ergonomic shape of the Airway Tool provides two major benefits: ease-of-use, and more importantly, patient comfort. In addition, a custom-designed disposable blade allows for easy placement of the tube into the patient's airway reducing the risks and complications of ETI.

The Airway Tools Company hopes to make the Airway Tool the device of choice in operating rooms and emergency rooms, and to create a new market opportunity in first-response practitioners. This would expand the market beyond Level 3 Emergency Medical Technicians (EMT). The Level 3 EMTs have additional training in airway management as opposed to EMT Level 1 and 2.

INTELLECTUAL PROPERTY (IP) PROTECTION AND LEGAL ISSUES

Airway has filed patents to protect its intellectual property (IP), but legal consultants and investors have raised concerns about what patents will issue and their potential degree of exclusivity in an intensely competitive marketplace. As their IP attorney reported:

> "...a number of pioneer patents and concepts applicable to laryngoscopes are now, or will shortly be, in the public domain. In particular, this prior art (in the public domain) shows the following major components: an anatomically conforming curved blade, having a lifter at the distal end for epiglottal engagement or control; a channel for an airtube; viewing means; and, illumination means... if a person of ordinary skill can implement a predictable variation or combination of these elements, this combination or variation may not be obvious (particularly in the U.S.). Further, these elements appear to be key to the current Airway Tools device or invention."

Another issue raised is the currently high rate of intubation failures, which creates the potential for liability claims and discourages the procedure. In fact, as a result of the potential for legal liability claims, many organizations employing paramedics now forbid the procedure. This situation can only be resolved by better design and extensive training for paramedics, but since each paramedic would likely perform few procedures during the course of their careers, training may not be cost effective.

Importantly it is not just the fear of liability in cases where incorrect procedures are used that causes concern, it is the cost and time of defending legal cases where the paramedic performed perfectly, but outcomes were unsuccessful (e.g., if a patient dies or has brain damage, it is difficult for the paramedic to prove they did nothing wrong). Nonetheless, ETI can be a life saving technique with no functional alternative.

CONCLUSION

Dr. Gross had completed his patient rounds and was leaving the hospital with his co-founders to meet Jim Monroe from Canaccord Adams, a Canadian investment bank specializing in healthcare. Canaccord thought Airway may be able to raise a venture capital Round B of $2-4 million, but also wanted to discuss the possibility of mergers and acquisitions (see Appendix G). Jim thought Dr. Gross, a country western fan, turned on his car to hear Taylor Swift's new song *Breathless* on the radio.

> *"When you, feel the world is crashing*
> *All around your feet*
> *Come running headlong into my arms*
> *Breathless I'll never judge you*
> *I can only love you"*

   "Well guys," Dr. Gross wryly remarked, "I don't know if we can raise money, but Taylor still loves us"!

APPENDICES

APPENDIX A: AIRWAY TOOLS ORGANIZATION CHART

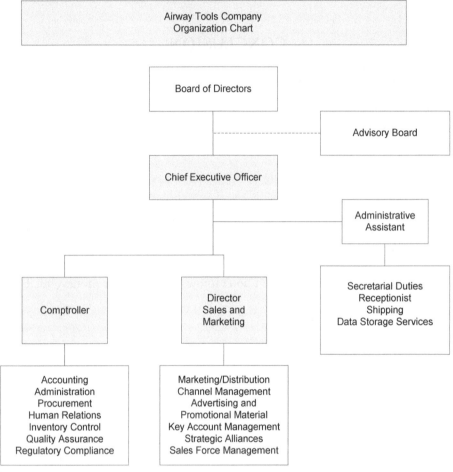

Source: Lumina

APPENDIX B: ESTIMATED INDUSTRY SALES NORTH AMERICAN AIRWAY DEVICES (2004-2005)

	Canada	United States	Total
Device Units Sold Annually			
Hospital OR's	70014	450736	520750
Hospital ER's	1255	7997	9252
Pre-Hospital	5803	36977	42780
Total	77072	495710	572783
Revenue Generated From Unit Sales			
Hospital OR's	$ 6,810,075	$ 43,841,750	$ 50,651,825
Hospital ER's	$ 2,241,280	$ 7,497,211	$ 9,738,491
Pre-Hospital	$ 10,365,920	$ 66,047,720	$ 76,413,640
Total	$ 19,417,275	$ 117,386,681	$ 136,803,956

ESTIMATED NORTH AMERICAN INTUBATION DEVICE MARKET SEGMENTS (2006)

Segment	Number
Existing North American Segments	
Number of Registered Hospitals	6,694
Estimated Number of Hospital Operating Rooms	43,803
Estimated Number of Hospital Emergency Rooms	9,252
Estimated Number of Hospital Crash Carts	27,135
Estimated Number of Level Three EMT's (Paramedics)	161,139
Estimated Number of Transporting Ambulance Services	18,700
Estimated Number of Ambulances	52,000
Estimated Number of United States Armed Forces Assigned Active Medical Personnel	5,000
Estimated Number of Canadian Armed Forces Assigned Active Medical Personnel	200
Total Existing	323,923
Potential New North American New Market Segments	
Estimated Number of North American Level One and Two EMT's	686,961

Source: Lumina

APPENDIX C – COMPETITOR ANALYSIS

MODEL	Blade Sizes	Disposable	Light Sources	Optics	Video	Re-chargeable Batteries	Price (Device) $USD	Price (Disposables) $USD
DUOSCOPE (PARKER)	Small, Medium, Large (Macintosh/Miller)	Yes	Yes	No	No	No	$33	N/A
WEST - TECH LARYNGOSCOPE	Two Sizes	Yes	Yes	No	No	Yes	N/A	N/A
XP LARYNGOSCOPE	Regular, Small, Pediatric, Infant	Yes	Yes	No	No	Yes, Green Standard ISO 7376	$578	$12/each
AIRTRAQ A-031 (PRODOL)	Regular, Small, Pediatric, Infant	Yes	Yes	Yes, Panoramic View Lens	Optional, Web-Cam Wireless Monitor	With the Web-Cam	N/A	$75 - $100
RUSH VIEWMAX (TELEFLEX MEDICAL)	Regular, Small, Pediatric, Infant (Macintosh/Miller)	Yes	Yes	Yes, Lens	No	Re-chargeable Batteries	$225	N/A
TRUVIEW EV02 (TRUPHATEK)	Regular, Small, Pediatric, Infant (Macintosh/Miller)	Yes	Yes	Yes, Panoramic View Lens	Optional, CCD camera digital LCD screen	Re-chargeable Batteries	$500	$75 - $100 (20 per box)
RES-Q-SCOPE II (RES-Q-TECH)	One Size	Yes	Yes	Yes, Fiber Optics	Embedded Camera and LCD Screen	Yes	$1,077	N/A

Source: Lumina

APPENDIX D: AIRWAY TOOLS PRO FORMA FINANCIALS

AIRWAY TOOLS COMPANY LIMITED
Forecasted Balance Sheet

	Pre-Market 2007	Total 2008	Total 2009	Total 2010	Total 2011	Total 2012
Current assets						
Cash	336,450	1,015,076	3,501,729	7,100,898	11,969,233	16,695,710
Accounts receivable		250,182	394,912	615,154	798,560	1,001,098
Inventory		152,458	177,543	230,749	289,591	318,550
	336,450	1,417,716	4,074,184	7,946,801	13,057,384	18,015,357
Deferred pre-market costs	1,563,550	1,250,840	938,130	625,420	312,710	-0
	1,900,000	2,668,556	5,012,314	8,572,221	13,370,094	18,015,357
Current liabilities						
Accounts payable		36,818	55,426	82,075	104,865	129,616
Accrued dividends	95,000	199,500	314,450	440,895	579,985	
	95,000	236,318	369,876	522,970	684,849	129,616
Debt	0	0	0	0	0	
	95,000	236,318	369,876	522,970	684,849	129,616
Shareholders equity						
Capital stock						
Common shares	950,000	950,000	950,000	950,000	950,000	950,000
Preferred shares	950,000	950,000	950,000	950,000	950,000	
	1,900,000	1,900,000	1,900,000	1,900,000	1,900,000	950,000
Retained earnings	-95,000	532,238	2,742,438	6,149,251	10,785,245	16,935,741
	1,805,000	2,432,238	4,642,438	8,049,251	12,685,245	17,885,741
	1,900,000	2,668,556	5,012,314	8,572,221	13,370,094	18,015,357
Number of common shares outstanding						
Opening balance	2,100,000					
Shares issued	950,000					
	3,050,000	3,050,000	3,050,000	3,050,000	3,050,000	3,050,000
Book value per common share	0.59	0.80	1.52	2.64	4.16	5.86

Source: Company reports

APPENDIX D: AIRWAY TOOLS PRO FORMA FINANCIALS

AIRWAY TOOLS COMPANY LIMITED
Forecasted Statement of Income and Retained Earnings

	Pre-Market 2007	Total 2008	Total 2009	Total 2010	Total 2011	Total 2012
Revenue	$0 $	2,001,456	$4,738,949	$7,381,846	$9,582,719	$12,013,171
Direct costs						
Airway tool	0	181,546	424,129	654,191	843,875	1,052,027
Disposable blades	0	0	2,622	8,746	18,193	30,379
Reusable blades	0	0	25,425	46,613	57,913	69,552
Packaging	0	13,108	30,623	47,234	60,930	75,959
Freight	0	6,554	15,312	23,617	30,465	37,979
	0	201,208	498,110	780,401	1,011,375	1,265,895
Gross margin	0	1,800,248	4,240,839	6,601,446	8,571,344	10,747,276
Overhead expenses	1,563,550	755,800	798,874	852,954	912,352	972,264
Amortization of deferred pre-market costs		312,710	312,710	312,710	312,710	312,710
Income (loss) before the undernoted	-1,563,550	731,738	3,129,255	5,435,782	7,346,282	9,462,302
Interest on debt						
Income (loss) before taxes	-1,563,550	731,738	3,129,255	5,435,782	7,346,282	9,462,302
Income taxes		0	804,105	1,902,524	2,571,199	3,311,806
Net income (loss)		731,738	2,325,150	3,533,258	4,775,083	6,150,496
Retained earnings beginning of period		-95,000	532,238	2,742,438	6,149,251	10,785,245
		636,738	2,857,388	6,275,696	10,924,335	16,935,741
Dividends	-95,000	-104,500	-114,950	-126,445	-139,090	0
Retained earnings end of period	-95,000	532,238	2,742,438	6,149,251	10,785,245	16,935,741

Source: Company reports

APPENDIX E: AIRWAY TOOLS PRO FORMA FINANCIALS

AIRWAY TOOLS COMPANY LIMITED
Forecasted Statement of Cash Flow

	Pre-Market 2007	Total 2008	Total 2009	Total 2010	Total 2011	Total 2012
Operations						
Net income (loss)	0	731,738	2,325,150	3,533,258	4,775,083	6,150,496
Dividends	-95,000	-104,500	-114,950	-126,445	-139,090	0
Amortization	0	312,710	312,710	312,710	312,710	312,710
Deferred pre-market costs	-1,563,550					
	-1,658,550	939,948	2,522,910	3,719,523	4,948,704	6,463,206
Change in non cash working capital						
Accounts receivable		-250,182	-144,730	-220,241	-183,406	-202,538
Inventory		-152,458	-25,085	-53,207	-58,842	-28,959
Accounts payable		36,818	18,608	26,649	22,790	24,752
Accrued dividends	95,000	104,500	114,950	126,445	139,090	-579,985
	95,000	-261,322	-36,257	-120,354	-80,369	-786,730
Financing						
Issue of common shares	950,000	0	0	0	0	0
Issue (redemption) of preferred shares	950,000	0	0	0	0	-950,000
	1,900,000	0	0	0	0	-950,000
Change in cash	336,450	678,626	2,486,653	3,599,169	4,868,335	4,726,477
Cash beginning of period	0	336,450	1,015,076	3,501,729	7,100,898	11,969,233
Cash end of period	336,450	1,015,076	3,501,729	7,100,898	11,969,233	16,695,710

Source: Company reports

APPENDIX F: MEDICAL DEVICE VENTURE CAPITAL

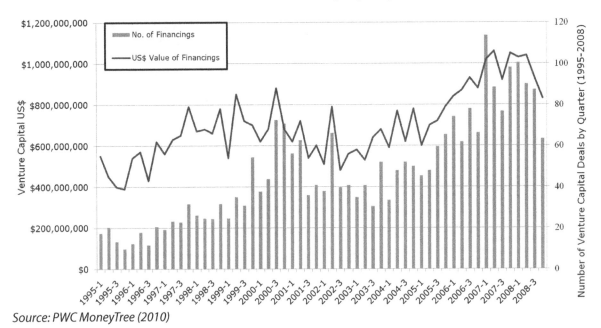

Source: PWC MoneyTree (2010)

APPENDIX G: MEDICAL DEVICE MERGERS & ACQUISITIONS 2008

	Amount US$ MM
Fresenius to acquire APP Pharmaceuticals for $3.7bn	$5,596.3
Primary Health buys Symbion for $A2.6bn	$2,400.6
Kinetic Concepts pays $1.7bn in cash for LifeCell	$1,744.4
J&J buys Mentor Corp. for $1.07bn in cash	$1,072.6
Getinge AB buys Datascope for $841mm in cash	$841.0
GE Healthcare buys research products firm Whatman for £357mm	$702.1
Cardinal Health buys Enturia for $490mm in cash	$490.0
J&J's Ethicon buys Omrix for $438mm in cash	$437.5
Medtronic buys CryoCath for $361mm	$361.1
Zimmer buys Abbott Spine for $360mm cash	$360.0
Medrad buys interventional cardiology firm Possis for $331mm	$331.3
St. Jude Medical buys MediGuide for $283mm plus liabilities	$300.0
St. Jude expands CV offerings in $250mm Radi Medical buy	$250.0
Lifecore Biomedical taken private through purchase by Warburg	$230.3
Datascope sells patient monitoring unit to Mindray for $209mm	$209.0
WuXi acquires AppTec Laboratory Services	$162.7
Novartis buys Nektar's pulmonary business for $115mm in cash	$115.0
NuVasive buys Osiris's Osteocel business	$102.6
St. Jude Medical buys EP MedSystems for $92.1mm	$91.2
Elliott Associates may buy the remaining 87.9% of Endologix	$86.7
Tissue Science to be acquired by Covidien for £36mm	$72.7
Opto Circuits tenders for monitoring device firm Criticare	$67.9
Boston Scientific divests endovascular assets	$65.0
Perrigo buys Unico Holdings for $49mm in cash	$49.0
Piramal builds critical care business in $40.1mm Minrad buy	$40.1
Wright Medical acquires Inbone for $24mm plus earn-outs	$37.7
Synovis sells interventional business to Heraeus Vadnais	$29.5
Medtronic buys Restore Medical for $29mm in cash/debt deal	$29.0
AngioDynamics buys Oncobionics for $25.4mm	$25.4
Elron ups its stake in Given Imaging by 5%	$24.2
St. Jude Medical buys Datascope's vascular closure business	$24.0
EUSA to buy Cytogen for $22mm in cash	$22.1
Volcano buys Axsun for $21.5mm in cash	$21.5
Moog buys AITECS for €15mm	$21.0
OccuLogix acquires 50% of OcuSense, now has full ownership	$19.0
Natus improves neurology business in $18mm NeuroCom buy	$18.0
Spectranetics buys Kensey Nash's endovascular unit	$18.0
Boston Scientific buys CryoCor for about $17.2mm in cash	$17.2
Volcano buys Novelis for $12mm in cash, plus earn-out	$15.0
Datascope buys Sorin Group's peripheral vascular stent unit	$14.3
AngioDynamics to buy Diomed's US and UK assets for $11mm	$11.0
Schwarzer sells neuro division to Natus Medical	$6.7
Diomed sells US operating assets to biolitec AG	$6.0
Mediscience merges with SensiVida Medical	$2.3

Source: Biocentury (2010)

REFERENCES

Wang, H.E., J.R. Lave, C.A. Sirio, and D.M. Yealy. 2006. *Paramedic Intubation Errors: Isolated Events Or Symptoms Of Larger Problems? Health Affairs*, 25(2): 501-509.

PART III: BUSINESS

Genentech Acquisition by Roche: Will Innovation Wither?

MARK J. AHN[1], ANNE S. YORK[2], DAVID ACKERLEY[3], HANNAH A. PEARCE[3], MARK J. CALCOTT[3], NATELLE C. QUEK[3], SONAI LIM[3], ROCHENE E. HIGGINSON[3], HANNAH D. HOANG[3], AND DAVID LEE[1]

[1]*Atkinson Graduate School of Management, Willamette University;* [2]*College of Business, Creighton University; and* [3]*School of Biological Sciences, Victoria University of Wellington*

- **Key Issue(s)/decision(s):** Genentech received an unsolicited offer from Roche, who already owned 56% of the company, to acquire the remaining 44% for US$89.00 per share or US$43.7 billion—thus making Genentech a wholly owned subsidiary of Roche. After seven months of unsuccessful negotiations with the Genentech board of directors, Roche lowered its original offer price from $89.00 to $86.50 per share—and launched a tender offer directly to shareholders. Thus, minority shareholders had to decide whether to tender their shares directly to Roche, reject the offer, or hold out for a higher bid price. Further, if the acquisition was successful, shareholders also had to evaluate the potential for converting their current Genentech holdings into Roche shares based on their assessment of whether integrating with Roche would hamper Genentech's edgy culture and innovation that has made them so successful to date.

- **Companies/institutions:** Genentech, Roche

- **Technology:** recombinant DNA, monoclonal antibodies

- **Stage of development:** Marketed biopharmaceuticals

- **Indication/therapeutic area:** oncology, immunology, and cardiology

- **Geography:** US, Switzerland

- **Keywords:** acquisition, biotechnology, cancer, culture, monoclonal antibodies

Acknowledgements: We would like to thank Michael Dothan for his thoughtful review and excellent suggestions, as well as our research assistants from BTEC 301. All views and errors are ours.

* *This case was prepared as a basis for class discussion rather than to illustrate either effective or ineffective handling of an administrative situation.*

INTRODUCTION

Genentech, Inc., derived from "Genetic Engineering Technology, Inc.", was founded in 1976 in South San Francisco, California and is considered the pioneer of the biotechnology industry. Over the ensuing three decades, researchers at Genentech have applied groundbreaking science to produce highly specialized and effective drugs that address unmet medical needs, particularly within the areas of oncology, immunology, neuroscience, infectious disease, and tissue growth and repair. Roche, who already owned 56% of Genentech's shares, launched an unsolicited offer for the remaining 44% of the company's shares in July 2008. After seven months of unsuccessful negotiations with the Genentech board of directors, Roche lowered its original offer price from $89.00 to $86.50 per share—and launched a tender offer directly to shareholders.[1] If successful, the proposed acquisition would result in Genentech becoming a wholly-owned subsidiary of Roche. The abrupt offer left shareholders to decide whether to tender their shares directly to Roche or hold out for a higher price.

One large institutional shareholder, Whitney Anderson, a hedge fund manager with the $2 billion Biofutures Fund, hung up the phone following an investment bank hosted teleconference briefing for clients who were Genentech shareholders. He asked his financial analyst, Piper Tomei, "Genentech is one of my oldest and largest share holdings, but I keep getting conflicting advice on whether we should tender our shares to the reduced Roche acquisition offer of $86.50 per share—or hold out for a higher bid price? Further, should we convert our share holdings to Roche stock or will they kill Genentech's edgy culture and innovation which has made them so successful?"

Piper noted, "we received a letter from the special committee, comprised of Genentech's independent board of directors, which advised:

> As you know, on February 9, 2009 Roche launched a unilateral tender offer to purchase all the common shares of Genentech that it does not already own for a price of $86.50 cash per share—a lower price than its original $89 proposal, which we continue to believe substantially undervalues Genentech. After exhaustive analysis, assisted by our independent financial and legal advisors, the Special Committee unanimously determined that Roche's tender offer price of $86.50 is inadequate and not in the best interest of Genentech stockholders, other than Roche and its affiliates. The Special Committee unanimously recommends that you reject Roche's offer and not tender your shares" (see full letter in Appendix 1).

However, Whitney noted, "DeutscheBank's biotechnology industry analyst published a report which calculated Genentech's intrinsic value at $85.52, indicating that Roche's offer is fair and reasonable—and that we should tender our shares. Let's do some more work before we make a final decision to make sure we don't leave any money on the table."

GENENTECH COMPANY HISTORY AND FINANCIALS

Genentech has been delivering on the promise of biotechnology for more than 30 years. Throughout Genentech's development, their goal has remained the same—to provide new generation therapeutics

---

1   *A tender offer is a proposal to buy shares of stock from the stockholders of a corporation, made by a group or company that desires to obtain control of the corporation. A tender offer to purchase may be for cash or some type of corporate security of the acquiring company—for example, stock, warrants, or debentures. Such an offer is sometimes subject to either a minimum or maximum that will be accepted, and is communicated to the stockholders through newspaper advertisements or a general mailing to the complete list of stockholders. Tender offers are subject to regulations by state and federal Securities laws, such as the Williams Act in the US.*

for patients with serious and life-threatening medical conditions, utilizing genetically engineered molecules. Today, Genentech continues to use genetic engineering techniques and advanced technologies to develop medicines, providing clinical benefits to millions of patients worldwide.

EARLY YEARS

In the early 1970s, biochemist Herbert Boyer and geneticist Stanley Cohen pioneered a new scientific field called recombinant DNA technology. The foundation of this field essentially used restriction enzymes with the ability to cut DNA fragments of interest, which were subsequently ligated into similarly cut plasmid vectors. Robert Swanson, then a 29-year old venture capitalist, learned of this development and contacted Boyer. During a now legendary three-hour barroom conversation, Swanson and Boyer concluded that this breakthrough technology had commercial potential. In the face of skepticism from both the academic and business community, Swanson and Boyer went forward with their idea—and Genentech was founded in 1976.

Initially, it was easy for the fledgling company to focus its scientific efforts. Genentech had one goal in mind—to clone a human protein. They were eager prove the scientific feasibility of their technology and to obtain funding for company growth. Success came in 1977 when Genentech's scientists cloned the brain hormone somastostatin in the bacterium *Escherichia coli*.[2] From this, Genentech was able to demonstrate that microorganisms could be used to produce and obtain products that would otherwise be extremely difficult or impossible to synthesize chemically.

ENTERING THE MARKET

Having achieved proof of concept, Genentech next pursued the development of their first marketable product. The company focused on proteins that would replace those that were either missing or were in short supply in certain medical conditions. In the summer of 1978, Genentech successfully cloned the human hormone insulin. This milestone in medical history was the most significant advancement in the treatment of diabetes since the development of animal insulin for use in humans in the 1920s. By 1982, Genentech had gained FDA approval and licensed commercial rights to Eli Lilly & Company, who marketed the hormone as Humulin® (recombinant insulin). Swanson knew that Genentech stood little chance of competing with Eli Lilly, who was the world's largest producer of synthetic insulin and commanded 75 percent of the U.S. insulin market. The licensing agreement also provided much needed research and development funding, and by 1987 the company was earning $5 million annually in licensing fees from Eli Lilly & Company.

Upon the introduction of Protropin®, a human growth hormone used to treat dwarfism in children, Genentech's became the first biotechnology company to independently market a product in 1985. The entry of Protropin into the market was facilitated by an FDA decision to ban the drug's predecessor due to viral contamination. Protropin generated $43.6 million in sales in 1986, growing to $155 million by 1991, thus establishing the foundation for Genentech as a fully integrated biopharmaceutical company.

Genentech's next achievement came in 1987 when the company introduced Activase®, a tissue plasmogen activator, indicated for the treatment of myocardial infarctions or heart attacks. Initially Genentech failed to provide the FDA with sufficient evidence that Activase prolonged the lives of heart attack victims, resulting in the delay of FDA approval until 1988. Nonetheless, upon approval sales in

---

2   E. coli *is a Gram negative bacterium that is commonly found in the lower intestine of warm-blooded organisms. Because of its long history of laboratory culture and ease of manipulation,* E. coli *also plays an important role in modern biological engineering and industrial microbiology. Due to the work of Cohen and Boyer with* E. coli, *the use of plasmids and restriction enzymes to create recombinant DNA became a foundation of biotechnology. Plasmids are small accessory DNA molecules that enable researchers to introduce new genes into microbes, allowing for the mass production of useful proteins in industrial fermentation processes. In this way,* E. coli *has been used to produce hormones such as insulin, vaccines, and a variety of enzymes.*

Activase sky rocketed, with the thrombolytic agent successfully generating almost half of the company's $400 million total operating revenue the following year (Rayasam, 2007).

However, success of the drug was not enjoyed for long as Activase was soon battered with legal and clinical setbacks. Genentech was challenged in court over patenting rights for Activase. More importantly, drug safety was questioned after subsequent analysis of clinical data revealed that Activase caused serious side effects, including severe internal bleeding. Moreover, a European study indicated that the drug was faster, but ultimately no more effective, than competitors costing just $200 per dose; whereas Activase sold at $2,200 per dose. Genentech's dilemma over Activase continued when The International Study of Infarct Survival (ISIS-3) performed a comparative study between Activase and two other thrombolytic agents Eminase, and streptokinase. Their results indicated that all three drugs were equally effective at keeping patients alive, which again reflected badly on Activase's high cost (Baxter-Jones *et al*, 1993). In an attempt to differentiate Activase, Genentech commissioned its own 41,000-patient comparative trial, which was completed in 1993 and demonstrated superiority of Activase over streptokinase. However, this success came at a steep price; Genentech's comparative study cost $55 million and did little to reverse the decline in Activase sales. By the mid-1990's, Genentech was selling just $300 million worth of Activase per year, a significant reduction from the $1 billion annual sales it had projected for the product in the late 1980's.

The regulatory, legal, and clinical roadblocks had significantly weakened Genentech financially. In addition, Genentech was spending more of its revenue in research and development (R&D), compared to other biotech companies (e.g., R&D spend was 40 percent of its revenue; 25 percent above the industry average) (Rebello, 1990). Combined with competition from large pharmaceutical and chemical companies that bought into biotechnology in the late 1980s, Genentech was facing financial duress. Ultimately, Genentech was rescued when Roche acquired 60 percent of Genentech's shares on September 7, 1990 for $2.1 billion. Additionally, Roche directly invested nearly $500 million into Genentech to maintain the company's research and development activities. After the first Roche acquisition, Genentech's cash position rose to $678.3 million, greatly enhancing their ability to develop new products and grow as an integrated biotechnology company.

GENENTECH REACHES NEW HEIGHTS

Genentech continued to successfully expand its product line during the early 1990s. In 1993, Genentech received regulatory approval to market Pulmozyme® (DNase) for the treatment of cystic fibrosis. Pulmozyme was considered the first major advance in the treatment of cystic fibrosis in 30 years. The company's relationship with Roche led to the establishment of a European subsidiary of Genentech to develop, register, and market pulmozyme in 17 European countries.

In the late 1990s, Genentech reached new heights with the company's revenue surpassing the $1 billion mark for the first time. In a move that later proved highly beneficial to the company, Arthur Levinson, upon his appointment as CEO, shifted Genentech's focus away from the marketing arena and back into research. The re-emphasis on research ultimately revitalized the company's product pipeline, which eventually led to a substantial increase in sales of products marketed by Genentech. The critical turning point for Genentech came with the validation of monoclonal antibodies as a technology platform with two new product approvals for Rituxan® and Herceptin®. In 1997, Rituxan was approved for the treatment of non-Hodgkin's lymphoma, making it the first monoclonal antibody approved for the treatment of cancer in the United States. In 1998, Genentech's next monoclonal antibody Herceptin was approved for the treatment of metastatic breast cancer.

The introduction of Rituxan® and Herceptin® provided an intoxicating start to the emphasis on cancer treatments, with each registering blockbuster success. Within its first year on the market, Rituxan® generated $162.6 million and Herceptin® generated $152 million in sales. Between 1999 and 2001, Rituxan's sales increased from $263 million to $818 million. Ultimately, Herceptin® and Rituxan®

became the two major sources of growth for Genentech during the first years of the decade and replaced the company's core business of growth hormones.

However, Genentech was yet again threatened by another legal battle. In 1990, the University of California, San Francisco sued Genentech for $400 million in compensation for alleged theft of the technology behind Protropin®. UCSF asserted that the technology was developed at the university and was covered by a 1982 patent. Genentech attempted to counteract the lawsuit, claiming the technology was developed independently after the university failed. After a nine-year long patent dispute, Genentech agreed to pay the university a $200 million settlement in 1999 (Albainy-Janei, 2008).

FINANCIAL PERFORMANCE

Genentech's financial and share price performance followed product successes and setbacks. Genentech revenues were $13.4 billion in 2008, up 14 percent from 2007; and net income was $3.4 billion, up 16 percent. Product sales were $10.5 billion in 2008, up 12 percent from the prior year. Royalty revenues were $2.5 billion in 2008, reflecting Roche's success in globalizing Genentech's products. Royalty revenues included licensing agreements related to Genentech's patents which include the Cabilly II patent covering monoclonal antibody production, which expires in 2018 and alone generated $237 million for the company in 2008 (Waltz, 2009). Genentech financial information is summarized in Appendix 8.

SHARE PRICE PERFORMANCE

Likewise, Genentech share performance (NYSE: DNA) has been linked to the progression of company product approvals (Figure 1). Share values rose rapidly in 1998-2000 when the company produced the first recombinant therapeutic antibodies indicated for treatment of cancer, Rituxan® and Herceptin®. Genentech shares saw another spike in the 2002-2004 period, with the approvals of Xolair®, Raptiva® and Avastin®. In the 2005-2008 period, preceding the offer by Roche, share values of Genentech have traded in the $80–90 per share range, with a high of $99.

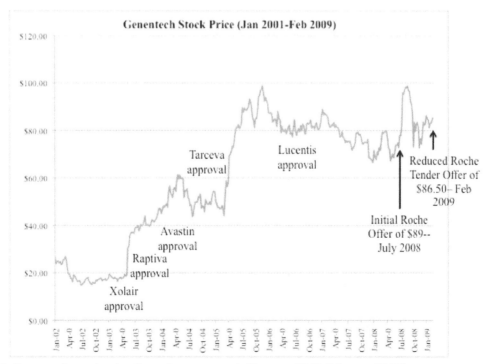

Figure 1: Genentech stock price & product approvals (January 2001-February 2009)
Source: Yahoo Finance

GENENTECH'S TECHNOLOGY, PRODUCTS, AND PIPELINE

As noted above, Genentech is a worldwide leader in the biotechnology industry, and focuses on discovering, developing, manufacturing, and commercializing multiple products in the areas of oncology, immunology, neuroscience, infectious disease, and tissue growth and repair. Recombinant DNA technology, the task of splicing genetic material derived from one organism into a new organism, enabling it to make proteins of potential therapeutic value, is the technology at the foundation of biotechnology and Genentech.

Following the success of recombinant proteins, therapeutic monoclonal antibodies (mAbs) represent the second wave of technological innovation created by the biotechnology industry during the past twenty years (Reichert *et al*, 2007). In recent years, the development of therapeutic antibodies has become the major focus of Genentech's research. With the successes of both Rituxan® and Herceptin®, Genentech is the leader in this field of research and technology. In 2008, over 95% of sales and nearly half of Genentech's current pipeline projects are evaluating new therapeutic antibodies or new uses for existing antibody-based products.

BACKGROUND ON MONOCLONAL ANTIBODIES

Emil von Behring and Shibasaburo Kitasato discovered an antitoxic component of a serum from animals immune to diphtheria or tetanus, over a century ago (Janeway, 2005). These components were later recognized as antibodies. Paul Ehrlich envisioned engineered antibodies as "magic bullets" to target disease (Adams & Weiner, 2005). The therapeutic potential of antibodies was tested as early as 1975 when British scientists Milstein and Kohler discovered hybridoma technology, for which they received the Nobel Prize in 1984. The pair had successfully generated large quantities of uniform (monoclonal) mouse antibodies designed to target specific proteins. However, despite their achievements, these early mouse (murine) antibodies were of limited therapeutic value primarily due to the patients' allergic response to them (Gura, 2002).

Although achieving limited success, Milstein and Kohler's discovery of hybridoma technology soon paved the way for the development of chimeric antibodies as therapeutics. To solve the problem of an allergic response, chimeric antibodies were specifically designed to incorporate only the front

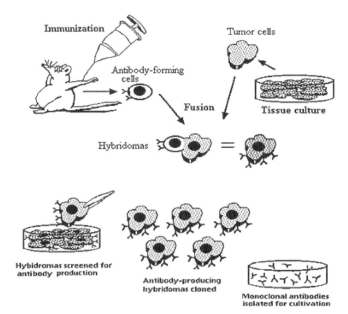

Figure 2: Monoclonal antibody production
Source: *www.accessexcellence.org*

half of the antibody binding site onto a human antibody. Although an improvement these antibodies, which have about 30% mouse and 70% human protein sequence, could still elicit an allergic response. To further optimize monoclonal antibodies for use in human therapeutics, scientists sought to make the chimeric antibody even more "human." Today, researchers at Genentech utilize genetic engineering to create humanized antibodies, an optimized form of antibody which contains over 90 percent human sequences. Humanized antibodies have demonstrated their efficacy and safety, exemplified by those already on the market such as Rituxan and Avastin.

Antibodies are highly specific for target molecules, which makes them ideal for drug therapy. They are Y shaped proteins secreted by B cells of the immune system that play an important role in the immune response, by eliminating foreign antigens through binding, thereby inhibiting these antigens from other deleterious actions. During their production, B cells subject antibodies to an intense selection process to eliminate self-recognition (Janeway, 2005). Monoclonal antibodies are even more specific than regular antibodies as they will recognise only a single unique region (or "epitope") on an antigen, enabling them to be precisely engineered for a specific target and response (Zangememeister-Wittke, 2005).

The key to Genentech's monoclonal antibody production is its patented method of using recombinant DNA technology to produce the monoclonal antibodies. Previous forms of monoclonal antibody production first require the generation and purification of antibodies from a living animal following injection of a target antigen. B cells producing antibodies that recognise that antigen are then recovered and fused with a tumor cell line to generate "hybridomas"—cell lines that combine the immortality of tumor cells with the specific antibody producing capabilities of the B cells. Effectively, these hybridomas are able to grow and divide indefinitely, allowing large quantities of monoclonal antibody production (Cabilly et al, 1989). This approach, however, is plagued by difficulties including contamination, expense and restrictions on modifications that can be made. Genentech's production method overcomes these challenges by using cell culture lines (primarily Chinese hamster ovary cells) which have been genetically engineered to produce the antibodies (Reff et al, 1994; Theil, 2004). The use of genetic recombination enables the antibody to be designed to produce either chimeric or humanized antibodies. This is important as it enables the production of antibodies for therapeutic use with a superior safety profile, as humanized antibodies do not invoke unwanted immune reactions. It has also led to the use of next generation antibody fragments in Genentech's pipeline. Genentech holds a patent for this technology (expiring in 2018) and, as such, is a powerhouse in the monoclonal antibody business across the industry (Cabilly et al, 1989).

Genentech's signature oncology treatments are therapeutic monoclonal antibodies, generated using recombinant DNA technology. They aim to design antibodies that are homogenous and familiar to the host's immune system. In this way, they have the ability to create an antibody against a foreign antigen (e.g., a specific cancerous molecule). Through these methods, Genentech has been able to create therapeutic antibodies which block cancerous growth by various approaches such as apoptosis and anti-angiogenesis.

An example is Rituxan, the first therapeutic antibody approved for the treatment of relapsed or refractory, CD20+, B cell low-grade non-Hodgkin's lymphoma (NHL). Rituxan acts by inhibiting the actions of CD20, a factor that is involved in the activation of cell cycle initiation and differentiation, a process that allows recognition by the immune system to ultimately mark that cell for destruction. The CD20 antigen was selected as the target as it is expressed on more than 90% of non-Hodgkin's lymphomas, and not on hematopoietic stem cells, pro-B cells, normal plasma cells or normal tissues, to induce B cell lysis. Rituxan can use three mechanisms that result in B cell lysis—antibody-mediated cytotoxicity, complement-dependent cytotoxicity, and apoptosis (Figure 3).

In 2008, Genentech had eleven product approvals in oncology, immunology and ophthalmology, and metabolism and primary care that generated product sales of $10.5 billion. By this time, Genentech's monoclonal antibodies comprised 95.7% of total product revenues. First, Genentech's leading oncol-

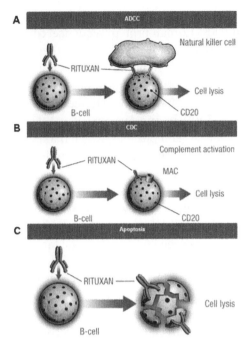

Figure 3: Proposed modes of action of Rituxan® resulting in B cell lysis. A) Antibody-dependent cell-mediated cytotoxicity. Macrophages, Natural killer cells, and T cells recognize bound Rituxan to CD20 resulting in cell lysis. B) Complement-dependent cytotoxicity. Upon Rituxan binding CD20 causes complement activation, resulting in over porous cell membrane disturbing the homeostatic state of the B cell causing the cell to fill and lysis. C) Apoptosis, whereby binding of the antibody results in signal transduction for self-destruction.

Source: Company reports

ogy division marketed Avastin® (bevacizumab), Rituxan® (rituximab), Herceptin® (trastuzumab), and Tarceva® (erlotinib). Second, the immunology and ophthalmology division marketed Lucentis® (ranibizumab), Xolair® (omalizumab), Rituxan® (rituximab), and Activase® (alteplase). Third, the metabolism and primary care division marketed Cathflo® Activase® (alteplase), TNKase® (tenecteplase), and Nutropin®/AQ®/Nutropin AQ Pen® (somatropin) (see Appendix 2 for product sales). In addition, Genentech received $2.5 billion in royalty revenues from licensing agreements with partners who sell and/or manufacture their products or use the company's intellectual property rights (see Appendix 3 for key patent expiry dates).

GENENTECH'S PIPELINE

Genentech's pipeline has been described as one of the "strongest, most well balanced pipelines in the industry" (Ransom, 2007). At the time of the Roche offer, Genentech's product pipeline was quite extensive, with 12 phase III, 16 phase II, and 23 phase I clinical trials ongoing. Three new molecular entities of note were Pertuzumab, Ocrelizumab and Tastuzumab-DM1. Pertuzumab is a monoclonal antibody that acts as a dimerization inhibitor for HER receptors, and represents a possible improvement on Herceptin and Tarceva (which bind to individual HER receptors). Pertuzumab has potential to treat breast, ovarian and lung cancers. Ocrelizumab is a second generation antibody targeting CD20 (similar to Rituxan®, but fully humanized), and is being studied against Lupus, Arthritis and Multiple Sclerosis. Trastuzumab-DM1 is a novel antibody-drug conjugate, targeted towards HER2 receptors (over expressed in certain breast cancers) by the Trastuzumab, which then allows the DM1 small molecule to kill selected cancer cells as a cytotoxic agent. This enhanced cell specific killing may enable trastuzumab-DM1 to be used in advanced and metastatic versions of breast cancer.

GENENTECH'S PEOPLE AND CULTURE

While there are many people responsible for the success of Genentech, there were three principal leaders who drove the company's success and serve as icons for corporate culture: Herbert Boyer, Robert Swanson, and Art Levinson. First, Herbert Boyer was born and raised in a corner of western Pennsylvania "where railroads and mines were the destiny of most young men" (Wiegand, 2009). The rags to riches story of Dr. Herbert Boyer started in the late 1960s when he was a low paid researcher at the University of California, San Francisco (UCSF). A decade later, Boyer became the first molecular biology multimillionaire, when he co-founded Genentech in 1976 (Golden, 2002). Dr. Boyer served as a director of Genentech since its founding in 1976, and served as a vice president of the company from 1976 to 1990.

A biochemist and a genetic engineer, Boyer first demonstrated the usefulness of recombinant DNA technology to produce commercial medicines, which ultimately laid the groundwork for Genentech's development. During his time as a researcher at UCSF, *Escherichia coli*, a common bacteria of the gut, had caught his attention. In particular, he had been studying restriction endonucleases, enzymes that possessed unique and useful properties—the capability of cutting DNA strands in a particular fashion, leaving what has become known as "sticky ends" that assist different DNA strands with compatible sticky ends in joining back together (Weigard, 2009). Recognition of this property enabled the technology of gene-splicing to take its next great leap. During a now-famous conversation over a midnight snack of corned beef sandwiches in 1972, Boyer and Stanley Cohen, a scientist at Stanford, agreed to collaborate on what was to become a critical breakthrough. Boyer's work on restriction enzymes complemented Cohen's goal of using bacterial plasmids to transfer DNA between organisms (Golden, 2002). Before this historical conversation, Cohen had developed a method to introduce antibiotic-carrying plasmids into certain bacteria, as well as a method of isolating and cloning genes carried by the plasmids.

Second, Robert A. Swanson, at the time of his premature death in December 1999 was already an icon of the biotechnology industry. Swanson was known widely as the co-founder of Genentech, as well as an innovative and far-sighted businessman and spokesman for the biotechnology industry (Campbell, 1999). The son of an Eastern Air Lines employee, Swanson grew up in Florida and earned degrees in chemistry and management from M.I.T. He joined Citicorp Venture Capital in New York and was transferred to San Francisco, then joined the venture capital firm now called Kleiner, Perkins, Caufield & Byers (Pollack, 1999). Following the publicity surrounding the revolutionary gene splicing work of Cohen and Boyer Swanson, now a 29-year old venture capitalist, approached Boyer about the possibility of developing biotechnology and commercializing useful products using recombinant DNA technology. Swanson served as director and chief executive officer of Genentech until 1990 when he was named chairman of the board, a position he held until his retirement (Pierce, 1999).

Third, Arthur D. Levinson, has been one of the most influential figures in the rise of Genentech. Levinson graduated with a Bachelor of Science degree in Molecular Biology from the University of Washington in 1972 and later completed his PhD at Princeton University in 1977. Subsequently he undertook a postdoctoral fellowship (1977-1980) in microbiology at the University of California, San Francisco (UCSF). Levinson joined Genentech as a senior scientist just before the company went public in 1980. Levinson held various executive positions before being promoted to vice president of research science and technology, and subsequently CEO in 1995.

As CEO after a difficult period at Genentech, Levinson brought a steadfast commitment to basic research, dedication to the company and nurturing its culture, and tight focus in key areas immunology, oncology and vascular biology. For example, Levinson played an important role in the start up of Genentech's post-doctoral research program, which recruits young scientists from throughout the world and encourages them to pursue research of their interest. His leadership revitalized the company's pipeline and by the late 1990s Genentech surpassed $1 billion in revenues in 1997 (Stipp, 2003).

GENENTECH'S CULTURE

Arguably, however, the unique culture of Genentech is the "secret sauce" of the company's success— and many would claim that this is what sets Genentech apart. Genentech's commitment to science, medicine and its patients enables it to undertake groundbreaking medical research. Yet this strong sense of purpose is complemented and balanced by a relaxed work environment, referred to as "casual intensity", generating an innovative staff culture that drives their success.

COMPANY PHILOSOPHY ON SCIENCE, BUSINESS AND PATIENTS

Genentech is a company driven by unmet clinical needs and translational research, with a strong focus on new science. It is a company where the people in charge have traditionally been interested in the science at least as much as the business. Art Levinson, Genentech's CEO, entered the company as a senior scientist and worked his way up the ranks to his current position. In his first two years as CEO, Levinson reinvested 50% of revenues straight back into research, resulting in a pipeline boom (Morris, 2006). Most top Genentech managers are from academia (Rayasam, 2007) and all the vice presidents at Genentech have their own lab, allowing them to keep themselves grounded in the research side of things (Schaeffer, 2009). A former CEO of Genentech, Kirk Raab, has been known to quote "The secret of research is that businessmen shouldn't mess with it" (Schaeffer, 2009). The strong science focus has lead many to comment that Genentech acts like a seemingly quixotic cross between academia and a pharmaceutical company (Bonnetta, 2009).

Genentech also places a strong emphasis on its patients: "What inspires them [staff] is being focused on improving patients' lives. That's what motivates us here", noted Mark Tessier-Lavigne, vice president, research (Bonnetta, 2009). The company's campus headquarters, annual report, and website is sprawled with pictures and stories of patient successes (Rayasam, 2007). Direct interactions with patients also help instill their strategic focus. Sue Desmond-Hellmann, Genentech's president, who was a practicing oncologist and clinical scientist, typifies the front line experience that maintains the grounded approach of the company (Morris, 2006). Amongst this dedication to science is a strong understanding that the staff holds the key to Genentech's innovative scientific breakthroughs. Genentech's co-founder Robert Swanson has been quoted saying "our most valuable assets go home in tennis shoes every night" (Bonnetta, 2009).

This is reportedly one of the main reasons that Genentech is frequently highly ranked in 'top company to work for' polls, both scientific and general. Genentech has been voted in *FORTUNE* magazine's 'top 100 companies to work for' list for the past 10 years, frequently appearing in the top ten. In 2006 it was ranked 1st, but has since dropped rankings to 2nd, 5th and finally 7th in 2009. In addition to this Genentech has also been ranked 'top biopharmaceutical company to work for' by Science magazine for the past two years. Indeed, one of the things that Genentech scientists say they value most about working for the company is that they feel they are doing meaningful work (Morris, 2006; Bonnetta, 2009). Besides offering meaningful work, other reasons for these consistently high rankings are onsite gyms, childcare, flexible job sharing and telecommuting, free coffee and food, carpool and public transportation subsidies, and stock options for all employees. In addition to these prerequisites, Genentech also keeps employees happy professionally and stimulated intellectually by providing researchers with advanced equipment, comfortable spaces to casually discuss ideas, and by encouraging scientific publications (Rayasam, 2007; Scheller, 2009).

Of note, Genentech is famous for allowing their researchers to work on 'pet projects' based on their own ideas (Rayasam, 2007). Giving researchers the freedom to 'play' allows Genentech the ability to harness the wildest, most creative ideas of its staff. It is generally expected that these projects won't work (Bonnetta, 2009), however it enables maintenance of staff morale and innovation, making them feel as though their ideas count. In some cases however these pet projects have really paid off. Napoleone Ferrara, whose research led to the development of Avastin®, Genentech's flagship product, was encouraged to follow up on observations that led to the discovery of VEGF and the idea of using angiogenesis as an anticancer treatment (Morris, 2006). Ferrara himself was unsure if the concept would work, yet the support of Genentech enabled the development of this fledging idea into a blockbuster product (Lawrence, 2007).

Another way in which Genentech nurtures innovation is by actively cultivating a fun, enthused and slightly mischievous environment. This effort is led from the top down by CEO Art Levinson. He closely monitors staff morale, and pounces when things seem to be going awry. Art has been known to dare scientists to see how many can fit into freezers, and has invented a game termed Buzz word bingo (Levinson decided that there were too many meaningless business buzz words flying around meetings and so designed bingo cards containing these words to be brought along to meetings. Anyone who could cross off a row would yell "BUZZ!" and silence the speaker). On one Halloween all the business directors, dressed up as Snow White and the Seven Dwarfs, snuck into the lobby of competitor Amgen to pose for photos, before being chased out by security guards (Morris, 2006). Genentech is also famous for its weekly Friday night keg parties, which have been a tradition since the company's inception (which started over a beer) (Rayasam, 2007). Concerts and costume parties are also used to reward successful work (Morris, 2006). Overall Genentech's environment is a 'blue jeans and tennis shoes with an attitude' culture (Ransom, 2007).

Amidst this casual style are high expectations for employee performance at Genentech. Science at Genentech runs more like a graduate program (e.g., Genentech was a pioneer in establishing a post-doctoral program within the company) and the intensity is similar. Scientists get grilled at research proposal presentations, where their science is stringently assessed, and vice president Vishva Dixit routinely works 100 hours a week as he 'rolls out of bed, waiting to get to the lab,' and expects similar dedication and passion from his researchers (Rayasam, 2007).

Obviously this type of environment is not for everyone, but Genentech makes an effort to make sure that the people they hire fit into the company's profile. While Genentech receives 15,000 unsolicited CVs a month, the company routinely keeps positions open for months in search of the right candidate for a job (Rayasam, 2007). As Art Levinson, CEO, noted, "I worry when people ask, 'How many stock options am I going to get?' We hire people who want to cure diseases" (Fortune, 2008). All this is done in an effort to maintain an enthusiastic innovative atmosphere, which is key to Genentech's ability to keep developing new innovative technologies.

ROCHE MOVES TO ACQUIRE GENENTECH

Headquartered in Basel, Switzerland, Roche has been a pioneer in the discovery, development and marketing of novel healthcare solutions for over 110 years (Roche, 2009). The founder of Roche, Fritz Hoffmann, was an entrepreneur who was convinced that the future belonged to branded pharmaceutical products. Fritz Hoffmann was among the first to recognize that the industrial manufacture of standardized medicines would be a major advance in the fight against disease. After gaining experience in pharmacy and the chemical trade, Fritz formed F. Hoffmann-La Roche & Company in 1896. Today, Roche is one of the world's leading research-focused healthcare groups in the field of pharmaceuticals and diagnostics. Their pharmaceutical division, for example, markets products in over 150 countries and has majority share holdings in Genentech in United States and Chugai in Japan.

Swiss pharmaceutical giant Roche has two research-intensive businesses. First, the pharmaceuticals division focuses on five disease areas: oncology, virology, inflammation, metabolism, and central nervous system. Second, the diagnostics division is a leader *in vitro* diagnostics, and also supplies instruments and reagents for life science research (e.g., applied science, molecular diagnostics, diabetes care, professional diagnostics, tissue diagnostics). Roche is currently the largest cancer drug manufacturer in the industry. In 2008, Roche group sales were US$42.6 billion in 2008, down 1.1% from 2007. Roche delivered US$8.4 billion of net income or 19.7% of sales, making the Swiss firm one of

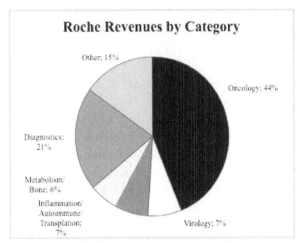

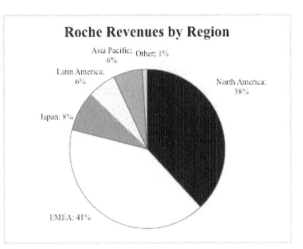

Figure 4: Roche financials and revenues by category and region

the most profitable companies in the world (Appendix 7). Four of Roche's top five selling products in 2008—Herceptin®, Avastin®, Rituxan®, and Tarceva®—were Genentech's.

In 1990, Roche acquired 60 percent of Genentech's shares, allowing a capital infusion into the American company and ultimately enabling Genentech to realize its potential while retaining relative independence. A minority of Genentech's shares was exchanged for special redeemable shares that continued to be traded publicly. On July 21, 2008, Roche made a move to acquire all outstanding shares in Genentech, offering $89 per share or $43.7 billion in cash for the remaining 44% of the company. Franz Humer, chairman of the board of Roche, stated:

> "Our long and successful participation in Genentech has provided great benefits to both of our companies and shareholders. It has resulted in one of the biggest success stories in the healthcare industry. Roche's significant investment in Genentech over many years has helped it to focus on innovation and long-term projects, leading to some of the most important breakthroughs in the treatment of cancer and other life-threatening diseases. The transaction will create a unique opportunity to evolve Roche's hub-and-spoke model into a structure that allows us to strengthen the focus on innovation and accelerate the search for new solutions for unmet medical needs. Combining the strengths of Roche and Genentech will create significant value and result in benefits for patients, employees and shareholders."

Genentech formed a Special Committee comprised of Genentech's independent board of directors—Herbert W. Boyer, Debra L. Reed, and Charles A. Sanders—and hired investment bank Goldman Sachs as an advisor. On August 13, 2008, the Special Committee rejected the offer as inadequate. On January 30, 2009, however, Roche's response was to lower the offer from $89.00 to $86.50 and launch an unsolicited cash tender offer for all remaining shares—thereby aggressively moving from negotiating with Genentech's special committee of the board of directors to placing the decision into the hands of individual shareholders. As Franz B. Humer, chairman of the Roche Group, noted:

> "We intend to create unrivaled benefits for our patients, employees and shareholders by combining Roche and Genentech. We are disappointed that the discussions over the last six months between Roche and the special committee of Genentech have not produced a negotiated agreement. We feel it is now time to give the Genentech minority shareholders the opportunity to decide on our offer. Especially in the current market environment the offer provides an opportunity for all public shareholders to achieve

liquidity and to receive a fair price for all their shares."

On February 23, 2009, the Genentech Special Committee again urged shareholders to take to reject the Roche tender:

> *"The Special Committee unanimously recommends that you reject Roche's offer and not tender your shares into the Roche offer... Over the past seven months, the Special Committee persistently attempted to work constructively with Roche and we were consistent in our stated willingness to negotiate toward a price that recognizes the full value of Genentech and reflects the significant benefits Roche would enjoy as a result of full ownership... Roche was consistently dismissive of these efforts and of the information we provided, and repeatedly refused to increase the price at which it sought to acquire the Shares... The Special Committee continues to believe that Roche's original $89 price substantially undervalues Genentech. We remain committed to considering a proposal that recognizes the full value of Genentech and reflects the significant benefits that Roche would enjoy as a result of full ownership."*

CONCLUSION

After the market close on February 24, Whitney Anderson and Piper Tomei met in the boardroom to review the team's analysis of Roche's reduced tender offer of $86.50 and decide how to respond, if at all. On one hand, Roche's tender offer was tempting. While Genentech had a strong track record of innovation, nothing guaranteed it would be sustainable. A comparison of Genentech's price-to-earnings ratio and the market's expectation for its pipeline to the rest of the biotech industry suggested that the company might already be overvalued (Appendix 5). In addition, Deutsche Bank's biotechnology industry analyst valued the company at $85.52 per share—below Roche's tender price. By selling now his fund stood to capture a healthy profit amidst the deepest recession in recent memory. Although a higher price might yet be negotiated, he feared his window of opportunity might be closing.

Still, Anderson was unsure. Genentech was less risky than most other biotech firms and the strength of its pipeline broadly recognized by the market (Appendix 4). Goldman Sachs deemed the $89 valuation inadequate, projecting substantial growth in both revenues and new drugs well into the future (Appendix 9). Anderson remembered well the huge value a "blockbuster drug" quickly creates, as was shown recently by Avastin.

As Anderson sat pondering, Tomei offered up yet another alternative, "you know, Whitney, if a deal goes through you'll be able roll your holdings into Roche. Effectively, Genentech is already a subsidiary of Roche and the merger simply increases operating efficiencies and tax benefits in achieving global scale. Roche has global reach, complimentary research interests and capabilities, and experienced management to take Genentech's technologies and products to the next level. Plus, Roche would cover some of the late stage trials, allowing Genentech to focus on what it does best: innovate."

Anderson considered her suggestion. Tomei made some good points and there was definite potential for synergies both in accelerating revenues and reducing costs, particularly in US operations. On the other hand, a buyout might kill the pioneering spirit and innovative culture that defined Genentech. Anderson also feared that rolling his holding into Roche shares would expose his fund to "big pharmaceuticals," a segment of the industry that was under increasing pressure from healthcare reform.

As the evening wore on, Whitney Anderson peered out the window of his office overlooking Battery Park and was increasingly aware that his next move could have a significant impact on his fund's future performance.

APPENDICES

APPENDIX 1: LETTER FROM SPECIAL BOARD COMMITTEE OF GENENTECH TO SHAREHOLDERS

February 23, 2009

Dear Fellow Stockholder:

We are writing to you as the Special Committee of Genentech's Board of Directors. Our committee exists for the sole purpose of representing your interests—the non-Roche owners of Genentech—and we take our responsibility seriously. As you know, on February 9, 2009 Roche launched a unilateral tender offer to purchase all the common shares of Genentech that it does not already own for a price of $86.50 cash per share – a lower price than its original $89 proposal, which we continue to believe substantially undervalues Genentech.

After exhaustive analysis, assisted by our independent financial and legal advisors, the Special Committee unanimously determined that Roche's tender offer price of $86.50 is inadequate and not in the best interest of Genentech stockholders, other than Roche and its affiliates.

The Special Committee unanimously recommends that you reject Roche's offer and not tender your shares into the Roche offer.

Below, we summarize a number of factors the Special Committee considered in reaching our conclusion that Roche's $86.50 price is inadequate. We also review the rigorous process we followed in a good-faith effort to work constructively with Roche to achieve a price for your shares that recognizes the full value of Genentech and reflects the substantial benefit Roche would enjoy if it were to acquire full ownership. The Special Committee continues to be willing to consider an offer from Roche that meets these objectives.

The Roche Offer Substantially Undervalues Genentech

The Special Committee, with its advisors and with Genentech's management team, considered many factors in reaching our conclusion that the $86.50 unsolicited tender offer price substantially undervalues the Company. We based our determination on numerous reasons:

- Strong Projected Financial Performance. Genentech's 2008 Financial Plan implies a valuation substantially in excess of Roche's tender offer price. The 2008 Financial Plan is based on the most current information available to Genentech. The 2008 Financial Plan was rigorously reviewed by Genentech's management and the Special Committee, and is believed by the Special Committee to be the best estimate of Genentech's prospects. For a detailed description of the 2008 Financial Plan, see Item 4 "—2008 Financial Plan" in the accompanying Schedule 14D-9. Our belief in the 2008 Financial Plan projections is also supported by Genentech's remarkable past achievements, including the strength of its unparalleled research and development organization, its robust and promising product pipeline, and its industry-leading commercial and financial success.

- Unparalleled Research Success. Genentech's tradition of, and commitment to, exceptional science serves as the foundation for the successful development and commercialization of new medicines. Genentech's scientists are among the world's most respected and they play an important role in the scientific discoveries upon which Genentech's products are based.

- Robust and Growing Product Pipeline. Genentech's integrated approach to research, development and commercialization has created one of the most robust and promising product pipelines in its industry, with Genentech's historical productivity significantly exceeding industry averages. Genentech has 25 new molecular entities in its clinical development pipeline. Many of these new molecular entities represent new scientific approaches that have the potential to significantly advance the standard of care in diseases with few currently effective therapies. Genentech has approximately 60 Phase II and Phase III clinical trials underway, including many new indications for our marketed products. Genentech had 15 consecutive positive Phase III studies between 2003 and 2006, and Genentech has obtained FDA approval for 15 biologics since 1985, more than any other biotechnology company.

- Industry-Leading Commercial Success. Genentech's scientific accomplishments continue to translate into product approvals, including five new medicines in the past six years, and strong market success. Since 1997, Genentech has experienced 11 consecutive years of double-digit revenue growth, and, based on sales, Genentech is the #1 oncology company in the United States.

- Extraordinary Financial Success. The strength of Genentech's research and development organization has enabled extraordinary financial success. Genentech's net income for 2008 was in excess of $3.4 billion, and for the period of 2003 through 2008, Genentech's compounded annual growth rate in GAAP earnings per share exceeded 40%.

- Significant Value to Post-2015 Ex-U.S. Commercialization. The exclusive option Genentech granted to Roche in the 1995 Commercialization Agreement to license, develop and commercialize outside of the United States new products that enter our clinical development pipeline expires in 2015. We are under no obligation to extend the term of the option contained in the Commercialization Agreement. The Special Committee believes that Genentech would receive significant value in agreeing to extend the term of the option with Roche or granting the option to a third party, through upfront payments, improvements in the royalty rates or other financial terms, or a combination thereof. Genentech could also choose not to license some or all of the rights and commercialize products outside the U.S. itself. This significant additional value is not accounted for in the 2008 Financial Plan or in the Roche tender offer price.

- The Offer Price Does Not Reflect the Substantial Benefits of a Business Combination to Roche or the Strategic Importance of the Transaction to Roche. The Special Committee believes there are other substantial benefits to be realized by Roche if it were to acquire full ownership of Genentech, including cost synergies, increased productivity and tax benefits. The Special Committee believes the unsolicited tender offer price does not reflect these significant additional benefits that would accrue to Roche as a result of a transaction. We also believe that Roche's tender offer price does not reflect its substantial and growing dependence on Genentech's pipeline and innovation. A substantial majority of Roche's product pipeline is composed of products under development in collaboration with Genentech.

Opinion of Goldman, Sachs & Co. As part of its full and diligent review process, the Special Committee engaged Goldman, Sachs & Co. as its independent financial advisor to evaluate the adequacy of the Roche unsolicited tender offer. Goldman Sachs rendered its opinion to the Special Committee that, as of February 22, 2009, and subject to the factors, assumptions and limitations set forth in Goldman Sachs' written opinion, the $86.50 price Roche is offering Genentech's public shareholders (other than Roche and its affiliates) pursuant to the tender offer is inadequate, from a financial point of view, to such holders. Please see the full text of Goldman Sachs' written opinion, which is set forth in Annex A of the Schedule 14D-9.

The Special Committee Engaged Actively and Openly with Roche to Achieve a Price That Recognizes the Full Value of Genentech and Reflects the Substantial Benefits to Roche

Over the past seven months, the Special Committee persistently attempted to work constructively with Roche and we were consistent in our stated willingness to negotiate toward a price that recognizes the full value of Genentech and reflects the significant benefits Roche would enjoy as a result of full ownership.

To this end, and to appropriately assess Roche's initial July 2008 proposal, we requested that management prepare the 2008 Financial Plan, incorporating updates of key assumptions from the 2007 LRP, prepared some 15 months ago, and reviewing in detail all critical business, financial, and scientific drivers of Genentech. We determined that the 2008 Financial Plan should be the Company's best estimate of its prospects and should be neither conservative nor aggressive, using assumptions where the probability-adjusted upsides and downsides are believed to be essentially equal. This approach differs from the annual LRPs, which, as Roche is aware, have a conservative bias because they are used for budgeting purposes, resulting in Genentech's actual financial results consistently surpassing the targets set forth in past LRPs.

In spite of:
- the significant and repeated efforts of the Special Committee, its advisors and Genentech's management to demonstrate that the value of Genentech exceeded the initial $89 proposal,

- the careful preparation and presentation of the 2008 Financial Plan,

- our attempts to bridge a portion of the valuation gap through the use of alternative consideration and transaction structures,

- our willingness to provide Roche with a price at which we would be willing to pursue a transaction, and

- our continued willingness to negotiate,

Roche was consistently dismissive of these efforts and of the information we provided, and repeatedly refused to increase the price at which it sought to acquire the Shares.

Even after all our efforts, and despite the expressed acknowledgement of additional value as reflected in its advisor's analyses, Roche not only refused to increase its original $89 proposal and to engage in productive negotiations with the Special Committee regarding a mutually acceptable valuation for the public Shares of Genentech – it reduced its offer price to $86.50.

Enhancing Shareholder Value

The Special Committee continues to believe that Roche's original $89 price substantially undervalues Genentech. We remain committed to considering a proposal that recognizes the full value of Genentech and reflects the significant benefits that Roche would enjoy as a result of full ownership.

At the same time, Genentech's uniquely productive ability to develop and commercialize breakthrough new medicines that improve and extend life for millions of patients will continue to fuel its momentum and financial success. The Special Committee is very confident in Genentech's ability to continue to create and deliver superior value for all of its stockholders.

The Schedule 14D-9 contains a more detailed explanation of the reasons for the Special Committee's recommendation. We urge you to read it carefully.

We appreciate your support as we work to protect and deliver the full value of your investment in Genentech.

Sincerely,

The Special Committee of the Genentech Board of Directors
Charles A. Sanders, M.D.
Herbert W. Boyer, Ph.D.
Debra L. Reed

This letter contains forward-looking statements. Please review the "Cautionary Note Regarding Forward-Looking Statements" in Item 8 of the Schedule 14D-9 for important information regarding these forward-looking statements.

APPENDIX 2: GENENTECH REVENUES BY PRODUCT

Year Approved		2006	2004	2004	2003	2003	1998	1997	1994	1994	1987	1986
Therapeutic Area		Opthamology	Cancer	Cancer	Immunology	Dermatology	Cancer	Cancer	Immunology	na	Cardiology	Endocronology
Technology Platform		monoclonal antibody	monoclonal antibody	small molecule	monoclonal antibody	monoclonal antibody	monoclonal antibody	monoclonal antibody				
Year	Total Revenues	Lucentis	Avastin	Tarceva	Xolair	Raptiva	Herceptin	Rituxan	Pulmozyme	Other	Thrombolytics	Nutropin
1985	$5											$5
1986	$44											$44
1987	$141										$56	$86
1988	$263										$151	$111
1989	$319										$196	$123
1990	$367										$210	$157
1991	$383										$197	$185
1992	$391										$182	$206
1993	$457										$236	$217
1994	$601								$88	$2	$281	$225
1995	$635								$111	$3	$301	$219
1996	$583								$76	$4	$284	$218
1997	$562							$6	$92	$6	$255	$224
1998	$686						$31	$163	$94	$4	$210	$214
1999	$995						$188	$279	$111	$5	$233	$221
2000	$1,207						$276	$444	$122	$3	$202	$227
2001	$1,641						$347	$819	$123	$4	$187	$250
2002	$2,018						$385	$1,163	$138	$3	$175	$297
2003	$2,437				$25	$1	$425	$1,489	$167	$4	$181	$322
2004	$3,551		$555	$13	$189	$52	$483	$1,711	$178	$7	$195	$354
2005	$5,162		$1,183	$275	$328	$79	$764	$1,989	$222	$0	$219	$375
2006	$7,169	$381	$1,853	$402	$428	$90	$1,330	$2,252	$245	$7	$243	$386
2007	$8,540	$825	$2,453	$417	$472	$107	$1,704	$2,515	$266	$11	$268	$383
2008	$9,503	$887	$2,908	$457	$517	$108	$1,819	$2,852	$305	$32	$275	$375

Source: Company reports

APPENDIX 3: GENENTECH PATENT EXPIRY FOR KEY PRODUCTS

Product	Lastest-to-Expire Patent	Date of Patent	Year of Expiration
Avastin	6 884 879	Apr-05	2017
	7 169 901	Jan-07	2019
Rituxan	5 677 180	Oct-97	2014
	5 736 137	Apr-98	2015
	7 381 560	Jun-08	2016
Herceptin	6 339 142	Jan-02	2019
	6 407 213	Jun-02	2019
	7 074 404	Jul-06	2019
Lucentis	6 884 879	Apr-05	2017
	7 169 901	Jan-07	2019
Xolair	6 329 509	Dec-01	2018

Source: Company reports

APPENDIX 4: PIPELINE VALUATION

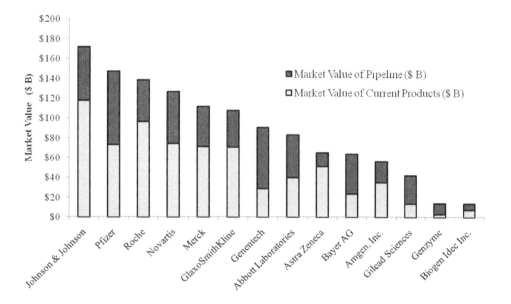

Leading Biopharmaceuticals: Market Valuation

■ Market Value of Pipeline ($ B)

□ Market Value of Current Products ($ B)

APPENDIX 5: PIPELINE VALUATION RELATIVE TO P/E

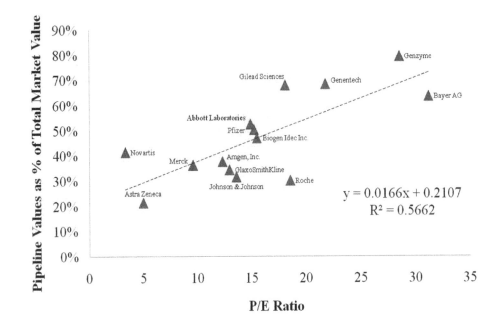

Leading Biopharmaceuticals: Pipeline Value as % of Total
Market Capitalization v. Price-to-Earnings Ratio

APPENDIX 6: GENENTECH FINANCIAL RATIOS VERSUS OTHER LARGE BIOTECHNOLOGY AND MULTINATIONAL PHARMACEUTICAL COMPANIES

	Company name	Share price	P/E ratio	Price-to-book value	Price-to-sales	Market value ($ billions)	Current ratio	Total debt to assets	Total debt to equity	Operating margin	Employees	Revenues ($ mms)	Net income ($mms)	Revenues per employee ($mms)	Net income per employee ($mms)
Large biotech companies	Genentech	$84.55	21.8	4.1	6.7	$90.2	3.3	28.1	39.0	39.7	11,229	$13,418.0	$3,427.0	$1.2	$0.3
	Amgen, Inc.	$55.38	12.3	2.8	3.7	$56.1	3.1	27.9	49.9	34.8	16,700	$15,003.0	$4,196.0	$0.9	$0.3
	Biogen Idec Inc.	$46.38	15.5	2.3	3.3	$13.4	2.7	13.1	19.2	28.2	4,700	$4,097.5	$783.2	$0.9	$0.2
	Genzyme	$50.05	28.5	1.9	2.9	$13.5	2.8	1.5	1.8	12.6	11,000	$4,605.0	$421.1	$0.4	$0.1
	Gilead Sciences	$46.39	18.1	10.2	7.8	$41.8	3.5	18.6	31.4	50.2	3,441	$5,335.8	$2,011.2	$1.6	$0.4
Multinational pharmaceutical companies	Abbott Laboratories	$53.64	14.9	4.8	2.8	$83.0	1.5	27.0	65.5	19.3	69,000	$29,527.6	$4,734.2	$0.4	$0.2
	Astra Zeneca	$34.33	5.0	4.1	2.1	$64.9	1.2	25.3	74.5	37.2	65,000	$31,601.0	$6,130.0	$0.5	$0.1
	Bayer AG	$76.81	31.2	2.4	1.3	$63.5	1.2	32.1	103.7	10.7	108,800	$48,871.7	$2,553.6	$0.4	$0.1
	GlaxoSmithKline	$41.53	13.0	9.0	2.7	$107.8	1.7	43.0	213.6	29.3	99,003	$40,213.4	$7,781.1	$0.4	$0.2
	Johnson & Johnson	$62.31	13.6	4.1	2.7	$171.9	1.7	14.0	27.9	26.6	117,000	$63,747.0	$12,949.0	$0.5	$0.2
	Merck	$36.46	9.6	4.1	4.7	$111.4	1.4	13.2	33.3	41.1	55,200	$23,850.3	$7,808.4	$0.4	$0.3
	Pfizer	$13.27	15.2	2.1	3.1	$147.2	1.6	15.6	30.0	29.0	80,025	$48,296.0	$8,049.0	$0.6	$0.1
	Roche	$40.10	18.6	3.2	3.1	$138.4	3.2	5.4	9.2	30.2	80,000	$44,742.3	$10,636.1	$0.6	$0.2
	Average	$49.32	16.7	4.2	3.6	$84.8	2.2	20.4	53.8	29.9	55,469	$28,716.0	$5,498.4	$0.7	$0.2
	Minimum	$13.27	5.0	1.9	1.3	$13.4	1.2	1.5	1.8	10.7	3,441	$4,097.5	$421.1	$0.4	$0.1
	Maximum	$84.55	31.2	10.2	7.8	$171.9	3.5	43.0	213.6	50.2	117,000	$63,747.0	$12,949.0	$1.6	$0.4
	Std Dev	$18.43	7.2	2.6	1.8	$49.4	0.9	11.5	55.4	11.2	41,560	$19,567.4	$3,816.4	$0.4	$0.1

APPENDIX 7: ROCHE FINANCIAL STATEMENTS

US$ Millions, except per share items	2008	08 v. '07 % Change	% of Revenues
Revenue	$42,633	-1.1%	100.0%
Cost of Revenue	$12,767	-0.6%	29.9%
Gross Profit	$29,865	-1.3%	70.1%
Selling/General/Admin. Expenses	$10,750	-2.4%	25.2%
Research & Development	$8,266	5.5%	19.4%
Unusual Expense (Income)	$131	35.9%	0.3%
Other Operating Expenses	-$2,137	2.0%	-5.0%
Total Operating Expense	$29,777	0.3%	69.8%
Operating Income	$12,856	-4.2%	30.2%
Other, Net	$623	11.0%	1.5%
Income Before Tax	$13,235	-7.5%	31.0%
Income After Tax	$10,135	-5.2%	23.8%
Minority Interest	-$1,752	11.9%	-4.1%
Net Income	$8,382	-8.1%	19.7%

Source: Company reports

APPENDIX 8: SELECT GENENTECH FINANCIAL INFORMATION

Income Statement

	1999	2000	2001	2002	2003	2004	2005	2006	2007	2008
Revenue	1,421.4	1,646.0	2,081.7	2,719.3	3,300.3	4,621.2	6,633.4	9,284.0	11,724.0	13,418.0
Cost of Goods Sold	285.6	364.9	354.4	441.6	480.1	672.5	1,011.1	1,181.0	1,571.0	1,744.0
Gross Profit	1,135.8	1,281.1	1,727.3	2,277.6	2,820.2	3,948.6	5,622.3	8,103.0	10,153.0	11,674.0
SG&A Expense	467.9	497.0	474.4	573.3	794.9	1,088.1	1,435.0	2,014.0	2,256.0	2,405.0
R&D Expense	367.3	489.9	526.2	623.5	722.0	947.5	1,261.8	1,773.0	2,446.0	2,800.0
Other Expense	1,350.6	375.3	568.5	1,050.4	498.7	776.2	1,003.6	1,164.0	1,222.0	1,140.0
Operating Income	(1,050.0)	(81.2)	158.2	30.5	804.7	1,136.8	1,921.9	3,152.0	4,229.0	5,329.0
Net Interest Income & Other	(290.9)	85.1	124.8	(0.8)	92.8	(0.5)	91.0	325.0	197.0	102.0
Earnings Before Taxes	(1,340.9)	4.0	283.0	29.8	897.5	1,219.4	2,012.9	3,403.0	4,426.0	5,431.0
Income Tax Expense	(196.4)	20.4	127.1	(34.0)	287.3	434.6	733.9	1,290.0	1,657.0	2,004.0
Earnings After Taxes	(1,144.5)	(16.4)	155.9	63.8	610.2	784.8	1,279.0	2,113.0	2,769.0	3,427.0
Accounting Changes	-	(57.8)	(5.6)	-	(47.7)	-	-	-	-	-
Net Income	(1,144.5)	(74.2)	150.2	63.8	562.5	784.8	1,279.0	2,113.0	2,769.0	3,427.0
Shares Outstanding	1,025	1,044	1,070	1,048	1,057	1,079	1,080	1,073	1,069	1,067

Balance Sheet

	1999	2000	2001	2002	2003	2004	2005	2006	2007	2008
Assets										
Cash and Equiv	337.7	551.4	395.2	208.1	372.2	270.1	1,225.5	1,250.0	2,514.0	4,533.0
Short-Term Investments	405.0	642.5	952.9	826.4	1,139.6	1,395.0	1,139.7	1,243.0	1,461.0	1,665.0
Accounts Receivable	214.8	261.7	303.3	521.6	612.6	957.4	1,083.4	1,666.0	1,766.0	1,941.0
Inventory	275.3	265.8	357.0	393.5	469.6	590.3	702.5	1,178.0	1,493.0	1,299.0
Other Current Assets	93.8	67.4	201.0	133.0	162.8	210.0	267.7	367.0	1,519.0	635.0
Total Current Assets	1,326.5	1,788.8	2,209.4	2,082.8	2,756.8	3,422.8	4,418.8	5,704.0	8,753.0	10,073.0
Property, Plant and Equiptment	730.1	752.9	865.7	1,068.7	1,617.9	2,091.4	3,349.4	4,173.0	4,986.0	5,404.0
Intangibles	3,082.0	2,736.1	2,415.8	2,261.8	2,125.8	1,983.4	1,888.8	1,791.0	2,745.0	2,598.0
Other Long-Term Assets	1,415.9	1,434.0	1,644.0	1,364.1	2,235.6	1,905.8	2,490.0	3,174.0	2,456.0	3,712.0
Total Assets	6,554.4	6,711.8	7,134.9	6,777.3	8,736.2	9,403.4	12,146.9	14,842.0	18,940.0	21,787.0

Liabilities and Stockholders' Equity

	1999	2000	2001	2002	2003	2004	2005	2006	2007	2008
Accounts Payable	33.1	34.5	33.4	51.4	59.7	104.8	339.0	346.0	420.0	228.0
Short-Term Debt	-	-	149.7	-	-	-	-	-	599.0	500.0
Taxes Payable	-	-	-	-	-	-	-	-	-	-
Accrued Liabilities	451.0	398.8	449.2	575.2	765.9	1,092.5	1,276.5	1,749.0	2,826.0	2,279.0
Other Short-Term Liabilities	-	15.4	19.6	20.0	47.5	46.0	44.3	62.0	73.0	88.0
Total Current Liabilities	484.1	448.7	651.8	646.7	873.0	1,243.3	1,659.8	2,157.0	3,918.0	3,095.0
Long-Term Debt	149.7	149.7	-	-	412.3	412.3	2,083.0	2,204.0	2,402.0	2,329.0
Other Long-Term Liabilities	637.8	439.2	563.3	791.8	930.6	965.7	934.5	1,003.0	715.0	692.0
Total Liabilities	1,271.6	1,037.6	1,215.0	1,438.4	2,215.9	2,621.2	4,677.3	5,364.0	7,035.0	6,116.0
Total Equity	5,282.8	5,674.2	5,919.8	5,338.9	6,520.3	6,782.2	7,469.6	9,478.0	11,905.0	15,671.0
Total Liabilities & Equity	6,554.5	6,711.8	7,134.9	6,777.3	8,736.2	9,403.4	12,146.9	14,842.0	18,940.0	21,787.0

Source: Morningstar

APPENDIX 9: GENENTECH PRO FORMA FINANCIAL FORECASTS (2009-2014)

Genentech Revenue Forecasts (US $ Millions)

	2009	2010	2011	2012	2013	2014	2015	2016	2017	2018	2019	2020	2021	2022	2023	2024
Revenues																
Product sales	11,244	12,345	13,690	14,860	16,083	17,349	18,061	19,004	20,764	23,408	24,279	24,995	26,230	28,732	31,551	33,803
Royalties	2,501	2,465	2,617	2,681	2,723	2,809	2,551	2,518	2,667	2,676	2,377	2,447	2,611	2,967	3,326	3,662
Contract & other	373	289	342	497	563	660	805	750	740	813	885	955	1,042	1,126	1,222	1,298
Total revenues	14,118	15,099	16,648	18,038	19,370	20,817	21,418	22,273	24,171	26,897	27,542	28,398	29,883	32,825	36,099	38,764
Expenses																
Cost of sales	1,541	1,794	1,872	1,760	1,784	1,947	1,969	1,948	2,022	2,288	2,412	2,742	3,048	3,369	3,800	4,235
R&D	2,824	3,020	3,330	3,608	3,874	4,163	4,284	4,400	4,649	4,951	5,273	5,658	5,977	6,488	6,967	7,395
MSG&A	2,233	2,183	2,321	2,432	2,495	2,692	2,828	3,157	3,484	3,867	4,077	4,444	4,933	5,472	6,027	6,448
Profit sharing	1,352	1,544	1,664	1,688	1,581	1,654	1,645	1,595	1,603	1,643	1,330	1,337	1,316	1,314	1,333	1,082
Total expense	7,950	8,541	9,187	9,488	9,734	10,456	10,726	11,100	11,757	12,749	13,092	14,180	15,274	16,644	18,128	19,160
Operating income	6,169	6,558	7,461	8,550	9,636	10,361	10,692	11,173	12,414	14,148	14,450	14,217	14,609	16,182	17,970	19,604
Tax expense	2,219	2,220	2,311	2,548	2,838	3,041	3,121	3,252	3,662	4,223	4,116	4,095	4,203	4,756	5,269	5,732
Post-tax operating income	3,950	4,339	5,150	6,002	6,798	7,320	7,570	7,920	8,752	9,925	10,334	10,122	10,406	11,426	12,701	13,872
Gross capital expenditure	(546)	(560)	(620)	(600)	(554)	(545)	(593)	(647)	(640)	(716)	(688)	(722)	(782)	(808)	(833)	(820)
Change in net working capital	(50)	(371)	125	(68)	(122)	(132)	(58)	(91)	(222)	(228)	(126)	7	(197)	(231)	(397)	(746)
Depreciation	507	586	559	575	577	585	594	593	594	606	613	638	625	622	635	661
Free cash flows	3,861	3,994	5,214	5,909	6,699	7,228	7,513	7,776	8,483	9,588	10,133	10,044	10,052	11,010	12,105	12,968

Source: Company reports

REFERENCES

Adams, G. and Weiner, L. 2005. Monoclonal antibody therapy of cancer. *Nature Biotechnology*. 23: 1147–1157.

Albainy-Janei, S. 2008. Can states have their (patent) and eat it, too?. *Drug Discovery News*. Old River Publications: Rocky River, Ohio.

Baxter-Jones, *et al*. 1993. An Overview of the Patency and Stroke Rates Following Thrombolysis with Streptokinase, Alteplase, and Anistreplase Used to Treat an Acute Myocardial Infarction, *Journal of Interventional Cardiology*. 6: 15 – 23.

Bonnetta, L. 2009. Innovation more critical than ever. *Science Careers*. http://sciencecareers.sciencemag.org/ career_magazine/previous_issues/articles/2009_10_02/science.opms.r0900079. Accessed October 2009.

Cabilly, S. Heyneker, H., Holmes, W., Riggs, A., Wetzel, R. 1989. Recombinant immunological preparations *Proc National Academy Science USA*. 1984 June; 81(11): 3273–3277

Campbell, K.D. 1999. *Robert Swanson, 52, alumnus who launched biotechnology industry*, Massachusetts Institute of Technology (Online). http://web.mit.edu/newsoffice/1999/swanson-1208.html. Accessed 12 October 2009.

Fortune Magazine. 2008. Power Point. *Fortune Magazine* (Online). 15 July 2008 www.opstcards.blogs.fortune.cnn.com.

Golden, F. 2002. The pioneers of molecular biology: Herb Boyer. *New York Times Magazine*.

Gura, T. 2002. Magic Bullets hit the target. *Nature*. 417: 548-587

Janeway *et al*. 2005. *Immunobiology, 6th Edition*. Garland Science Publishing.

Lawrence, S. 2007. Billion dollar babies - Biotech drugs as blockbusters. *Nature Biotechnology*. 25: 4.

Levinson. A. D. 1998. For success, focus your strengths. *Nature Biotechnology*. 16: 45-46.

Morris, B. 2006. The 100 best companies to work for 2006. *Fortune Magazine*. 2006.

Pierce-Wright, Swanson, R. 1999. Entrepreneur of the genetic engineering evolution. *The Guardian*. http://www.guardian.co.uk/theguardian. Accessed 2009.

Pollack, A. 1990. Genenetch-Roche deal may spur similar ties. *Special New York Times*.

Ransom, J. 2007. Genentech faces the consequences of growth. *Nature Biotechnology*. 25: 4.

Rayasam, R. 2007. Scientific methods (Genentech). *U.S. News and World Report*. 142: 6.

Rebello, K. 1990. Genentech's well-bred deal; Roche pact is likely to spawn others. *USA TODAY*.

Reff, ME, Carner, K., Chambers, KS, Chinn, PC, Leonard, JE, Raab, R., Newman, RA, Hanna, N. and Anderson, DR. 1984. Depletion of B cells *in vivo* by a chimeric mouse human monoclonal antibody to CD20. *Blood*. 83: 435-445.

Reichert *et al*. 2007. Development trends for monoclonal antibody cancer therapeutics. *Nature Reviews Drug Discovery*. 6, 349–356

Roche, 2009. *Biotechnology: New Directions in Medicine*. Roche Web Site. http://www.roche.com/biotechnology_new_directions_in_medicine-2.pdf. Accessed 13 November 2009.

Schaeffer, S., Lawrence, S. 2009. Genentech's staying power. *Biocentury*. May: 2009.

Scheller, R. 2009. *Our research vision*. Genentech Web Site. www.gene.com/gene/research /researchvision.html. Accessed October 2009.

Stipp, D. 2003. Biotech: How Genentech Got It. *Fortune*. May 27.

Thiel, K. 2004. Biomanufacturing from boom to bust... To bubble? *Nature Biotechnology*. 22: 1365-1372.

Thrive. 2009. *Madison Region Economic Development Enterprises*. http://www.thrivehere.org/media/documents/Internal/Definition%20of%20Biotechnology.pdf. Accessed 3 October 2009.

Waltz, E. 2009. Genentech's cabilly victory. *Nature Biotechnology*. 27: 4.

Weingard, Susan and Boyer, Herbert. 2009. *National Health Museum: AccessExcellence*. http://www.accessexcellence.org/RC/AB/BC/Herbert_Boyer.php. Accessed 12 October 2009.

Zangememeister-Wittke, U. 2005. Antibodies for targeted cancer therapy -- technical aspects and clinical perspectives. *Pathobiology*. 72(6): 279-86.

Myogen: Are We There Yet?

MARK J. AHN AND TRAVIS COOK

Atkinson Graduate School of Management, Willamette University

- **Summary and key issue/decision:** In 2006, Myogen, Inc., a ten-year old Boulder, Colorado-based development-stage biotechnology company with two late stage product candidates, received an acquisition offer from Gilead. Myogen and multinational pharmaceutical GlaxoSmithKline had already entered into a global collaboration in which Myogen received marketing and distribution rights to GlaxoSmithKline's Flolan® (epoprostenol sodium) in the United States and GlaxoSmithKline in exchange licensed Letairis® (ambrisentan) from Myogen for all territories outside of the United States. This decision-based case evaluates the strategic options of Dr. William Freytag, president and chief executive officer and the board of directors of whether to sell Myogen to Gilead, vertically integrate and establish commercial operations, or seek other approaches to maximize value and manage risk from its late-stage product pipeline.

- **Companies/institutions**: Myogen, Gilead, GlaxoSmithKline, University of Colorado-Boulder

- **Technology**: Endothelin receptor antagonists

- **Stage of development at time of issue/decision:** Myogen's lead product, Letairis® (ambrisentan) for the treatment of pulmonary arterial hypertension (PAH), had recently completed two Phase III clinical trials and was about to submit a New Drug Application (NDA) to the US FDA (Food and Drug Administration). Myogen was also developing darusentan for the treatment of patients with resistant hypertension.

- **Indication/therapeutic area:** Treatment of pulmonary arterial hypertension (WHO Group I) in patients with WHO class II or III symptoms to improve exercise capacity and delay clinical worsening. Pulmonary arterial hypertension (PAH) is a disease of the pulmonary vasculature characterized by vasoconstriction and vascular proliferation, which leads to right heart failure and death.

- **Geography:** US

- **Keywords:** Letairis® (ambrisentan), pulmonary arterial hypertension, acquisition

---

* *This case was prepared as a basis for class discussion rather than to illustrate either effective or ineffective handling of an administrative situation.*

INTRODUCTION

Myogen, Inc. is a Westminister, Colorado-based biopharmaceutical company founded in 1996. The company's name, Myogen, is a term that describes the process of protein extraction from muscle tissue, which became the foundation of their cardiac tissue bank. Myogen's focus from inception was the development of small molecule therapeutics for the leading killer of Americans—cardiovascular disease. In pursuing this strategy, Myogen developed several discovery partnerships with large multinational pharmaceuticals including GlaxoSmithKline and Novartis. Dr. William Freytag, CEO of Myogen, recognized that the firm's novel discovery-stage research platform would not produce a clinical stage candidate for several years and began actively seeking ways to leverage the company's expertise through business development.

After a careful analysis of the pharmaceutical industry, Myogen developed a strategy to in-license patent-protected clinical stage compounds no longer being pursued by other pharmaceutical companies. However, Myogen's license of ambrisentan and darusentan from Abbott Laboratories, both of which treatment of pulmonary arterial hypertension and were effectively discarded, accelerated their progress into clinical trials and raised their ability to obtain financing. After promising Phase II clinical results with ambrisentan, Myogen commenced two Phase III clinical trials to seek commercial approval by the US FDA (Food and Drug Administration).

After a successful IPO in 2003 and raising a total of $210.9 million in funding to date, Myogen was at a crossroads and started to evaluate its strategic options. In 2006, based on Myogen's success with ambrisentan and a promising pipeline the company received an acquisition offer from Gilead, a Foster City, California-based biotechnology leader focused on infectious diseases and actively seeking diversification. Myogen's management and board of directors met to consider if they should reject Gilead's offer and continue to focus on their strategy to "create an integrated biopharmaceutical company focused on the discovery, development, and commercialization of novel therapies that address the fundamental mechanisms involved in cardiovascular disease." If Myogen rejected the acquisition offer, however, they would need to raise significant additional, dilutive equity capital to complete clinical trials, further advance other pipeline drugs, and establish commercial operations. On one hand, Freytag reflected: "Becoming the next Amgen or Genentech would be difficult and risky, but it would be worth it." On the other hand, "Gilead's offer represents a significant premium for shareholders who have funded our efforts to date."

MYOGEN, INC. BACKGROUND

Myogen's primary focus was the discovery, development, and commercialization of small molecule therapeutics for cardiovascular disorders that were not adequately treated with existing therapies. Their drug discovery programs were based on the discoveries of three prominent academic scientists who are recognized experts in the field of cardiac hypertrophy and heart failure: Dr. Michael Bristow, professor of cardiology at the University of Colorado Health Sciences Center, Dr. Leslie Leinwand, chairperson of molecular, cellular and developmental biology at the University of Colorado, and Dr. Eric Olson, chairman of molecular biology at the University of Texas Southwestern Medical Center.

Bill Freytag became CEO and Director of Myogen in 1998, and brought a diverse range of experiences. Freytag served as an independent consultant to the healthcare industry, senior vice president at Somatogen, Inc., a biopharmaceutical company; president of research and development at Boehringer Mannheim Corporation; and DuPont Medical Products in various research and business positions. Dr. Freytag was also a director of Immunicon Corporation. Dr. Freytag received a Ph.D. in biochemistry from the University of Kansas Medical Center. While he was excited about the novelty and long-term potential of Myogen's research platform, Myogen aggressively sought to use its scientific insights to in-license products that other companies had shelved to unlock value and accelerate the company's pipeline development.

In 1998, Myogen obtained a license to Perfan® I.V. enoximone for the treatment and prevention of certain forms of heart disease in humans from Aventis Pharmaceuticals, Inc. In 1999, Myogen established a wholly owned subsidiary located in Germany, through which Perfan® was sold in eight European countries. In June 2000, Myogen initiated two Phase 3 trials with enoximone capsules, both of which yielded negative results, and the company discontinued development of enoximone capsules in June 2005. In December 2005, Myogen exited the business and sublicensed the rights to Perfan® I.V. in markets outside North America to Wülfing Holding, GmbH.

Despite these setbacks Myogen continued to investigate opportunities to fill its pipeline, eventually in-licensing rights to ambrisentan in 2001, and darusentan in 2003, from Abbott Laboratories. In 2002, Myogen initiated a Phase 2 clinical trial of ambrisentan for pulmonary arterial hypertension (PAH), which was completed at the end of 2003. Myogen then proceeded to initiate two pivotal Phase 3 trials for ambrisentan, ARIES-1 and -2 in January 2004. In 2005, Myogen completed patient enrollment in ARIES-1 and -2 and announced positive "top line" summary results in the ARIES-2 trial. During this time Myogen also continued to develop darusentan, initiating a Phase 2b clinical trial in resistant systolic hypertension in 2004. In August 2005, Myogen reported positive top line results of the darusentan trial.

Myogen's primary source of financing has been equity stock offerings. Myogen raised $395.9 million in capital in six financings between 1998 and 2005. Management successfully completed its first round of venture capital in 1998, raising $6 million for operations in a Series A financing. In 1999, the company raised $18 million in a Series C financing activity; and raised $106.4 million in a Series D financing activity in 2001. In 2003, Myogen crossed over from a private to a public company and raised $80.5 million in their initial public offering. Myogen continued to seek additional financing from the public capital markets and in 2004 they raised $60 million in a follow on offering; and $125 million in 2005 (Table 1).

In addition to Myogen's equity financing activities, the company's management explored and assessed strategic alternatives for accelerating growth and value. These alternatives included strategies to grow Myogen's business operations through partnerships and licensing agreements with large pharmaceutical and biotechnology companies. In 2003, Myogen entered into a research collaboration with Novartis Institutes for BioMedical Research, Inc. to conduct a research program to discover and develop disease-modifying drugs for chronic heart failure and related disorders. The partnership also

Table 1: Myogen financing history

Financing Type	Date Completed	Amount Raised	Investors	Underwriters
Follow-on	9/16/05	$125,000,000		Goldman Sachs, CIBC World Markets, First Albany, Lazard
Common and wts (other)	9/27/04	$60,000,000	New Enterprise Associates, InterWest Partners, Perseus-Soros BioPharmaceutical Fund, Sequel Venture Partners	
IPO	10/29/03	$80,500,000		CS First Boston, JPMorgan, CIBC World Markets, Lazard
Venture (Series D financing)	8/28/01	$106,400,000	JPMorgan Partners, New Enterprise Associates, InterWest Partners, Adams Street Partners, Pacific Rim Ventures, Sequel Venture Partners, CMEA Ventures, Montagu Newhall Assoc., Perseus-Soros BioPharmaceutical Fund, and others	
Venture (Series C financing)	12/15/99	$18,000,000	New Enterprise Associates, Sequel Venture Partners, Crosspoint Venture Partners, InterWest Partners, New Venture Partners, CMEA Ventures	
Venture (Series A financing)	11/1/98	$6,000,000	Sequel Venture Partners, Crosspoint Venture Partners, InterWest Partners	

Source: Biocentury (2009)

sought to discover and develop novel therapeutics to slow or reverse the progression of cardiovascular disease. Ultimately, the collaboration would provide opportunity for both organizations to identify and acquire additional preclinical and clinical-stage compounds for development.

As Myogen matured into a development stage company, they began to incur greater operational losses. In December 2005, Myogen had an accumulated deficit of $239.2 million due to expenses associated with research and development and operational activities—and having no revenues (see Figure 1; and Appendices A and B). These expenses represented both the clinical development costs and the costs associated with non-clinical support activities such as toxicological testing, manufacturing process development, and regulatory consulting services.

Due to the magnitude of the company's burn rate, Myogen entered into a broad collaboration with GlaxoSmithKline in 2006. The terms of the collaboration provided GlaxoSmithKline the license to manufacture, develop and commercialize ambrisentan in all territories outside of the United States. In addition to the agreement, Myogen received rights to market and distribute Flolan® (epoprostenol sodium) for a three year period in the United States. Flolan®, a long term intravenous treatment for primary pulmonary hyptertension and pulmonary hypertension associated with the scleroderma spectrum of disease, was approved by the FDA in 1995.

Like all biopharmaceutical companies, Myogen's share price was dependent on the company's financing, product pipeline, clinical advancements and probability of FDA approval for product launch. In 2005, Myogen's shares epitomized the ups and downs of biotechnology investments with a trading range of $5 to $38 per share. The volatility of the stock in 2005 was the result of the clinical performance of Myogen's lead drug candidates: Enoximone, Darusentan, and Ambrisentan. During the first two quarters of operations in 2005, enoximone, a drug for treatment of chronic heart failure, performed poorly in phase 3 clinical trials, resulting in the observed decrease in stock price to a low of $5.21. At the end of the second quarter, Myogen released news of darusentan's quality phase 2 performances for treatment-resistant hypertension. This news inflated the stock price to a new high, reaching $24 at the close of 2005. At the start of the first quarter in 2006, news of abrisentan's positive phase 3 studies were released, reporting fewer side-effects than existing treatments, lifting the stock price to $38.

Market analysts compared preliminary clinical results of ambrisentan to the industry benchmark, Tracleer®, a PAH drug developed and commercialized by Swiss-based Acetlion. From the phase 3 re-

Product	Years Ended December 31,		
	2005	2004	2003
Development fees (in thousands):			
Enoximone capsules	$18,902	$26,757	$19,718
Ambrisentan	21,150	16,255	7,962
Darusentan	5,329	4,250	149
Total development	45,381	47,262	27,829
License fees:			
Enoximone	—	—	—
Ambrisentan	—	1,500	1,000
Darusentan	—	—	5,000
Other	106	117	188
Total license fees	106	1,617	6,188
Discovery research	7,115	5,245	3,348
Total research & development	**$52,602**	**$54,124**	**$37,365**

Figure 1: Product development fees
Source: Company reports (2006)

sults, ambrisentan was thought to be the best-in-class for treatment of PAH, and biotechnology analysts estimated peak sales to be between $750 million and $2.1 billion. Analysts forecast sales of ambrisentan based, in part, on Acetlion's $350 million in revenue for Tracleer during the first nine months of 2005. Overall, combining Myogen's two most promising developmental drugs, the company's target market value was estimated to be between $2.1 billion and $3.3 billion—or between $48 to $79 per share (Ransom, 2006).

MYOGEN'S TECHNOLOGY, PRODUCTS, AND PIPELINE

Myogen has focused on the research and development of therapeutics based on signaling pathways in cardiac, hypertrophy, myosin heavy chain research and gene/protein expression profiling using its human cardiac tissue bank. With a focus on developing receptor antagonists, medicinal chemists developed structure activity relationships (SAR) to identify receptor antagonists for cardiovascular related diseases.

The term cardiovascular disease is used to describe a continuum of clinical conditions resulting primarily from three underlying chronic diseases: atherosclerosis, hypertension and diabetes. These underlying diseases cause permanent damage to the heart, blood vessels and kidneys, leading to progressively debilitating clinical conditions such as chronic heart failure, hypertension, chronic renal disease, heart attack and stroke.

Cardiovascular disease is the second leading cause of death and disability in the United States, accounting for 19% of all hospitalizations in short-stay, non-federal hospitals and over 58% of total mortality in 2003. The American Heart Association estimates that the total direct and indirect costs of cardiovascular disease in the United States will be approximately $403 billion in 2006, including $50 billion in costs for drugs and related medical durables and $157 billion in hospitalization and nursing home costs. Despite improved treatments and increased awareness of preventative measures, approximately 71 million people in the United States currently suffer from one or more types of cardiovascular disease.

Over the past 25 years, a variety of drug classes such as beta-blockers, calcium channel blockers, diuretics, angiotensin converting enzyme inhibitors, endothelin receptor antagonists and angiotensin receptor blockers have been used to treat various cardiovascular diseases. Several of these agents have helped to increase the survival times of patients who suffer from cardiovascular diseases. However, many current therapies do not adequately address the underlying molecular mechanisms of cardiovascular disease. Cardiovascular disease remains progressive in a large portion of patients, many of whom continue to deteriorate even when treated with multiple drugs simultaneously. Myogen believes that recent advances in the understanding of the molecular biology of cardiovascular diseases provide an opportunity to improve on existing therapies and to discover and develop new therapeutics to ameliorate the symptoms and perhaps to slow or reverse the progression of these diseases.

DRUG DISCOVERY STRATEGY

Many patients with chronic heart failure develop an abnormal enlargement of the heart called cardiac hypertrophy. The causes and effects of cardiac hypertrophy have been extensively documented, but the underlying molecular mechanisms that link the molecular signals to cell changes, or cardiac signaling pathways, remain poorly understood.

Myogen believes that the fundamental drivers of pathological remodeling of the heart (abnormal growth, shape and function of the heart) are increases in ventricular wall stress and neurohormonal and growth factor stimulation of cardiac muscle. These processes are set in motion by primary insults to the heart, including myocardial infarction (heart attack) and chronic high blood pressure. Wall stress and associated growth promoting stimuli lead to changes in cardiomyocyte signaling pathways that ultimately produce pathological changes in gene expression in the heart.

One of the characteristic changes that occur in a failing heart is a change in gene expression wherein fetal genes that were turned off shortly after birth are reactivated in the disease process. Although this response may initially be beneficial to a patient with chronic heart failure, it becomes harmful as the disease progresses. Myogen scientists and academic collaborators at the University of Colorado and the University of Texas are focused on identifying the set of fetal genes that are reactivated in chronic heart failure, understanding the consequences of their reactivation and discovering the means to control their expression. This work has led to the discovery of several signaling pathways that appear to control the reexpression of fetal genes, down-regulation of adult genes, cardiac hypertrophy and its progression to dilated cardiomyopathy.

DISCOVERY RESEARCH

Through a license agreement with the University of Colorado, Myogen gained access to the human cardiac tissue library which consists of hundreds of failing and non-failing human hearts that they use to discover and validate targets for drug discovery. Myogen have shared rights to the new discoveries through agreements with the investigators' academic institutions, creating a source of novel molecular mechanisms and targets for their drug discovery operations.

In addition, Myogen built a drug discovery research team and infrastructure, which includes a compound library and high-throughput screening robotics. This high-throughput platform made it possible for Myogen to advance several targets through screening rapidly, ultimately leading to the identification of a series of promising lead structures.

MYOGEN AND PULMONARY ARTERIAL HYPERTENSION (PAH)

Pulmonary arterial hypertension is defined as a blood pressure elevation in the vessels supplying the lung. It is characterized by vasoconstriction and vascular obstruction that eventually lead to increased pulmonary vascular resistance and right heart failure (Newman, *et al*, 2004). According to its etiology, PAH may be classified as idiopathic PAH, familial PAH, or PAH caused by pulmonary veno-occlusive disease and pulmonary capillary hemangiomatosis, collagen vascular disease, portal hypertension, HIV infection, drugs and toxins, or congenital systemic-to-pulmonary shunts, as well as persistent PAH of the newborn. There are an estimated 200,000 patients with PAH in the US and EU. The World Health Organization (WHO) classification of PAH is based on patients' function (Simonneau, *et al*, 2004) as shown in Appendix C.

Treatment options focus on symptomatic management of the disease, including improving breathlessness on exertion, chest pain and syncope. Prior to the availability of specific disease related therapies, the median life expectancy of patients was 2.8 years with 1-, 3-, and 5-year survival rates of 68%, 48%, and 34% respectively (Hoeper, 2005). However, recent advances in understanding the role of the vascular endothelium in the pathogenesis of PAH have led to the development of new pharmacological approaches to managing these diseases (D'Alonzo, 1991). The control of three key mediators is thought to support the pathogenesis of PAH: prostacyclin, nitric oxide, and endothelin-1 (ET-1).

Typical treatment of PAH includes lifestyle modifications, conventional treatments, and disease-specific treatments. Lifestyle modifications include low-level graded exercise, such as walking, as tolerated. Conventional treatments include diuretics to manage volume overload due to right ventricular failure. An example of such an agent is the anticoagulant, warfarin, a controversial treatment due to the associated risk of gastrointestinal bleeding. Current disease specific therapies approved for the treatment of PAH in the United States and/or Europe includes prostacyclin analogues such as, intravenous epoprostenol, treprostinil, and iloprost. Other approved treatments include oral endothelin-receptor antagonist bosentan and sitaxsentan, and oral phophodiesterase type 5 inhibitor sildenafil. Although these agents are efficacious, each has safety-, tolerability-, or drug delivery-related adverse effects (McLaughlin, 2006). Due to the lack of efficacious therapies and the adverse side effects associated with PAH treatment, Myogen chose to apply their technology to treatment of PAH.

PATHOBIOLOGY OF PAH

PAH is known to be associated with the perturbation of three major mechanistic pathways. The first major mechanism is the Nitric Oxide (NO) pathway. In this process, NO is produced in endothelial cells by type III NO synthase. NO induces conversion of guanosine triphosphate (GTP) to cyclic-guanosine monophosphate (cGMP) by the enzyme guanylate cyclase (GC). cGMP is a secondary messenger that manages pulmonary artery smooth muscle cell relaxation (PASMC). In other words, an endothelial cell deficient in cGMP will contribute to hypertension. The second major mechanism is the endothelin (ET) pathway. In this process, endothelin is converted to ET-1 (21-amino acid residues) by endothelin-converting enzyme (ECE) in endothelial cells. In the PASMC's, ET-1 binds to the ET_A and ET_B receptors, leading to contractions, proliferation, and hypertrophy. The third major mechanism is the prostacyclin pathway. In this pathway, prostacyclin synthase (PS) catalyzes production of prostacyclin in endothelin cells. Prostacylin (PGI2) perturbs adenylate cyclase (AC), which induces cAMP to produce ATP, maintaining PASMC relaxation and inhibiting proliferation.

Modulation of these pathways may be a result of perturbation of any one mechanism in the endothelial cells. These pathways will also respond to signals from transmitters and other stimuli that act on cell membrane receptors. Modulation of these pathways may be a result of perturbation of any one mechanism in the endothelial cells. These pathways will also respond to signals from transmitters and other stimuli that act on cell membrane receptors. For example, thrombin, bradykinin, arginine, vasopressin (AVP), vessel-wall shear stress, angiotensin II (Ang II), cytokines, reactive oxygen species (ROS), and others, will perturb the system. Response to the signaling transmitter depends on whether the interaction was with ET_A, ET_B, or both. Marked by arrows in Figure 2 are the aberrations observed among patients with PAH. Agents found to provide clinical benefit include dual and ET_A selective endothelin receptor antagonists, such as the bosentan, ambrisentan and sitaxsentan, and PDE-5 inhibitor sildenafil. Others include prostaniods epoprostenol, treprostinil, and iloprost that supplement levels of PGI2.

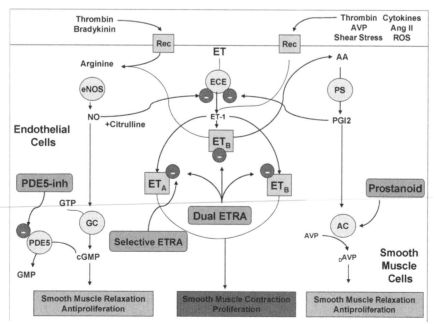

Figure 2: Three major biochemical mechanisms of PAH.
Source: Adapted from McLaughlin and McGoon (2009)

MYOGEN TARGETS ENDOTHELIN (ET) FAMILY OF NEUROHORMONES

Myogen began studying potential drug candidates that mediate plasma levels of ET-1 that are elevated in patients with PAH. Over expression of ET-1 is found in the muscular pulmonary arteries of patients with idiopathic PAH, and over expression of ET-1 is strongly correlated with increased mean right arterial pressure and severity of the disease (McLaughlin, 2006). The ET family of neurohormones consists of three poly peptides with similar structures: ET-1, ET-2, and ET-3. ET-1, the principle isoform in disease states, is synthesized and secreted from the vascular endothelium in response to numerous stimuli, including hypoxia, ischemia, shear stress, and growth factors and other neurohormones (Levin, 1995). ET-1 is a 21 amino acid peptide that plays a role in the pathobiology of PAH, exerting vasoconstrictor and mitogenic effects by binding to two distinct receptor isoforms in the pulmonary vascular smooth muscle cells, endothelin A (ET_A) and B (ET_B) receptors. ET_B receptors are also present in endothelial cells, and their activation leads to release of vasodilators and antiproliferative substances such as nitric oxide and prostacyclin that may counterbalance the harmful effects of over expressed ET-1 (Cheng, 2008).

AMBRISENTAN

Ambrisentan is a small chiral organic molecule with a molecular weight of 378 g/mol (Figure 3). It is a selective ET_A receptor antagonist with a measured IC_{50} of 0.011 nM. Ambrisentan's affinity for the ET_A receptor is >4000-fold its affinity for the ET_B receptor (Vatter *et al*, 2002). Bosentan, Myogen's competitors selective ET-1 antagonist, has ~100 times greater affinity for the ET_A receptor than for the ET_B receptor (Cheng, *et al*, 2008). This data suggests that's ambrisentan is a selective antagonist for the ET_A-receptor, and provides Myogen the opportunity to be the best in class ET_A selective endothelin molecule for treatment of PAH.

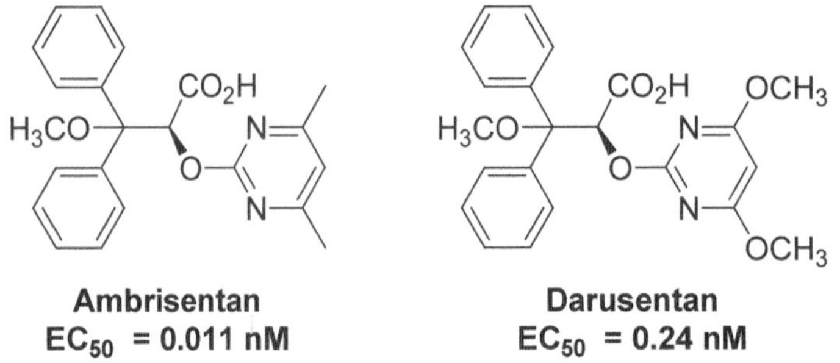

Ambrisentan
EC_{50} = 0.011 nM

Darusentan
EC_{50} = 0.24 nM

Figure 3: Myogen's endothelin antagonists ambrisentan and darusentan. Ambrisentan is an orally active, highly selective ETA receptor antagonist with >4000-fold higher selectivity over the ETB receptor.
Source: Author's elaboration

AMBRISENTAN PHARMACOKINETICS

After oral administration, single doses of ambrisentan 1, 5, 10, 15, 20, 100 mg were readily absorbed, with the maximum drug concentration observed in the blood stream attained in 1.7 to 3.3 hours in both healthy subjects and patients with PAH. Ambrisentan is a substrate for P-glycoprotein and is ~99% bound to plasma protein. It is primarily eliminated by non-renal pathways (77%), mostly by the phase II metabolism via glucuronidation. *In vitro* data indicate that the metabolism of ambrisentan is affected by cytochrome P450 (CYP) 3A4 and 2C19 isozymes, as well as by strong inhibitors of p-glycoprotein. The mean oral clearance of ambrisentan in healthy subjects and patients with PAH is reported to be 38 and 19 mL/min. The mean elimination half-life of ambrisentan after repeated dosing ranged from 13.6 to 16.5 hours across all doses between 5 and10 mg per day. The oral bioavailability and the half-life pharmacokinetic data for ambrisentan allows for once-daily dosing in clinical studies (Cheng, 2008).

AMBRISENTAN CLINICAL DATA

To date, randomized controlled clinical trials performed in pulmonary arterial hypertension (PAH) have been relatively short-term studies involving mainly patients with advanced disease. The primary end points in these trials have addressed exercise capacity, usually by using the 6-minute walk distance (6MWD) test. The 6MWD measures the change in distance walked in meters after treatment, relative to the patient's baseline at initiation of the clinical study. Myogen carried out three randomized, double-blind studies of ambrisentan in patients with PAH. In a Phase II double-blind, dose ranging study, 64 patients aged 18 years and older with symptomatic idiopathic PAH or PAH associated with collagen vascular disease, use of weight-loss agents, or HIV infection were randomized to receive oral doses of ambrisentan 1, 2.5, 5, or 10 mg per day for 12 weeks, followed by 12 weeks of open-label ambrisentan. The primary efficacy end point was exercise capacity, measured in terms of change in the 6MWD. After 12 weeks of treatment, this end point was significantly improved from baseline in all ambrisentan individual dose groups, with increases from baseline of 33.9, 37.1, 38.1, and 35.1 meters. Patients in WHO functional class II showed improvements of 37.7 meters and those in WHO class III improved 35.2 meters. Improvements from baseline continued to increase up to 24 weeks for the 10 mg group (50.2 m), and up to 48 weeks (54.5 m). The one-year survival was 92%, compared with the 68% predicted by the National Institutes of Health registry formula for untreated PAH subjects. In addition, improvements in patients between WHO functional classes were observed, with 36.2% of patients improving ≥ 1 functional class, and the mean pulmonary arterial pressure was reduced from baseline by 5.2 mm Hg after 12 weeks of dosing (Cheng, 2008).

The most common adverse effects associated with ambrisentan in clinical trials were peripheral edema, nasal congestion, palpitation, constipation, flushing, abdominal pain, nasopharyngitis, and sinusitis (see Figure 4).

Following the positive phase II results, two large phase III double-blind, dose ranging studies were conducted; ARIES-1 and ARIES-2 in patients with PAH in WHO group I. Patients were administered ambrisentan dosed at 2.5-10 mg once daily. Significant increases in the 6-minute walk distance ranging from 31-59 m from baseline versus placebo group were observed (see Table 2 and Figure 5). It was determined at week 12 that WHO functional class distribution had significantly improved for the once-daily

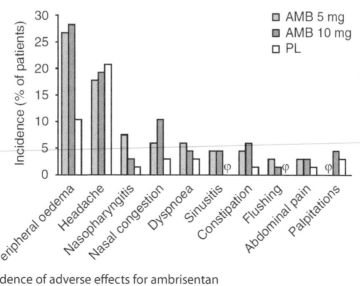

Figure 4: Relative incidence of adverse effects for ambrisentan
Source: Croxtall (2008)

ambrisentan 5 mg group. The beneficial effects of ambrisentan on exercise capacity were demonstrated over time, and WHO functional class scores seen at 12 weeks were maintained at 48 weeks in the ARIES-E phase III extension trial with 361 patients. One year survival rates with ambrisentan were up to 95-97% (Cheng, 2008; Galiè, 2008).

The clinical data demonstrates that ambrisentan improved exercise capacity and decreased levels of brain natriuretic peptide. Ambrisentan was well tolerated and was not associated with any clinically significant serum aminotransferase abnormalities, a marker of potential liver damage, and a major side effect of ET-1 receptor antagonists. The percentage of subjects with elevations of serum aminotransferase concentrations > 3 times the upper limit of normal appears to be lowest for patients treated with ambrisentan (2.8%) than for patients treated with sitaxsentan (3 to 5%) and bosentan (11 to 12%). These

Table 2: Ambrisentan clinical data, 6MWD

	ARIES-1			ARIES-2		
	Placebo (N=67)	5 mg (N=67)	10 mg (N=67)	Placebo (N=65)	2.5 mg (N=64)	5 mg (N=63)
Baseline	342 ± 73	340± 77	342 ± 78	343 ± 86	347± 84	355 ± 84
Mean change from baseline	-8 ± 79	23 ± 83	44 ± 63	-10 ± 94	22 ± 83	49 ± 75
Placebo-adjusted mean change from baseline		31	51		32	59
Placebo-adjusted median change from baseline		27	39		30	45
p-value†		0.008	<0.001		0.022	<0.001

Source: Ambrisentan (Letairis®) Company Full Prescribing Information

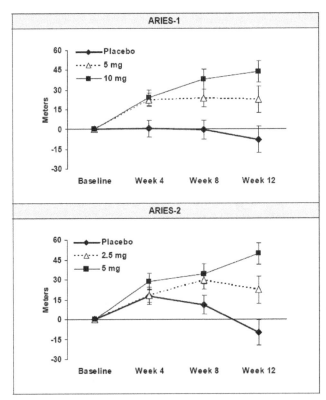

Figure 5: Mean change in 6 minute walking distance
Source: Galiè (2008)

findings suggest significant advantages in treatment of PAH patients with ambrisentan as opposed to sitaxsentan and bosentan (MacLaughlin and McGoon 2006).

CARDIOVASCULAR MARKETS ARE HIGHLY COMPETITIVE

Many pharmaceutical and biotechnology companies have developed or are developing products that will compete with the products Myogen distributes or the product candidates Myogen is developing. Several significant competitors of Myogen are working on, some already have approval for, drugs for the same indications as ambrisentan and darusentan. It is also possible that competitors will develop and market products that are less expensive, more effective or safer than Myogen's future products, potentially rendering Myogen's products obsolete.

In the coming years, competition from pharmaceutical and biotechnology companies, universities and public and private research institutions will increase. Many of these competitors have substantially greater financial, technical, research and other resources. For example, Actelion, Ltd. received FDA approval in December 2001 for bosentan (Tracleer®), an endothelin receptor antagonist for the treatment of Pulmonary Arterial Hypertension (PAH). United Therapeutics Corp. received FDA approval in May 2002 for treprostinil (Remodulin®) for the treatment of PAH. In addition, Schering AG and CoTherix, Inc. market iloprost (Ventavis®) for the treatment of PAH. Encysive Pharmaceuticals, Inc. is developing sitaxsentan (Thelin™), an endothelin A receptor selective Endothelin Receptor Antagonist which has demonstrated efficacy in a Phase 3 study and may be approved for the treatment of PAH earlier than ambrisentan. Pfizer, Inc. received approval for the use of sildenafil (Revatio®) for the treatment of PAH in the United States and Europe, and if ambrisentan is approved, sildenafil is likely to be a major competitor. ICOS Corporation and Eli Lilly and Co. have initiated clinical trials to evaluate oral tadalafil (Cialis®) for the treatment of PAH. In addition, Predix Pharmaceuticals Holdings, Inc. is in early stages of clinical development of a compound with a novel mechanism of action for the possible treatment of PAH. A number of other companies, including Abbott Laboratories and Speedel, have ET_A selective ET receptor antagonists in late-stage clinical development that could compete with ambrisentan and darusentan.

Finally, Myogen could experience pressures on the pricing of their product, and if approved, their product candidates. Competition from manufacturers of competing drugs and generic drugs is a major challenge in the United States and is increasing internationally. Upon the expiration or loss of patent protection for a product, or upon the "at-risk" launch (while patent infringement litigation against the generic product is pending) by a generic manufacturer of a generic version of a product, Myogen could lose the major portion of sales of that product in a very short period of time.

GILEAD MOVES TO ACQUIRE MYOGEN

Gilead Sciences, Inc. is a global biopharmaceutical company that discovers, develops and commercializes therapeutics to advance the care of patients suffering from life-threatening diseases. Gilead was founded in 1987, becoming incorporated in Delaware on June 22, 1997, Their worldwide headquarters are located in Foster City, California. Gilead is a multinational company, with revenues from nine approved products and marketing operations in twelve countries. Gilead focuses their research and clinical programs on anti-infectives. From inception, Gilead pursued partnerships to advance their product pipeline.

In August of 1987, "Oligogen, Inc.", later changed to "Gilead Sciences" in 1988, was founded by Michael Riordan. Dr. Riordan, a young medical doctor aged 29 at the time of inception, sought mentorship from Donald Rumsfeld, a professional that understood the political landscape of the business world. The company initially focused its discovery efforts on gene therapy, lead by Scientist Mark Matteucci, designing DNA oligomers that regulate genetic code blockers. Later in 1992, John Martin led a licensed program to develop small molecule antiviral therapeutics, discovered at two European universities.

Further development of this technology coined "antisense" resulted in a research collaboration with GlaxoSmithKline in 1990. The GSK collaboration was terminated in 1998, and ownership of the company's antisense technology was sold to Isis Pharmaceuticals.

At the start of 1992, Gilead raised $86.25 million in its IPO on the NASDAQ. Nine years after inception in 1996, Gilead commercialized Vistide® (cidofovir injection) for treatment of cytomegalovirus (CMV) retinitis in AIDS patients, marketed in cooperation with Pharmacia & Upjohn outside the United States. Following two years of negations, Gilead acquired the Colorado-based company NeXstar Pharmaceuticals for their $130 million revenue-generating drugs, AmBisome and DaunoXome. AmBisome is an injectable fungal treatment, and DaunoXome, an anti-cancer drug used to treat HIV patients. In 1999, the Gilead's influenza drug Tamiflu® (oseltamivir) licensed to Roche for late-phase development was approved for treatment. Gilead's Viread® (tenofovir) achieved first approval in 2001 for the treatment of HIV. Gilead continued to grow horizontally, acquiring Triangle Pharmaceuticals in January 2003, and was able to announce its first year of profitability. During this time Hepsera® (adefovir) was approved for the treatment of chronic hepatitis B, and Emtriva® (emtricitabine) for the treatment of HIV. In 2004, Gilead launched Truvada®, a fixed-dose combination of tenofovir and emtricitabine.

The U.S. Food and Drug Administration (FDA) approved Atripla® in 2006, a once-a-day single tablet regimen for HIV, combining Sustiva® (efavirenz), a Bristol-Myers Squibb product, and Truvada® (emtricitabine and tenofovir disoproxil fumarate), a Gilead product. The same year Gilead purchased Alberta-based Raylo Chemicals, Inc., a wholly-owned subsidiary of Degussa AG, for $133.3 million. Raylo Chemical custom-manufactured active pharmaceutical ingredients and advanced intermediates for the pharmaceutical and biopharmaceutical industries. In effort to diversify their historical antiviral franchise, Gilead acquired two companies in the cardiovascular and respiratory therapeutic arenas. One of those companies was Corus Pharma, acquired for their lead Phase 3 candidate for cystic fibrosis patients with a lung infection caused by *Pseudomonas aeruginosa*. In addition to the late-stage product Gilead obtained an inhalation formulation of two antibiotics for treatment of respiratory infections from the acquisition (see Figure 6 and Figure 7).

Collaboration Partner	Program Area	Year
Emory	Emtricitabine	2005
Japan Tobacco	Viread, Truvada and Emtriva	2003
GSK	Hepsera	2002
Pfizer	Macugen and Vistide	2002; 1996
OSI	Liposome products and Macugen	2001; 2000
Roche	Tamiflu	1996
Sumitomo	AmBisome	1996
M.D. Anderson Cancer Center	Hepsera	1994
IOCB/REGA	Viread, Truvada, Hepsera and Vistide	1991
Astellas	AmBisome	1991
ULEHI	SELEX	1991

Figure 6: Gilead collaborations
Source: Company reports (2009)

Year	Company	Price	Comments
1999	Nexstar Pharmaceuticals	$550 million	Nexstar had two drugs (AmBisome and DaunoXome). Nexstar also provided Gilead with a sales force and commercialization team in Europe and Australia, and a manufacturing plant in San Dimas, California.
2003	Triangle Pharmaceuticals	$464 million	Triangle owned the development and commercialization rights to emtricitabine
2006	Corus Pharma, Inc.	$365 million	The acquisition of Corus signaled Gilead's entry into the respiratory arena. Corus was developing aztreonam lysine for the treatment of patients with cystic fibrosis who are infected with Pseudomonas aeruginosa

Figure 7: Gilead mergers & acquisitions history
Source: Company reports (2009)

POTENTIAL FIT WITH GILEAD PRODUCTS

Gilead has U.S. and international commercial sales operations, with marketing subsidiaries in Australia, Canada, France, Germany, Greece, Ireland, Italy, New Zealand, Portugal, Spain, the United Kingdom and the United States. Their commercial teams promote HIV and HBV products, Viread®, Truvada®, Emtriva® and Hepsera®, through direct field contact with physicians, hospitals, clinics and other health-care providers that are involved in the treatment of patients with HIV or chronic hepatitis B. The HIV products and Hepsera® are sold in the United States exclusively through their wholesale channel. Gilead's corporate partner, Astellas, promotes and sells AmBisome® in the United States. Gilead sells their HIV products, Hepsera® and AmBisome® in major European countries through their European commercial team and distributors. In addition, Gilead promotes, sells and distributes their products in countries outside of the United States and Europe, including countries in Asia, Latin America, the Middle East and Africa. In certain territories, Gilead negotiated agreements with third party distributors granting them the exclusive right to sell Gilead products in a particular territory for a specified period of time. Most of these agreements provide for collaborative efforts between the distributor and Gilead for obtaining regulatory approval for the product in the specified territory.

Gilead revenues were $2,588 million in 2006, comprised of 82% HIV therapeutics (Truvad® $1,194 million; Viread® $689.4 million; Atripla® $205.7 million; Emtriva® $36.4 million), and the remainder through additional products as follows: Hepsera® $230.5 million; AmBisome® $223.0 million; and other $8.9 million (see Figure 8 and Appendix F).

Of note, the HIV landscape is becoming more competitive and complex for Gilead as treatment trends continue to evolve which made diversification an attractive strategic alternative. A growing number of anti-HIV drugs are getting approved for treatment or are in advanced stages of clinical development. Of the approximately 25 branded drugs available in the United States, Gilead's HIV products primarily compete with the fixed-dose combination products in the nucleotide/nucleoside reverse transcriptase inhibitors (NRTI) class, including Combivir® (AZT and 3TC); Epzicom® (3TC and ABC) and Trizivir® (AZT/3TC/ABC), each sold by GSK. Other companies with HIV products competing in the same NRTI class include Bristol-Myers Squibb Company (BMS) and Roche, while Gileads HIV products also compete broadly with HIV products from Boehringer Ingelheim, Merck & Co., Inc. (Merck) and Abbott Laboratories. In addition to new therapeutic discoveries, BMS's Videx EC® (didanosine) became the first generic HIV product in the United States in 2004. GSK's Retrovir® (AZT) also now faces generic competition in the United States as a result of the launch of generic AZT in 2005. There has been little impact from generic didanosine or generic AZT on the price of Gilead's HIV products, but price decreases for all HIV products may result.

Product	Product sales (2005)	Percent of sales	U.S. patent expiration	European patent expiration
DaunoXome	$1,287	---	2009	2008
Vistide	$6,629	1%	2010	2012
Hepsera	$186,532	10%	2014	2011
AmBisome	$220,753	12%	2016	2008
Tamiflu[1]			2016	2016
Macugen[2]			2017	2017
HIV products:				
Viread	$778,783	43%	2017	2018
Emtriva	$47,486	3%	2021	2016
Truvada	$567,829	31%	2021	2018
Total HIV products	$1,394,098	77%		
Total product sales	$1,809,299	100%		

Figure 8: Gilead revenues and patent status in 2005

Source: Company reports (2009)

[1] *Roche maintains exclusive rights to manufacture and distribute Tamiflu*

[2] *FDA Approved in 2005, no product sales data available for Macugen*

ARE WE THERE YET?

On November 4, 2005, the Myogen board held a meeting at which they appointed three members of the board, including Dr. Freytag, to an engagement committee who was authorized to retain Goldman Sachs & Co. to act as the financial advisor. At that meeting, representatives of Goldman Sachs presented a review of the current biotechnology mergers and acquisitions environment, a preliminary analysis of Myogen's business and financial profile, a preliminary financial analysis of Myogen and a discussion of the potential impact of the upcoming Phase 3 clinical trial of ambrisentan. Representatives of Goldman Sachs also discussed potential strategic alternatives available to Myogen, which included selling the company, undertaking acquisitions, or licensing arrangements and pursuing additional equity financing in order to remain independent.

After receiving approval from the Myogen board and the engagement committee, the company requested that Goldman Sachs, on behalf of the company, conduct a limited "market check." During a two-week period in late November and early December 2005, representatives of Goldman Sachs contacted approximately fifteen parties that the company and Goldman Sachs believed could potentially be interested in pursuing a business combination transaction with Myogen, including a potential partner (see Appendix E).

Commencing in December 2005 and continuing through February 2006, Myogen allowed four parties that had expressed an interest in entering into a collaboration or licensing arrangement regarding ambrisentan, to conduct a due diligence investigation relating to ambrisentan—and began preliminary negotiations with each of these four parties regarding a possible collaboration arrangement. On December 12, 2005, Myogen publicly announced favorable results from the first Phase 3 clinical trial of ambrisentan. On December 14, 2005, the Myogen board held a regularly scheduled meeting in Westminster, Colorado. At that meeting, the company's management team reviewed the status of the licensing and collaboration discussions relating to ambrisentan. In addition, representatives of Goldman Sachs reviewed the results of the market check and indicated that five parties had expressed an interest in reviewing additional information regarding Myogen and its products with a view toward a potential business combination if Myogen were to undertake a formal auction or sale process.

While the Myogen board determined that there was moderate strategic interest on the part of certain companies in Myogen and its products, they concluded that it was premature to undertake an auction or sale process; results of the second pivotal Phase 3 trial of ambrisentan (ARIES-1) were not expected to be received and reported until April 2006, discussions were still underway with the U.S. FDA and the EMEA (European Agency for the Evaluation of Medicinal Products) regarding the design of Phase 3 studies relating to darusentan, and any significant delay in ongoing collaboration discussions with GlaxoSmithKline could result in the company losing an opportunity to obtain distribution or license rights to Flolan®.

In February 2006, Myogen received written non-binding proposals or expressions of interest from four potential partners that had expressed an interest in a possible collaboration or licensing arrangement regarding ambrisentan. After several months of negotiations, Gilead emerged as a potential suitor and offered $2.5 billion or $52.50 per share to fully acquire Myogen.

On a fall evening in Denver, Colorado, Bill Freytag reflected on Myogen's key milestones and potential. Tonight, the board of directors will be making a critical decision: continue Myogen's strategy to "create an integrated biopharmaceutical company" or accept Gilead's merger acquisition offer.

On the one hand, Bill thought: "Gilead is the ideal partner with multiple sources of operating leverage. They are well funded, experienced in making acquisitions work, and have an outstanding commercial organization. In addition, if we remain independent we're going to need a significant financing to establish commercial operations which will be dilutive to existing shareholders—who have stuck with us in the ups and downs of building our company." On the other hand, "Are we selling too early? Why couldn't we build the next Gilead, Genentech, or Amgen ourselves?"

APPENDICES

APPENDIX A: MYOGEN, INC. BALANCE SHEET (US$ 000s)

	2005	2004
	(in thousands, except share data)	
ASSETS		
Current assets:		
Cash and cash equivalents	$138,380	$70,669
Short-term investments	38,575	48,331
Research and development contract amounts due within one year	-	300
Prepaid expenses, accrued interest receivable and other current assets	2,752	1,772
Assets of discontinued operations	1,289	2,052
Total current assets	180,996	123,124
Long-term investments	5,362	-
Property and equipment, net	2,622	2,452
Other assets	27	27
Total assets	$189,007	$125,603
LIABILITIES AND STOCKHOLDERS' EQUITY		
Current liabilities:		
Accounts payable	$10,345	$10,510
Accrued liabilities	2,797	1,893
Current portion of deferred revenue	1,187	1,823
Current portion of other liabilities	142	119
Current portion of notes payable, net of discount	172	1,822
Liabilities of discontinued operations	264	220
Total current liabilities	14,907	16,387
Deferred revenue, net of current portion	1,656	1,399
Other long term liabilities, net of current portion	220	331
Notes payable, net of current portion and discount	-	172
Commitments and contingencies (see Note 8 and Note 13)		
Stockholders' equity:		
Preferred Stock, $0.001 par value; 5,000,000 shares authorized at December 31, 2005 and 2004, no shares issued or outstanding	-	-
Common stock, $0.001 par value; 100,000,000 shares authorized and 41,962,587 and 35,731,581 shares issued and outstanding as of December 31, 2005 and 2004, respectively	42	36
Additional paid-in capital	412,862	286,017
Deferred stock-based compensation	(1,406)	(2,535)
Other comprehensive loss	(88)	(42)
Deficit accumulated during the development stage	(239,186)	(176,162)
Total stockholders' equity	172,224	107,314
Total liabilities and stockholders' equity	$189,007	$125,603

Source: Company reports (2009)

APPENDIX B: MYOGEN, INC. INCOME STATEMENT (US$ 000s)

	2005	2004
Revenues — research and development contracts	6,963	6,606
Costs and expenses:		
Research and development (excluding stock-based compensation expense of $4,111, $1,971, $2,373, and $8,938 respectively)	52,602	54,124
Selling, general and administrative (excluding stock-based compensation expense of $4,776, $1,977, $1,819 and $8,822, respectively)	13,141	8,358
Stock-based compensation	8,887	3,948
	74,630	66,430
Loss from operations	-67,667	-59,824
Interest income (expense), net	3,661	821
Loss from continuing operations	-64,006	-59,003
Discontinued operations, net of income taxes	982	1,318
Net loss	-63,024	-57,685
Accretion of mandatorily redeemable convertible preferred stock	—	—
Deemed dividend related to beneficial conversion feature of preferred stock	—	—
Net loss attributable to common stockholders	-63,024	-57,685
Basic and diluted net loss per common share attributable to common stockholders:		
Continuing operations	-1.71	-2.05
Discontinued operations, net of income taxes	0	0
	-1.68	-2
Weighted average common shares outstanding	37,416,368	28,839,076

Source: Company reports (2009)

APPENDIX C: WHO FUNCTIONAL CLASSIFICATION OF PAH
Source: WHO

Class	Description
I	Patients with PH in whom there is no limitation of usual physical activity; ordinary physical activity does not cause increased dyspnea, fatigue, chest pain, or presyncope.
II	Patients with PH who have mild limitation of physical activity. There is no discomfort at rest, but normal physical activity causes increased dyspnea, fatigue, chest pain, or presyncope.
III	Patients with PH who have a marked limitation of physical activity. There is no discomfort at rest, but less than ordinary activity causes increased dyspnea, fatigue, chest pain, or presyncope.
IV	Patients with PH who are unable to perform any physical activity at rest and who may have signs of right ventricular failure. Dyspnea and/or fatigue may be present at rest, and symptoms are increased by almost any physical activity.

APPENDIX D: MYOGEN'S SENIOR MANAGEMENT & BOARD OF DIRECTORS

J. William Freytag, Ph.D. has served as our president and chief executive officer and as a director since July 1998 and as chairman of our board since December 2000. From May 1998 to July 1998, Dr. Freytag was an independent consultant to the healthcare industry. From October 1994 to May 1998, Dr. Freytag was a senior vice president at Somatogen, Inc., a biopharmaceutical company, where he was responsible for corporate and commercial development. Prior to Somatogen, he was president of research and development at Boehringer Mannheim Corporation, an international healthcare company, from May 1990 to September 1994. Previously, Dr. Freytag spent ten years with DuPont Medical Products in various research and business positions. Dr. Freytag is a director of Immunicon Corporation. Dr. Freytag received a Ph.D. in biochemistry from the University of Kansas Medical Center.

Michael R. Bristow, M.D., Ph.D. was the founder of Myogen and has served as a director since October 1996. Dr. Bristow has served as a scientific advisor to Myogen since February 2006 and served as our chief science and medical officer from October 1996 to February 2006. Dr. Bristow founded and is the chief executive officer of ARCA Discovery, Inc., a pharmocogenomic drug development and discovery company in the field of cardiovascular disease. Dr. Bristow is currently professor of medicine at the University of Colorado and co-director of the University of Colorado Cardiovascular Institute. He is the author of over 300 peer-reviewed publications and chapters on heart failure or cardiomyopathy for cardiology textbooks, including Braunwald's "Heart Disease: A Textbook of Cardiovascular Medicine" and Hurst's "The Heart." Dr. Bristow holds a M.D. and Ph.D. from the University of Illinois.

Kirk K. Calhoun has served as a Director since January 2004. Mr. Calhoun joined Ernst & Young, LLP, a public accounting firm, in 1965 and served as a partner of the firm from 1975 until his retirement in 2002. His responsibilities included both area management and serving clients in a variety of industries, including biotechnology. Mr. Calhoun is a certified public accountant with a background in auditing and accounting. He is currently on the board of directors of Abraxis Bioscience, Inc., Adams Respiratory Therapeutics, Inc., Aspreva Pharmaceuticals Corporation and Replidyne, Inc. Mr. Calhoun received a B.S. in accounting from the University of Southern California.

Judith A. Hemberger, Ph.D. has served as a director since December 2005. Dr. Hemberger is the chief executive officer of the Macroflux Corporation, a medical device life sciences company. Dr. Hemberger was a co-founder of and has served as executive vice president and chief operating officer and a director of Pharmion Corporation from its inception until her retirement on April 1, 2006. From 1997 to 1999, she worked as a consultant to various healthcare companies. During this period she also served as a senior vice president of business development at AVAX Technologies, Inc., a vaccine technology company. From 1979 to 1997, Dr. Hemberger worked at Marion Laboratories and successor companies Marion Merrell Dow and Hoechst Marion Roussel. She led a number of strategic functions including professional education, global regulatory affairs, global medical affairs and commercial development. Her final role in the company was senior vice president of global drug regulatory affairs. Dr. Hemberger currently serves on the board of directors of PRA International, Zymogenetics, Inc. and Renovis. Dr. Hemberger holds a Ph.D. in pharmacology from the University of Missouri and a M.B.A. from Rockhurst College.

Jerry T. Jackson has served as a director since September 2002. Mr. Jackson was employed by Merck & Co., Inc., a pharmaceutical company, from 1965 until his retirement in 1995. During this time, he had extensive experience in sales, marketing and corporate management, including joint ventures. From 1993 until retirement, Mr. Jackson served as executive vice president of Merck with broad responsibilities for numerous operating groups, including being president of Merck's worldwide human health division in 1993. Mr. Jackson was senior vice president responsible for Merck's specialty chemicals business from 1991 to 1992. Previously, he was president of Merck's international division from 1988 to 1991. Mr. Jackson has served on the board of directors of several biotech pharmaceutical companies and currently is on the board of directors of Langford I.C. Systems Inc., a privately-held company. Mr. Jackson holds a B.A. in education from the University of New Mexico.

Daniel J. Mitchell has served as a Director since May 1998. Mr. Mitchell founded and is a manager of Sequel Venture Partners, L.L.C., a venture capital firm formed in January 1997. Mr. Mitchell was a founder of Capital Health Venture Partners, a health care focused venture capital firm, in October 1986 and was a general partner.

He is currently on the board of directors of Replidyne, Inc. Mr. Mitchell holds a M.B.A. from the University of California at Berkeley.

Arnold L. Oronsky, Ph.D. has served as a director since October 1998. Dr. Oronsky is general partner with InterWest Partners, a venture capital firm focusing on investments in medical and information technology. Dr. Oronsky joined InterWest in a full-time capacity in 1994 after serving as a special limited partner since 1989. In addition to the position of general partner at InterWest, he also serves as a senior lecturer in the department of medicine at Johns Hopkins Medical School. From 1980 to 1993, Dr. Oronsky was the vice president for discovery research at the Lederle Laboratories division of American Cyanamid Company, a pharmaceutical company. From 1973 to 1976, Dr. Oronsky was the head of the inflammation, allergy, and immunology research program for Ciba-Geigy Pharmaceutical Company. From 1970 to 1972, Dr. Oronsky was assistant professor at Harvard Medical School, where he also served as a research fellow from 1968 to 1970. Dr. Oronsky is currently on the board of directors of Anesiva Inc., Aspreva Pharmaceuticals Corporation, Dynavax Technologies Corporation and Metabasis Therapeutics, Inc. Dr. Oronsky holds a Ph.D. from Columbia University's College of Physicians & Surgeons.

Michael J. Valentino has served as a director since March 2005. Mr. Valentino has thirty years of experience in the prescription and consumer pharmaceuticals industry. He currently serves as the president and chief executive officer and a director of Adams Respiratory Therapeutics, Inc. During 2003, Mr. Valentino served as president and chief operating officer of the global human pharmaceutical division of Alpharma, Inc. From 1999 to 2002, he served in senior management roles at Novartis AG, including as the executive vice president, global head consumer pharmaceuticals and the president and chief operating officer of Novartis, North America. From 1996 to 1998, Mr. Valentino served the president of Pharmacia & Upjohn's consumer products division. Prior to 1996, Mr. Valentino spent 12 years at Warner Lambert in a variety of senior level roles. Mr. Valentino holds a B.A. in psychology from State University of New York at Stony Brook.

Robert Caspari, M.D. has served as our senior vice president of commercial operations and medical affairs since March 2006. Prior to joining us, Dr. Caspari worked as an independent consultant to the biotechnology, pharmaceutical and venture capital industries. From August 2000 to April 2004, Dr. Caspari served as the vice president and general manager, biopharmaceuticals of Novo Nordisk Pharmaceuticals, Inc. From April 1999 to July 2000, Dr. Caspari was vice president of medical affairs at NeoRX Corporation. From 1982 until April 1999, Dr. Caspari held a variety of management positions at Baxter International, Inc., Somatogen, Inc., Boehringer Mannheim Corporation, Gynopharma, Inc., Schering-Plough and Lederle Laboratories. Dr. Caspari entered the pharmaceutical industry in 1982 after practicing internal medicine for four years. Dr. Caspari received a B.A. in psychology from UCLA and obtained his medical degree from Georgetown University. He has more than twenty-four years experience in drug development and commercialization.

Andrew D. Dickinson has served as our vice president, general counsel and secretary since December 2004. Since March 6, 2006, Mr. Dickinson has served as our vice president of corporate development. From January 2003 to December 2004, Mr. Dickinson was a partner at Kendall, Dickinson & Koenig PC, a corporate and securities law firm. From May 1999 to December 2002, Mr. Dickinson was a business associate at Cooley Godward LLP. Previously, Mr. Dickinson was a corporate and securities associate at Kirkland & Ellis LLP. Mr. Dickinson holds a J.D. from the Loyola University of Chicago School of Law.

Michael J. Gerber, M.D. has served as our senior vice president of clinical development and regulatory affairs since February 2002. From October 2001 to February 2002, Dr. Gerber was an independent consultant to the healthcare industry. From July 1999 until October 2001, Dr. Gerber was senior vice president, clinical development and regulatory affairs of Allos Therapeutics, Inc., a pharmaceutical company. Dr. Gerber also served as vice president, medical affairs of Allos from November 1994 until July 1999. From 1991 to 1994, Dr. Gerber was executive director, clinical sciences and medical affairs at Somatogen Inc., where he directed nonclinical and clinical development. Prior to joining Somatogen, Dr. Gerber was in private practice since 1987 and directed the pulmonary drug evaluation program subsidiary of Pulmonary Consultants, Inc. Dr. Gerber is board certified in internal, pulmonary and critical care medicine, and is clinical assistant professor of medicine at the University of Colorado Health Sciences Center. Dr. Gerber received his M.D. from the University of Colorado School of Medicine.

Richard J. Gorczynski, Ph.D. has served as our senior vice president of research and development since December 2003, and prior to that served as our vice president of research and development since December 1998. From December 1994 to November 1998, Dr. Gorczynski was vice president of R&D for Somatogen, Inc. From September 1985 to November 1994, Dr. Gorczynski served as executive director of cardiovascular diseases research for Monsanto-Searle Pharmaceuticals. From September 1976 to August 1985, Dr. Gorczynski was head of pharmacology at American Critical Care, a pharmaceutical division of American Hospital Supply Corporation. Dr. Gorczynski holds a Ph.D. in physiology from the University of Virginia School of Medicine.

Joseph L. Turner has served as our senior vice president of finance and administration and chief financial officer since December 2003. He initially joined us on a part-time basis in December 1999 as our acting chief financial officer and joined us full-time as our vice president of finance and administration and chief financial officer in September 2000. From July 1999 to May 2000, Mr. Turner was an independent strategic financial consultant to emerging companies. From November 1997 to June 1999, Mr. Turner worked at Centaur Pharmaceuticals, a biopharmaceutical company, where he served in several positions, including vice president, finance and chief financial officer. From March 1992 to October 1997, Mr. Turner served as vice president, finance and chief financial officer of Cortech, Inc., a biopharmaceutical company. Previously, Mr. Turner spent twelve years with Eli Lilly and Company, where he held a variety of financial management positions both within the United States and abroad. Mr. Turner holds a M.A. in molecular, cellular and developmental biology from the University of Colorado and a M.B.A. from the University of North Carolina.

APPENDIX E: VALUATION

Goldman Sachs analyzed the consideration to be received by holders of shares of Common Stock pursuant to the Merger Agreement in relation to the closing market price or average market price of the shares of Common Stock for various periods since the initial public offering of Myogen. The following bar graph represents the results of this analysis:

Gilead Offer v. Historical Share Value 2005-2006

Category	Price Per Share
Gilead Offer Per Share	$52.50
Prior to Close Sept. 29, 2006	$35.08
One Year High	$41.31
One Week Prior	$34.43
One Month Prior	$34.88
Three Months Prior	$28.83
One Year Prior	$23.99
One Week Average	$34.87
One Month Average	$34.63
Three Month Average	$31.77
One Year Average	$30.49
Since IPO Average	$17.46

Price Per Share

Source: Company reports

APPENDIX F: GILEAD, INC. FINANCIAL STATEMENTS

Income Statement (US$ Millions)	2006	2005
Revenue	4,230.1	3,026.1
Cost of Revenue, Total	768.8	433.3
Gross Profit	3,461.3	2,592.8
Selling/General/Admin. Expenses, Total	705.7	573.7
Research & Development	591.0	383.9
Unusual Expense (Income)	-	2,394.1
Total Operating Expense	2,065.5	3,784.9
Operating Income	2,164.5	-758.8
Income Before Tax	2,270.3	-638.2
Net Income	1,615.3	-1,190.0

Balance Sheet (US$ Millions)	2006	2005
Cash and Short Term Investments	1,172.0	936.9
Accounts Receivable - Trade, Net	795.1	609.3
Total Receivables, Net	795.1	609.3
Total Inventory	600.0	564.1
Prepaid Expenses	273.5	50.1
Other Current Assets, Total	187.8	268.8
Total Current Assets	3,028.3	2,429.2
Property/Plant/Equipment, Total - Gross	649.0	522.0
Long Term Investments	1,550.4	452.7
Other Long Term Assets, Total	808.3	842.8
Total Assets	5,834.7	4,086.0
Accounts Payable	290.3	367.0
Accrued Expenses	414.9	334.1
Current Port. of LT Debt/Capital Leases	0.3	18.8
Other Current liabilities, Total	30.8	44.4
Total Current Liabilities	736.3	764.3
Long Term Debt	1,300.0	1,300.0
Total Long Term Debt	1,300.0	1,300.0
Total Debt	1,300.3	1,318.8
Minority Interest	140.3	53.1
Other Liabilities, Total	198.2	152.9
Total Liabilities	2,374.7	2,270.3
Common Stock, Total	0.9	0.9
Additional Paid-In Capital	3,214.3	2,703.9
Retained Earnings (Accumulated Deficit)	249.1	-891.4
Other Equity, Total	-4.4	2.2
Total Equity	3,460.0	1,815.7
Total Liabilities & Shareholders' Equity	5,834.7	4,086.0
Total Common Shares Outstanding	932.5	922.3

Source: Company reports

REFERENCES

Cheng, J. W. M. 2008. Ambrisentan for the Management of Pulmonary Arterial Hypertension. *Clinical Therapeutics*, 30: 825-833.

Croxtall, J. D., Keam S. J. 2008. Ambrisentan. *Drugs*, 68: 2195-2204.

D'Alonzo G, Barst R, Ayres S, *et al.* 1991. Survival in patients with primary pulmonary hypertension: results from a national prospective registry. *Ann Intern Med*, 115: 343-349.

Farber H. W., Loscalzo J. 2004. Pulmonary arterial hypertension. *New England Journal of Medicine*, 351: 1655-1665.

Galie *et al* 2008. Ambrisentan in Pulmonary Arterial Hypertension. *Circulation*, 117: 3010-3019.

Hoeper, M. M. 2005. Drug treatment of pulmonary arterial hypertension: current and future agents. *Drugs*, 65:1337-54.

Levin, E. R. 1995. Endothelins. *New England Journal of Medicine*, 333: 356-363.

McLaughlin, V. V., McGoon M. 2006. Pulmonary Arterial Hypertension. *Circulation*, 114: 1417-1431.

NEA, (Online), 2009, Available at http://www.gene.com/gene/news/press-releases/display. do?method=detail&id=4160 Accessed 05.12.2009

Newman, J. H., Fanburg, B. L., Archer, S. L., *et al*, for the National Heart, Lung, and Blood Institute/Office of Rare Diseases. 2004. Pulmonary arterial hypertension: Future directions: Report of a National Heart, Lung, and Blood Institute/Office of Rare Diseases workshop. *Circulation*, 109: 2947-2952.

Simonneau G, Galiè N, Rubin, LJ, *et al.* 2004. Clinical classification of pulmonary hypertension. *Journal of American College of Cardiology*, 43(Suppl S): 5S-12S.

Traupe, T., Ortmann, J., Haas, E., Münter, K., Parekh, N., Hofmann-Lehmann, R., Baumann, K., Barton, M. 2006. Endothelin ETA Receptor Blockade With Darusentan Increases Sodium and Potassium Excretion in Aging Rats. *Journal of Cardiovascular Pharmacology*, 47: 456-462.

U.S. Securities and Exchange Commission. (2006). Myogen, Inc. 10-K Form: http://www.sec.gov/Archives/edgar/data/1101052/000095013406005189/d33922e10vk.htm Accessed 15.12.2009.

U.S. Securities and Exchange Commission. (2006). Myogen, Inc. Schedule 14D-9 Form: http://www.sec.gov/Archives/edgar/data/1101052/000119312506207895/dsc14d9.htm Accessed 15.12.2009.

U.S. Securities and Exchange Commission. (2006). Gilead, Inc. 10-K Form: http://www.sec.gov/Archives/edgar/data/882095/000119312506045128/d10k.htm Accessed 01.12.2010.

Vatter H, Zimmermann M, Jung C, *et al.* 2002. Effect of the novel endothelin(A) receptor antagonist LU 208075 on contraction and relaxation of isolated rat basilar artery. *Clinical Science (London)*, 103 (Suppl 48): 408S-413S.

Compression Dynamics: In Search of Sales

ANNE S. YORK[1] AND MARTIN WINKLER[2]

[1]*College of Business, Creighton University;* [2]*University of Nebraska Medical Center and Creighton University Medical Center*

- **Summary and key issue/decision:** The Compression Dynamics case describes an inventor faced with developing a startup after several rejected licensing offers. Determined to bootstrap, vascular surgeon Dr. Martin Winkler identified target markets for his EdemaWear® compression stockinet. Over the first two years, a lack of sales through his primary marketing channel, an internet website, surprised Winkler. The case addresses licensing and startup differences, bootstrapping pros and cons, and marketing channels required to bring a medical product to a range of non-medical target markets.

- **Companies/institutions**: Compression Dynamics, Inc.

- **Technology:** EdemaWear is the name for a class of compression products which utilize a patented elastic yarn which can be made into various compression garments.

- **Stage of development at time of issue/decision:** The product, for which the patent has issues, falls into the FDA classification of Class I and II devices, that is either not regulated or required to show substantial equivalence. Surgical applications (kidney dialysis and post-vascular or breast surgery), because of the sterilization requirements, fall into Class II, while non-surgical applications (distance running, cellulite appearance reduction), because they are not used in a medical setting involving open wounds, fall into Class I.

- **Indication/therapeutic area:** Class I and II medical devices for athletic muscular recover, cellulite appearance reduction, post-stint surgery, and post-vascular or breast surgery to reduce swelling, promote more rapid healing, and lessen the chance of recurrence.

- **Geography:** US, with potential international markets.

- **Keywords:** medical devices, vascular compression, market segmentation, marketing and selling channels, edema, product line extensions

---

* *This case was prepared as a basis for class discussion rather than to illustrate either effective or ineffective handling of an administrative situation.*

INTRODUCTION

Dr. Martin Winkler, a vascular and general surgeon, and his wife, were celebrating their daughter Sally's acceptance to Stanford in December, 2009. Sally was interested in following in her parents' footsteps to become a physician but opted for engineering as her major. While Winkler was proud of her accomplishments and future goals, he also wondered aloud how he and his wife, a breast cancer survivor, would manage to pay for four years at Stanford and, possibly, medical school.

A long-time inventor, in 1992 Winkler licensed and patented a unique gastro-jejunostomy feeding tube that minimized patient injury for patients who had been tube fed for long periods of time. The jejunostomy system was developed after ten years of "jerry rigging" gastro-jejunal tubes for his patients who required small bowel feeding. The patent license for the jejunostomy system was sold to Kimberly Clark Corporation. Winkler has received royalties over the 20-year life of the patent. However, in 2012 the jejunal tube will come off patent, meaning that royalty checks would soon cease. To replace that soon-to-be lost income, Winkler turned to another invention that he had been working on since 1999.

At the same time that Winkler was treating a patient who had massive swelling of both legs, he was developing a new compression textile to control swelling. The textile was knitted by a family-owned textile factory near Chattanooga. Winkler admitted that from the beginning in 1999 he was hoping that his research with the innovative compression textile would produce profit. This would enable him, among other things, to help his daughter Sally to fulfill her dreams of attending Stanford University as an undergraduate and, following that, medical school. By providing his patient with free compression stockinets, he was able to follow this patient's remarkable recovery, which Winkler believed to be aided by the use of EdemaWear® (please see Figures 1 and 2 for photos of the product). He then began treating patients with pressure ulcers with EdemaWear®, again witnessing a rapid reduction in healing time (see Figures 3 and 4).

In the midst of toasting his daughter's admission to Stanford, several questions about this new invention kept running through his head. First, this new invention, a compression stocking made from a lycra textile, seemed to have a positive effect on patients with pressure sores and wounds. Yet, he knew that there were other products on the market that claimed similar results. Was his compression prod-

Figure 1: Sample EdemaWear® Stockinet	Figure 2: Sample EdemaWear® Stockinet

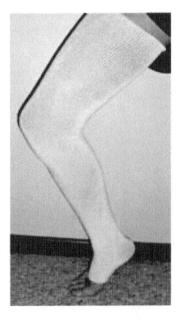

uct really superior to other products on the market and, if so, how would he prove it? Should he try to license or sell the product, for which he had a pending patent, to a larger firm, as he had done with his feeding tube? Should he continue to try to market, sell and distribute it himself? He had paid a student group to develop a website, but to date that passive marketing channel had produced few sales. Sales so far had mostly come from word of mouth to hospitals and pharmacies in the Omaha area.

While Winkler thought that there might be non-medical markets for his textile that controls swelling and helps wounds to heal, he had little idea how to research and identify market opportunities. He had prevailed on family and friends to invest small amounts of capital to get his bootstrap venture, Compression Dynamics LLC, up and running, but he had worries about the future. If textile sales were to take off, he would need a significant amount of capital to scale up production capacity, fund inventory and hire an implementation team.

THE ELASTIC COMPRESSION GARMENT MARKET

Elastic compression therapy involves increasing the external pressure on vascular elements under the skin to assist with the pumping of venous fluid return to the heart. Compression therapy is the mainstay of treatment for venous stasis and venous ulcers. By compressing skin and subcutaneous fat with and an elastic textile, the calf muscles and other lower extremity skeletal muscles have help in pumping venous blood upward to the heart.

Existing compression therapy uses both static and dynamic compression systems to treat venous stasis and venous stasis ulcers. Static compression is characterized by graduated pressure from the ankle proceeding up the lower leg and ending just below the knee. The greatest level of elastic compression pressure is at the ankle and averages between 30 mmHg and 40 mmHg. The exact pressure can vary depending on the type of products utilized, including hosiery, self-adhering, long and short-stretch, and tubular. Dynamic compression products tend to be mechanical pneumatic compression devices, with gauntlets and booties that assist in pumping venous fluid towards the heart

According to a Frost and Sullivan industry report (2005), the existing compression hosiery market is divided into four segments: circular knit, flat-bed knit, net, and one-way stretch. Circular knit products are made of cotton or elastane, and tend not to stretch, which makes them difficult to put on. Flat-bed knit products are made of nylon or cotton and are more flexible. Net products are seamed, less cosmetically accepted stockings and must be custom fitted. The one-way stretch products have very heavy circular machine knitting and also must be custom fitted. Pressures tend to fall into four ranges: 8-15 mmHg, 15-20 mmHg, 20-30 mmHg, and 30-40 mmHg.

Compression products promote healing by increasing: blood flow velocity in deep veins, local capillary clearance, transcutaneous oxygen pressure, and lymphatic clearance of water outside of the vascular system. Elastic compression textiles also decrease capacity and pressure in the veins, decrease visible superficial varicose veins, decrease edema (swelling), and decrease ambulatory venous pressure. Essentially, compression controls swelling of the upper and lower extremity. Dozens of research articles on clinical trials support the effectiveness of using compression garments to reduce deep vein thrombosis, pulmonary embolisms, and edema.

COMPANY HISTORY

Dr. Martin Winkler completed a vascular fellowship at Baylor College of Medicine in Houston. Dr. Michael Debakey, a pioneer of open heart surgery, was the program director. DeBakey invented prolifically. His products included the roller pump used for cardio-pulmonary bypass that allows the heart to stop beating while surgeons operate, a left ventricle assist device for patients who would otherwise need a heart transplant, a "cell saver" that cleans and returns blood lost during surgery, vascular clamps, and a Dacron textile tube that is still used to repair aneurismal and occluded blood vessels.

DeBakey encouraged, by example, his surgical residents and fellows to develop new surgical technology. While Winkler worked in Houston, he envisioned a new stainless steel table-mounted retractor for aortic surgery. Much like his mentor Debakey, Winkler's inventions stemmed from his need for a solution to a surgery-related problem. Winkler chose to stick with surgical innovation projects that he could develop himself without a large budget. Winkler licensed the jejunal feeding to a company, and he paid the legal fees required to file the patent with an "advance" on royalty payments.

Since 1999, Winkler has developed six prototypes of the lycra and nylon tubular compression stockinet. Hoping to land another product license for the compression textile, Winkler approached four compression textile corporate manufacturers. BSN-Jobst, the largest manufacture of compression textiles in the world, then listed on Frankfurt Germany stock exchange, was interested. Jobst sent a representative to Omaha to review clinical results. However, in January 2006, Jobst was taken private, and plans for collaboration fell apart. With a license not forthcoming, and with a cadre of doctors who wanted longitudinal yarn compression stockinet for their patients, the time was right to start a business to sell the new product. The first pair of EdemaWear® were sold in July, 2007; Winkler's wife named the product.

EDEMAWEAR

What differentiates EdemaWear® from the competition? Most standard compression stockings are just very tight elastic socks. EdemaWear® has fuzzy longitudinal yarns with elastic compression provided by transverse medical lycra spandex fibers. The yarns go up and down the limb, forming a tubular compression stockinet. Yarns under tension form furrows in the subcutaneous fat. Between each yarn furrow is an "island" of non-compressed skin. These islands, or skin ridges between "cornrow furrows," according to Winkler, create areas of high and low pressure in the skin. Fluid in the compressed skin tissue flows away in lymphatic vessel in the non-compressed fat between yarns. The physiologic effect on swelling can be dramatic. Reduction of swelling increases blood flow, which in turn enables needed oxygen and nutrients to promote healing and recovery.

While working with a local attorney on a patent application, Winkler contracted with IT students from an area university to design a website for a company that he incorporated as Compression Dynamics, LLC. He named his new product EdemaWear®, the website went live (see www.edemawear.com), and he waited for sales inquiries to roll in. In the beginning, EdemaWear® stockinets came in two sizes, regular and large, and retailed for $19.99 per pair. Winkler touted the products as a low cost compression option for consumers. In fact, Winkler wasn't sure how to accurately calculate his costs, especially when it came to management, sales and marketing. He spent time out of his regular practice and the operating room working on the business and marketing the product to fellow surgeons. His surgical office staff also assisted with the business. Annual sales were: 2007, $38K; 2008, $73K; and 2009, $124K. Winkler noted that 2007/08 were lean "proof of concept" years for the start up.

Winkler realized that when he traveled to conferences or other medical centers, he was the best salesperson for his product. He was relying on word of mouth among the Omaha medical community, and the website as his primary channel to drive sales. Thus, by the end of year three, Compression Dynamics had sales of only $124K per year and had obtained orders from only 20 medical centers in the Omaha area. Winkler realized that something had to change.

In 2008, he was approached by Dr. Anne York, director of Creighton's Entrepreneurship programs, who had been referred to him by a mutual friend and investor in Compression Dynamics LLC. She was seeking a project for students in Creighton University's new NSF-funded Bioscience Entrepreneurship program. He was intrigued because the final output of the course would be a commercialization plan researched by four student teams consisting of business, science, health science, and law students, supervised by Creighton faculty and technology transfer directors. All it would cost him was time and some sample products. Winkler signed on.

POTENTIAL TARGET MARKETS AND MARKETING CHANNELS

Winkler began targeting the medical/surgical market for his fuzzy yarn elastic compression stockinet, because he personally had tested the compression stockinets on patients and had testimonial research from patients to back up his claims that EdemaWear® was "magic" for swelling. However, he was sure that other markets existed, including hemodialysis patients with kidney failure, athletes such as distance runners, and women with cellulite. Winkler had some research data on these applications (markets) for the textile. However, Winkler felt that more research needed to be done. He had barely scratched the surface of potential applications, especially in the medical arena. The following sections briefly define compression products and describe the market opportunities in the areas mentioned above.

POST-SURGICAL AND OTHER MEDICAL MARKETS

According to a 2005 Frost and Sullivan compression products industry report, the medical use market for compression products in 2004 was estimated to be around $460 million and was expected to grow at a compounded average rate of 7.5 percent per year, reaching $760 million by the year 2011. This market was split into static and dynamic products. Frost and Sullivan went on to say that growth in the compression hosiery segment was likely to drive market growth.

However, due to changes in Medicare reimbursement policies, in which hospitals and nursing homes no longer qualified for reimbursement when pressure ulcers developed during the patients' stay, interest in the prevention and inexpensive treatment of these ulcers had increased significantly, as the average cost to treat a single ulcer could be as high as $40,000. Thus, that market was estimated to be much larger than the surgical market and much more of a growth opportunity. Estimates of the cost to manage pressure ulcers in these two settings were over $10 billion dollars per year in the U.S. alone (Aronovitch, 2008). Winkler, however, felt that any surgeries involving grafts and implants followed by swelling, such as those experienced by kidney dialysis patients, could be potential markets for his products.

EdemaWear's® largest competitors in the medical/surgical elastic compression textile markets included BSN-Jobst, Kendall, and MedLine, followed by Juzo, Sigvaris, and Mediven. Contour MD provided custom vs. pre-sized compression stocking. While Jobst produced products primarily for support wear, varicose veins, plastic/post surgical, and lymphedema, Kendall provided products primarily for anti-embolism, travel, and medical. The compression ranges of these products was 8mmHG to 40mmHG and were knee high or calf high.

One challenge in this relatively slow growing market was increased competition from contract manufacturing and imports (there were currently over 100 products on the market), especially in low-end product segments. Another was low clinician awareness, which resulted in misdiagnosis and non-compliance. Firm strategies to combat these challenges include focusing on niche marketing techniques by adding new features or creating new designs, improving efficiencies in production, and achieving economies of scale by expanding into new and international markets (Frost & Sullivan, 2005). To combat lack of awareness, firms were encouraged to train clinicians on proper diagnosis, educate them on products and application techniques, and to provide visual guides for caregivers.

DISTANCE RUNNING RECOVERY

Because EdemaWear® stockinets increase the re-absorption of lymph and edema (swelling) in limbs that builds up during and after strenuous physical activity, Winkler believed that athletes who wore his product would result in decreased recovery time for muscles, which would translate into the ability to train harder, longer, and more often. Winkler, who had completed the San Francisco marathon three times himself, had first-hand experience which suggested that his longitudinal yarn compression textile would be an excellent product to market to runners in any sport, whether competitive, fitness or recreational.

The potential market size for such a product is huge, with 22 million "active" runners who ran 50+ times a year. The demographics of this market were men and women with a net worth of nearly one million dollars, an average annual income of $135,000 and a median age of 44 years old. Roughly 74% were college graduates. The market was increasing at about 10% a year, which had huge implications for new entrants. For example, Under Armour had revenues of $50 million in 2002; those grew to $600 million by 2007. In the active running market alone, with a pricing point of $19.99 per pair, the potential revenue for compression products was $440 million. Future possibilities for growth were in higher performance fabrics such as those providing compression, moisture management, and temperature control, as well as more stylish designs, which helped make sports apparel fashionable for everyday wear (*Running USA*, 2008).

Competition in the running compression products market to date was stiff, largely because giants like Nike and Under Armour were entering the fray. But probably the closest competitor to EdemaWear® was Zensah, although they did not advertise any patented technology to support their products. The only proprietary element of the product was the ratio of antimicrobial silver ions in their products. All of these competitors priced at a considerably higher point than EdemaWear, with Zensah at $39.99, and others even higher.

Research on the effectiveness of compression garments on muscle recovery was mixed. An article in *Running Times* (2008) reported that a South African study published by the American College of Sports Medicine suggested that while maximum oxygen consumption, heart rate, and minute ventilation were not improved by wearing the compression stockings, there did appear to be a statistically significant effect of a faster lactate recovery rate post-exercising for those who wore the stockings after running. Several years earlier, Winkler had approached Nike with an offer to sell or license EdemaWear® but was summarily rebuffed.

Winkler had offered the position of president of EdemaWear's® sports division to a Creighton bioscience entrepreneurship student but had not figured out how to fund the position or what sort of budget, incentives, or marketing channels made sense.

COSMETIC/CELLULITE REDUCTION

Capturing the cellulite market for women was a greater stretch. Despite a constellation of products on the cellulite treatment market, ranging from creams to surgery, and despite many references to clinical trials that proved them effective (none of which could be found searching the literature or FDA site), no reputable firm claimed that their products would completely eliminate cellulite. Most claims, in fine print, admitted that cellulite couldn't be "cured" but its appearance could be reduced or improved. Although Winkler believed that EdemaWear® could be a much less costly and equally effective method of eliminating the unsightly appearance of cellulite, no "proof of concept" (wide scale use by aging females) yet existed to support this claim.

As mentioned above, the presence of so many cellulite-related products on the market demonstrated how big a problem this issue was perceived to be, primarily among women. According to the *Journal of Cosmetic and Laser Therapy* (Avram, 2004), cellulite affects 85-98% of post-pubertal females of all races. With approximately 91 million women in the 20-60 age range, who might be likely to purchase such products, and assuming that the retail price of Winkler's product was set at $39.00, the potential market size was over $3 trillion dollars. As yet, Winkler had not focused on this potential market and was unsure how to start in terms of costs and marketing channels.

So far, Winkler had given away 4,400 pairs of longitudinal yarn compression textile, for "research" between 1999 (prototype 001) and July of 2007, when the first sale of prototype 006, now called EdemaWear®, occurred. In 2008, he partnered with a seamstress in the economically blighted North Omaha area who was interested in fashion design and helped to create new fashionable and customized products from the compression material.

CHOOSING AND FUNDING A SALES GROWTH STRATEGY

In a 2009 report to his board and investors, Winkler had touted the growth of sales over the last year with an average sales of $11,500 per month totaling about $124,000 for the 2009 year. He set a goal of doubling that amount over the next calendar year to $250,000. Of current sales, 5% were from the website channel, with the remainder (95%) from direct word of mouth selling. All sales were in the original post-surgical and pressure ulcer market niche. Without addressing the small volume of website sales, he attributed the recent past growth in direct sales to the hiring of a summer marketing intern whom he then hired as a full-time MBA level marketing director for a year, until she left to enter medical school. Winkler had come to realize that given his and his staff's regular work schedules, he would need some help. In January, 2009 Compression Dynamics was moved into its own office space in the Old Market section of Omaha.

By this time, it was clear to Winkler that growing a startup was not going to be the same as licensing a technology and waiting for the quarterly royalty checks to roll in. He also realized that the product would not sell itself through the internet channel and that direct selling or finding the right distributor and employees would be key, particularly in the medical and surgical markets. Additionally, while he believed that non-medical niches potentially existed for his product, he wasn't sure that the same channels would work for reaching those customers. With little cash available to invest in his company, Winkler was searching for the best way to spend whatever funds might be freed up for marketing. Clearly, getting the word out about his product was the most daunting sales growth challenge for 2010 and beyond.

REFERENCES

Aronovitch, Sharon A. 2008. Intra-operatively acquired pressure ulcers: Are there common risk factors? *Ostomy Wound Management.* http://www.o-wm.com/article/6776. March 11.

Avram, Mathew M. 2004. Cellulite: A review of its physiology and treatment. *Journal of Cosmetic & Laser Therapy,* December. 6(4): 181-185.

Frost and Sullivan. 2005. F333-54. *The U.S. Compression Therapy Market.* www.frost.com.

Running USA. 2009. *RRIC: Trends and Demographics.* www.runningusa.org/cgi/trends.pl, May 2.

iKaryos Diagnostics: The Rocky Road from Concept to Startup

ANNE S. YORK

College of Business, Creighton University

- **Summary and key issue/decision:** iKaryos was the brainchild of Dr. Jill Hagenkord, a cancer researcher who brought her lab and grants to Creighton University to begin developing a diagnostic-based business around the concept of virtual karyotyping. The procedure designed by Hagenkord, based on trade secrets, has been able to identify aggressive vs. non-aggressive cancers 80% more accurately than existing karyotyping processes. While the concept of virtual karyotyping had been widely written about in the scientific literature, only one other lab in the country had commercialized the concept, although for use in the developmental diagnostic market, rather than for cancer applications. Unfortunately, Hagenkord and her partners had little training in business or technology commercialization, nor did they have contacts in the industry to help them develop a company and market their product. This case follows Hagenkord and her iKaryos team's decisions to incorporate, seek equity financing for their new company, develop a viable organization structure, and ultimately market their product before their competition beat them to it.

- **Companies/institutions**: iKaryos, Creighton University

- **Technology:** Recently, platforms for generating high-resolution karyotypes *in silico* from disrupted DNA have emerged, such as array comparative genomic hybridization (arrayCGH) and SNP arrays. Conceptually, the arrays are composed of hundreds to millions of probes which are complementary to a region of interest in the genome. The disrupted DNA from the test sample is fragmented, labeled, and hybridized to the array. The hybridization signal intensities for each probe are used by specialized software to generate a log2 ratio of test/normal for each probe on the array. Knowing the address of each probe on the array and the address of each probe in the genome, the software lines up the probes in chromosomal order and reconstructs the genome *in silico* (Fig 2 and 3). Virtual karyotypes have dramatically higher resolution than conventional cytogenetics, currently up to a 1000-fold greater resolution than karyotypes obtained from conventional cytogenetics.

- **Stage of development at time of issue/decision:** The process of conducting virtual karyotyping is based on trade secret IP, because the concept itself is not patentable. At the time of the case, iKaryos was performing diagnostics for a small but growing group of customers from a functioning, CLIA certified lab. However, current capacity was limited and competitors in other virtual karyotyping niches were moving into the cancer domain. Brand recognition and customer rela-

---

*   *This case was prepared as a basis for class discussion rather than to illustrate either effective or ineffective handling of an administrative situation.*

tionship development are important in gaining first to market advantage, which is why the firm needed to move quickly to gain equity for expansion and market penetration.

- **Indication/therapeutic area:** cancer tumor classification, although other applications are possible, including developmental diagnostics
- **Geography:** US
- **Keywords:** virtual karyotyping, cancer diagnostics, venture team formation, early stage valuation, first to market strategy

INTRODUCTION

Dr. Jill Hagenkord sat alone in her lab, worried that she'd be late picking up her two pre-schoolers. iKaryos Diagnostics, her newest "baby," was at risk of being still-born. She had arrived at Creighton University with a grant to pay for her lab equipment and part of her salary, and full of hope that with one day a week allowed for developing her virtual karyotyping technology she would be able to move it forward from concept to start-up. Early interest from prominent venture capital firms evaporated in the worst economic climate in decades and the nagging lack of intellectual property—a standard ingredient in the VC recipe for diagnostic companies. Her project had languished for two years with no clear strategic direction and no time to spend on anything other than doing the work at hand. As a single parent trying to make ends meet on an academic salary, she could really use the extra money, but more than that, she wanted to make this new type of testing available to the world. She believed in her science and believed that it could make a difference.

A native of Altoona, Iowa (population 14,000), Hagenkord was a self-described "troublemaker" in high school who served more than a few hours in detention for skipping class. Her high school science teacher bet her $5 that she would fail out of college in one semester. But when Hagenkord got to the University of Iowa, where she earned a degree in history—with honors—she found that she loved learning and that she was smart. Because she felt there was so much more to learn, she spent another four years at Iowa as an undergraduate, this time studying biochemistry. "I always worked 2-3 jobs to support myself and pay my tuition, but I still had a lot of fun in college. I enjoyed being an undergrad", she says. "I didn't really even think of becoming a doctor until I was 23, when I took my first science class," she says. "My boyfriend, who was in med school, told me that I only got good grades because I took humanities classes and that I wouldn't get such good grades in pre-med classes. So, we made a bet. I signed up for the hardest class he ever took as an undergrad, and whichever one of us got the lower grade had to take the other one to Red Lobster. I got an A+, crushing his A-, and discovered that I am really good at science." This bet ultimately led her to Stanford Medical School.

Her time at Stanford impacted her in two very important ways. First, early array technology was emerging from research labs during this time at Stanford. Hagenkord realized that this massively parallel testing technology could change the way we practice medicine. She wanted to be one of the pioneers ushering the technology into the clinical laboratory. More importantly perhaps, Hagenkord would often cry when she was with cancer patients who were drawing their last breaths. "When I was in the room with the family and the patient died, I would always tear up and sometimes I would even just sob... I also cried when I had to tell someone that their cancer had come back and that there was nothing more we could do. Pediatric oncology was the hardest."

Never afraid to take the road less traveled, Hagenkord aborted her pathology residency at University of California San Francisco to be a mouse pathologist at a Silicon Valley start up company, Deltagen, Inc., whose mission what the high-throughput development and phenotypic analysis of genetically engineered mice for the purpose of drug discovery and development. "Everyone warned me that once you leave academia, you can never come back," she explains. "I have never been one to stay inside the box, and it was too good of an opportunity to pass up. So, I went for it. And it turned out to be one of the best professional decisions of my life." Hagenkord says that in the start-up environment she met some of the smartest and most inspired people she has every worked with. "They were so passionate and believed that their science could make a difference. They pushed me to play at the top of my game in a way that medical training never did." And again, although she didn't realize it at the time, Hagenkord was making the connections with the people she would approach a decade later when she wanted to start her own company.

Hagenkord was the only physician with fellowship training in both molecular genetic pathology and pathology/oncology informatics, which made her uniquely trained to design, validate, and implement array-based testing in the clinical laboratory. She was considered by many to be the foremost expert in the field. Those deeply poignant encounters with cancer patients and their families would help start her down a path to founding a company to help those patients find the right treatment for their diseases—helping people deal with the uncertainty of a cancer diagnosis would be central to the corporate identity of her company, which was reflected in the initial advertising campaign (see Appendix 1).

Wanting to get closer to her family in the Midwest, Hagenkord took a faculty position in the school of medicine at Creighton University in Omaha, NE. As an assistant professor of pathology and medical director of molecular pathology and clinical genomics, she taught medical students and residents, signed out molecular pathology, and was given ample time for research. Hagenkord chuckles when she admits that coming to Creighton also turned out to be a brilliant decision despite many people telling her that, with her credentials, she should work at a more prominent, "coastal" institution. Hagenkord explains that her chairman was unbelievably supportive in enabling her to accomplish her goal of being the first clinical lab world-wide to offer SNP array karyotyping for cancer. "Creighton is a relatively small institution. They are able to be nimble, when needed. They knew when they hired me that I had a vision for iKaryos and they worked with me to find a way to make it happen."

Hagenkord's working name for her company was iKaryos. *Karyo* means nucleus in Greek and "i" connotes personalized—capturing the essence of personalized molecular diagnostics. The mission of iKaryos would be to provide detailed and personalized DNA analysis cancer samples using cutting edge technologies like SNP array karyotyping. One of Hagenkord's favorite mantras was "knowledge is power in the fight against cancer." Using her skills as a physician, a scientist, and a bioinformatician, she wanted to look into the DNA of each patient's tumor and provide more information about that patient's type of cancer, the aggressiveness of the cancer, or the responsiveness to specific drugs. That is, to take a personalized look into the nucleus of each patient's cancer. (see Appendix 1 for the home page of www.ikaryosdiagnostics.com).

Hagenkord had kept her nose to the grindstone for a year at Creighton—setting up the laboratory, doing extensive clinical validations of the SNP array technology, publishing papers, and launching the assay. She hadn't given up on iKaryos, but had kind of lost track of it in the daily grind. Local physicians had started ordering the test, and Hagenkord had tangible evidence of how it was providing better patient care. More than ever, she wanted to advertise widely that this test was available for clinical use. She also knew that time was of the essence for iKaryos. Array-based karyotyping had moved quickly into the clinical laboratory for testing the DNA of children with mental retardation and other constitutional disorders. Most people in the field conceded that the next major market for the technology was cancer. She was several years ahead of the pack, but knew that there was no patent protection for her processes. The only barrier to entry for other commercial laboratories was know-how.

They were now a functional lab making a profit—one of the biggest milestones for a diagnostic start up. Angel investors were approaching the team and a locally-based beauty products marketing entrepreneur offered to provide assistance, but the cost could be giving up more than 50% of the equity in her company to partners who provided a relatively small percentage of the capital needed to commercialize her diagnostic tool. The beauty products entrepreneur brought little relevant experience to the table. She was at a loss for objective guidance in making these difficult decisions. Omaha had so few bioscience startups, and the discipline was so technical and specialized. Most firms doing this kind of work were either on the West Coast or in the Boston area. Creighton's technology transfer office had little interest in the venture as there was no intellectual property connected to it, and their expertise lay more in the realm of licensing as opposed to startups.

Before turning off the lights in her office, Hagenkord checked her phone messages one last time, quickly listening to a request from the Creighton's professional science masters program director to guest speak to their current issues in bioscience class. While she loved teaching and talking about her work, how could she justify the time with everything else on her plate? This almost chance meeting with Anne York, Ph.D., director of entrepreneurship in the Creighton University School of Business changed everything. With her encouragement and connections, iKaryos began to rise from the ashes.

VIRTUAL KARYOTYPING TECHNOLOGY: A BREAKTHROUGH IN CANCER DIAGNOSTIC

To understand virtual karyotyping, it is useful to contrast it to traditional karyotyping, also called conventional metaphase cytogenetics. Essentially, traditional karyotyping involves examining intact chromosomes under a microscope on a glass slide (*in vitro*, or literally, "in the glass"), while virtual karyotyping physically disrupts the chromosomes, applying the fragments to an array, scanning the signals into a computer, and reconstructing the karyotype inside the computer (*in silico*). Wikipedia provides basic explanations of both traditional and virtual karyotyping. A karyotype is typically presented as an image of the chromosomes from a single cell arranged from largest (chromosome 1) to smallest (chromosome 22), with the sex chromosomes (X and/or Y) shown last. Karyotypes have been used for several decades to identify chromosomal abnormalities in both germline (inherited genetic information that is present in all cells from conception) and cancer cells. Conventional karyotypes can assess the entire genome for changes in chromosome structure and number, but it requires tissue culture and the resolution is relatively coarse with a detection limit of 5-10Mb.

Recently, computerized platforms for generating much higher resolution karyotypes from disrupted DNA have emerged, such as comparative genomic hybridization (CGH) and single nucleotide polymorphism (SNP) arrays. SNPs are single nucleotide sequence variants which are often used for whole genome association or linkage studies. SNP arrays have a SNP embedded in each probe on the array, so they can provide both copy number information at that locus of the genome and the genotype of the embedded SNP. The use of these diagnostic platforms has come to be known as virtual karyotyping.

Conceptually, the arrays are composed of hundreds to millions of probes which are complementary to a region of interest in the genome. The disrupted DNA from the test sample is fragmented, labeled, and hybridized to the array. Knowing the address of each probe on the array and the address of each probe in the genome, the software lines up the probes in chromosomal order and reconstructs the genome on the computer. Analyzing virtual karyotype data for tumors requires not only an understanding of tumor biology and morphology, but also the molecular biology of the platform and the assumptions underlying every single step in every single algorithm used to generate a single virtual karyotype. Each tumor is unique, and not all assumptions hold true for every tumor.

Virtual karyotypes have dramatically higher resolution than conventional karyotypes. The actual resolution will depend on the density of probes on the array. Currently, the Affymetrix Cytogenetics Array is the highest density commercially available array for virtual karyotyping applications. It contains 2.7 million polymorphic and non-polymorphic markers, with a median intermarker distance of 700bp and denser coverage of known genes, thus providing exon level resolution. This is approximately 1,000-fold greater resolution than karyotypes obtained from conventional cytogenetics. (see Appendices 2a-b for figures that depict differences in traditional karyotyping and virtual karyotyping).

The key value in karyotyping, both traditional and virtual, is used to detect copy number changes in germline or tumor samples. In humans, a normal copy number is always two for the non-sex chromosomes. A deletion is the loss of genetic material. The deletion can be heterozygous (copy number of 1) or homozygous (copy number of 0). Microdeletion syndromes, such as DiGeorge Syndrome, are examples of constitutional disorders due to small deletions in germline DNA. Deletions in tumor cells may represent the inactivation of a tumor suppressor gene and may have diagnostic, prognostic, or therapeutic implications. A copy number gain represents the gain of genetic material. If the gain is of just one additional copy of a segment of DNA, it may be called a duplication. If there is one extra copy of an entire chromosome, it may be called a trisomy. Copy number gains in germline samples may be disease-associated or may be a benign copy number variant. When seen in tumor cells, copy gains also may have diagnostic, prognostic, or therapeutic implications. Technically, an amplification is a type of copy number gain in which there is a copy number >10. In the context of cancer biology, amplifications are often seen in oncogenes. This could indicate a worse prognosis, help categorize the tumor, or indicate drug eligibility. (see Appendices 3a-d for graphical representations of the virtual SNP-based array karyotyping).

While a virtual karyotype can be generated from nearly any tumor, the clinical meaning of the genomic aberrations identified is different for each tumor type. Clinical utility varies, and appropriateness is best determined by an oncologist or pathologist in consultation with the laboratory director of the lab performing the virtual karyotyping. Examples of types of cancers where the clinical implications of specific genomic aberrations are well established include neuroblastoma, Wilm's tumors, renal cell carcinoma, chronic lymphocytic leukemia, multiple myeloma, myodysplastic syndrome, and colorectal cancer. In conjunction with disease stage, virtual karyotyping can provide a markedly improved ability to not only distinguish between rapidly growing vs. slow growing tumors, but also to target intensity of treatment, select appropriate therapeutics, and to identify risk of treatment failure. As such, it can help clinicians and patients make better treatment decisions. In certain cases, virtual karyotyping can replace existing testing strategies. In other cases, it complements existing methods, providing a more complete picture of the cancer genome, but adding to the cost of standard test panels. However, because it may create a better match between treatment and disease stage of specific cancers, virtual karyotyping can reduce unnecessary therapeutic costs and undesirable side effects.

Hagenkord began using SNP arrays to create virtual karyotypes of cancer genomes in 2006, while a fellow at the University of Pittsburgh working with Federico Monzon, M.D. (who is a co-founder in iKaryos). "When we got the first cancer genome to appear inside the computer, both Federico and I knew we were onto something amazing, something that would change the way we practice medicine," says Hagenkord. "But when I tried to explain to other physicians what we were doing, "the cricket would chirp." And there was very little in the medical literature about using SNP arrays to karyotype cancer.

In 2008, Karen Lusky wrote in *CAP Today*, that SNP arrays might be the newest molecular sleuth poised for a prime-time position in the unfolding oncology drama of "what is it and how can we best treat it?" While many times pathologists might report that they favored a benign diagnosis, there was a chance that the tumor just looked benign but would behave in a malignant fashion; with virtual karyotyping, the oncologist could decide how to proceed with that caveat in mind. In May 2009, the *Omaha World Herald* published a feature story on Dr. Hagenkord's work. They contacted experts from around the country as part of their reporting, one of whom was Dr. Jared Schwartz, the president of the College of American Pathologists. When Dr. Schwartz was asked what he thought the impact of SNP array karyotyping would be on the practice of medicine, he said "This is the beginning of the end—the beginning of the end—of trial and error medicine." The March, 2010 issue of the *Journal of Molecular Diagostics* used Hagenkord's work as the cover art. A commentary on the Hagenkord, *et al* manuscript written by Shelley Gunn, M.D., Ph.D. touts that "In an age when oncology patients expecting a more personalized approach to their individual disease, array-based karyotyping will become an indispensible tool in the clinical laboratory..." The title of Gunn's commentary is *The Vanguard Has Arrived in the Clinical Laboratory*.

By this time, crickets were no longer chirping when Dr. Hagenkord spoke about her work. In the first few months of 2010, she gave invited talks in Dubai, Saudi Arabia, Jordan, Japan, and Korea. There was an international ground swell of interest in using SNP array karyotyping for cancer testing, and iKaryos was standing up on the board, ready to ride the wave, while other labs were looking on from the beach.

THE MARKET FOR VIRTUAL KARYOTYPING

The most common applications for SNP-related research tools are gene disease association studies, drug target validation, disease susceptibility studies, diagnostics, drug target screening and pharmacogenomics for clinical trials. SNPs can be used to help determine how a person responds to therapeutic treatments or can act as markers for populations at risk for developing a disease. Examples of the variety of disease applications of SNPs include sickle-cell anemia, susceptibility to late-onset Alzheimer's disease, hereditary deep-vein thrombosis and drug metabolism. Inherited differences in DNA sequences contribute to phenotypic variation, an individual's susceptibility to disease and response to therapy.

But Hagenkord did not use SNP array in this typical application. Instead, she used it as a cytogenetics tool—to recreate the chromosomes inside the computer. The fact that a SNP is present in the probes merely allowed her to detect loss-of-heterozygosity. The SNPs on the array are not clinically meaningful point mutations, like the sequence variants that cause sickle cell disease or hereditary deep vein thrombosis. Hagenkord's work never considered the genotype of one specific probe in isolation, but rather the behavior of dozens of consecutive probes as an indicator of chromosomal structure. Thus, her analysis did not provide any information about specific sequence variants that would impact drug metabolism.

Market projections suggested that the cancer diagnostics market in the U.S. was estimated to expand from $73 million in 2004 to $2.5 billion by 2014; a cumulative average growth rate of more than 30%. With breast, colon, and prostate diagnostics as the top three segments, the "all other" segement was small ($60 million in 2008) but the most rapidly growing, with a CAGR projected at 48%. From Hagenkord's perspective, however, the market for virtual karyotyping is essentially all of human cancer, a market estimated at 1.5 million new cases diagnosed each year (American Cancer Society), with cancer being the second greatest cause of death in the U.S., following heart disease. Currently, the majority of the market was being served by related but "old" technology—conventional cytogenetics, which is currently being performed by most cytogenetics lab in the world. Tier 1 of iKaryos's business model was just using a better technology to detect known biomarkers. A key industry challenge was to educate physicians about the usefulness of such tools over and above traditional karyotyping.

Factors favoring the development of this technology included the decreased costs associated with clinical trials (possibly a savings of as much as $80 million per drug), the possibility of targeting drugs to the more specific populations who could be helped by the drugs, and limiting the powerful side effects of such drugs to those who would be unlikely to realize the benefits. However, narrowing the market would result in decreased drug sales. Many in the industry felt that changes in FDA regulations would be required to provide incentives for companion drug development once diagnostics were developed to be used as screening to screen for prospective patients who would benefit from the new, expensive, drugs.

A final issue facing molecular diagnostics firms was the possibility of proposed legislation which would change the way that diagnostics were regulated. Currently, new diagnostics could be introduced to the market as long as they were performed in existing CLIA certified labs. Most could be reimbursed under current insurance codes. These labs were mainly required to produce tests with reliability and validity, but did not require diagnostic specific clinical trials for efficacy. Firms were using CLIA certified products to enter the market quickly and provide cash flow, estimated at $5-30 million, for funding the trials needed to obtain FDA approval. The benefit of FDA certification would be to achieve a new, higher-level reimbursement code. As the molecular diagnostic market became more competitive, it was thought that the average price point for cytogenomic array CGH-based tests would fall between $1,200 and $1,500, although prices could be as low as $75 to detect point-based mutations to as high as $10,000 for a multi-gene SCA mutation panel.

THE COMPETITIVE LANDSCAPE

Signature Genomic Laboratories is a company that Hagenkord has been watching since its inception. Signature Genomics was the first clinical laboratory to provide microarray-based diagnostic testing of individuals with mental retardation and birth defects. In their first three years, Signature captured the market, defined themselves as the experts, and saw an 1800% growth. Watching Signature succeed without being located in a major biotech or cancer research cluster gave Hagenkord confidence that iKaryos could do the same in Omaha. Signature gave her a company to emulate—the first tier of the iKaryos business model was to become the Signature Genomics for cancer applications.

Signature Genomics was founded in 2003 by Lisa Shaffer, Ph.D. (president and CEO) and Bassem Bejjani, M.D. (CMO) as a partnership between Signature Genomic Services, Pathology Associate Medical Laboratories, and Sacred Heart Medical Center. Opening its doors with three employees, Signature began offering testing in March of 2004, and by August of that year was making a profit. In 2007, Ampersand Ventures, a private equity firm focusing on investments in the health care and industrial technology sectors, and in 2008, jVen Capital, a life sciences investment company with a focus in the areas of biotechnology, diagnostics, and medical devices, acquired stakes in Signature Genomics. By that time, Signature Genomics had grown to 107 employees and had processed over 40,000 cases since its inception. It offers a full test menu of cytogenetic diagnostic procedures, including array CGH, FISH, and G-banded karyotype analysis, along with a genomics genome browser and a proprietary web-based visualization software for laboratory use.

While Hagenkord admired Signature and hoped to follow in its footsteps, she also was aware that Signature was making a move into cancer diagnostics, recruiting researchers, and laboratory staff to prepare for the new market segment. However, Signature Genomics is built on array CGH technology—not SNP arrays. Currently, Signature is in a contractual relationship to use arrays designed by Roche-Nimblegen. Because these arrays do not contain SNPs in their probes, they are considered 'copy number only' arrays. Although copy number only arrays may be sufficient for constitutional applications, they are not appropriate for cancer diagnostics. Although unsuspected by the cytogenetic community a few years ago, the advent of SNP array karyotyping has revealed that it is very common for cancers to have a previously under-appreciated genetic lesion called acquired uniparental disomy (aUPD). aUPD is known to be clinically relevant, and the cytogenetics community has conceded that array-based karyotyping of cancer must be done on SNP-based platforms. When Signature established a relationship with Roche-Nimblegen, the critical importance of aUPD detection for cancer was not fully established. Because of this, Signature Genomics went from being one of iKaryos' potential key competitors to being several years from being in the game, unless they were to buy a start-up lab which already has the experience, expertise, and market share.

Another firm in the market space of which Hagenkord was aware was CombiMatrix Molecular Diagnostics (www.cmdiagnostics.com). CombiMatrix, a firm headquartered near Seattle, Washington, had three divisions: diagnostic testing services, clinical trials, and manufacturing. The CLIA-certified and CAP-accredited diagnostic lab which opened in Irvine, California in 2006, offered genomic testing services in the area of cancer diagnostics. The clinical trials unit was partnering with pharmaceutical companies to identify patients who stood to receive the greatest benefit from currently available drugs, as well as those under development. The manufacturing unit, CMDX, sold custom CGH microarrays and reagents to researchers worldwide. Among CombiMatrix's management team were president and CEO Dr. Mansoor Mohammed, Ph.D., a specialist in molecular immunology, molecular genetics, and cytogenetics; chief operating officer Chris Emery, MBA and former sales and marketing manager in a variety of pharmaceutical and biotechnology firms; medical director Shelly Gunn, M.D., Ph.D., a geneticist specializing in the detection of cryptic chromosomal abnormalities; and VP of Operations Dr. Lony Lim, Ph.D., who had years of experience running a variety of clinical labs. CombiMatrix also had vice presidents of sales and marketing and of scientific affairs.

Like Signature Genomics, Combimatrix Diagnostics committed to a copy number-only platform for array based karyotyping. They use custom BAC arrays, which are rarely used clinically because they have been replaced by the more reliably manufactured synthetic oligonucleotide arrays. Combimatrix was poised to be a competitor in the cancer market, but the use of copy number only arrays renders them almost irrelevant.

Hagenkord felt that Lab Corp (www.labcorp.com) was probably the biggest threat. Lab Corp came into existence in 1995, when National Health Laboratories and Roche Biomedical Laboratories merged to become one of the largest clinical lab providers in the world. Initially led by Dr. James Powell, the new company—Laboratory Corporation of America—was headquartered in Burlington, NC. Over the last 10 years, LabCorp continued to expand specialty laboratory testing capabilities by acquiring scientific leaders in the genomics, anatomic pathology, and personalized medicine arenas. They had Affymetrix instrumentation in-house and currently used the SNP 6.0 arrays clinically for constitutional testing. While they lacked in-house expertise in cancer applications, they had gone on record as having an active interest in moving into the cancer market. LabCorp operated a sophisticated laboratory network and logistics infrastructure, with more than 28,000 employees worldwide and 220,000 clients, including physician offices, hospitals, managed care organizations, and biotechnology and pharmaceutical companies. They were currently examining more than 10 million cytology and 2 million surgical pathology samples each year.

There are two key features that distinguish iKaryos from their rivals. One is that iKaryos used subspecialty trained molecular pathologists—M.D.'s—to manage the laboratory, interpret the data, and generate a value-added report with clinic-pathologic correlation. Constitutional testing is done in cytogenetics laboratories by Ph.D. cytogeneticists. The technical aspects of array-based karyotyping for constitutional samples were relatively straightforward compared to cancer applications. For cancer applications, an in-depth knowledge of tumor biology and morphology, appreciation of the clinical impact of abnormal findings, the ability to synthesize clinical data from other laboratories, and a solid foundation in molecular biology and bioinformatics all were needed. In short, the process required a molecular pathologist with informatics training, and there were few physicians, other than Drs. Hagenkord and Monzon, with this constellation of skill sets. In addition, M.D.'s could bill using the array codes on the Professional Fee Schedule, while Ph.D.'s could only charge for the technical component, a differential of at least $1000 per test. Another key feature that distinguished iKaryos from their rivals was their ability to handle both fresh tumors and formalin fixed paraffin embedded (FFPE) tumors. While fresh tumors would always produce a cleaner SNP array virtual karyotype, it was an advantage to be able to use FFPE tumor samples as well, since it is standard protocol for solid tumors to be fixed in formalin immediately after being removed from a patient in the operating room.

The product of iKaryos' first tier is a personalized report with clinic-pathologic correlation, including concise results for the tumor being tested, an analytical interpretation of the results, pertinent details of the procedure, and a list of references (see Appendix 4 for a sample of a report).

TEAM iKARYOS

The skills needed to develop SNP-based virtual karyotyping technology were complex. Much like the scientists and founders of Signature Genomics and CombiMatrix, Hagenkord had acquired them in a roundabout way over time, primarily by working with scientists doing related research, both in university settings and at innovative Silicon Valley biotechnology startups such as Deltagen, Inc.

In 2006, when the idea of making personalized cancer diagnostics company based on SNP array karyotyping solidified in her head, Hagenkord asked herself, "Who would be the best people to work with? Who could make this happen?" She immediately thought of Robert Klein, Ph.D., former VP of technology development at Deltagen, and Shera Kash, Ph.D., former VP of operations at Deltagen. "The three of us had that special chemistry when we worked at Deltagen. Our skill sets are very complementary, and they are two of the most brilliant people I have ever met," said Hagenkord. Klein's extensive experience and connections in Silicon Valley biotechnology start-ups got the team in front of several prominent VC firms, but the time wasn't right. Hagenkord explained, "The economy was in a tail spin, we had more potential than accomplishments or proof, and we didn't fit into the IP-central paradigm of VC investment in diagnostics. They tend to focus almost exclusively on IP and very little on clinical utility. iKaryos was just the opposite—loads of clinical utility and very little IP." Despite this, she refused to give up, and convinced Dr. Kash to relocate from Silicon Valley, where she had lived for 15 years, to Omaha in order to keep the dream alive.

She needed a CEO. Robert Klein had been focusing on other projects while Hagenkord and Kash were setting up iKaryos in Omaha. He wasn't as naturally connected to the project as he had been a few years earlier. Several people, including Klein, were suddenly interested in taking the executive helm at iKaryos. Klein was also wanted to participate in series A as an angel investor, making him inherently conflicted in valuation discussions. Hagenkord described this time period as her crash-course MBA. "People were quite passionately giving us differing advice. It was hard to know who to trust, so I had to try to learn as much as I could about cap tables, pre-money valuation, LCC versus C-corp—things I had quite frankly never heard of before," she said.

"We were worried about Robert being a long-distance CEO of iKaryos from California. He had also kind of checked out of iKaryos after the initial VC tour. People told us that we needed a local CEO or even that I should be the CEO." It came down to the trust and confidence that Hagenkord and Kash had established with Klein when they were co-workers at the Silicon Valley start up 10 years earlier. "We all knew that we worked together well. We all knew that we worked hard. We all trust each other personally and professionally. Once Robert made it clear that he was fully committed to iKaryos, it was a no-brainer to have him be the CEO, even though it meant he was long distance," explained Hagenkord. "Besides, now we were building a new paradigm for diagnostic companies—one where the company doesn't have to be in California or Boston, and where clinical utility matters more than IP. And in the age of Skype, do you really need your CEO in the office next to you? With Robert in Silicon Valley, he is close to the money, which is more important than having him be close to the lab."

Klein negotiated a creative arrangement with Creighton University where all samples would be processed in Creighton Medical Laboratories CLIA-certified laboratory on a fee-for-service basis, and then signed out by Dr. Hagenkord while she is consulting for iKaryos. Maintenance of equipment and laboratory personnel was Creighton's responsibility, while iKaryos focused on marketing and informatics. iKaryos had no actual overhead, allowing them to run lean while they built value and moved toward their series B financing round. Creighton University was a small equity holder in the company. Said Hagenkord, "It is a win, win, win situation. I know I could not have done something this non-traditional at a larger institution. I am so grateful for the institution-wide support I received from Creighton to make iKaryos happen."

A SHORT BIO OF EACH KEY IKARYOS FOUNDING TEAM MEMBER

Hagenkord was a board-certified molecular genetic pathologist and a founder and chief medical officer of iKaryos. She also was an assistant professor of pathology at Creighton University School of Medicine and served as the medical director of molecular pathology and clinical genomics at Creighton Medical Laboratories. She obtained her M.D. from Stanford University School of Medicine in 1999. After residency training in pathology at UCSF and University of Iowa, Hagenkord completed fellowships in pathology/oncology informatics and molecular genetic pathology at the University of Pittsburgh, making her uniquely trained to design, validate, and perform array-based testing in the clinical laboratory. Hagenkord had been performing virtual karyotyping of human tumors, hematologic and solid, since 2006.

Described by Hagenkord as her long-time good friend, Dr. Robert Klein, CEO and founder of iKaryos, had nearly 20 years biotechnology experience at multiple companies including Rinat Neuroscience, Genentech, and Deltagen. Klein's resume included senior positions as vice president and CSO in both the scientific and business areas with a focus in new technology development, pre–clinical research, and mergers and acquisitions. He received his bachelor's degree in biochemistry from UC Berkeley and his Ph.D. in Biology from MIT.

Dr. Shera Kash, chief operating officer and founder of iKaryos, obtained her Ph.D. from Baylor College of Medicine in molecular genetics and did post-doctoral training at UCSF. Prior to coming to iKaryos, Kash was vice president of operations at Deltagen, Inc, where Kash worked with both Klein and Hagenkord. Kash had 10 years of management experience in biotechnology, experience establishing CLIA-certified clinical laboratories, and dozens of scientific publications in peer reviewed journals.

Dr. Federico Monzon served as head of the scientific advisory board and was a founder of iKaryos. A board-certified molecular genetic pathologist, Monzon had extensive experience in clinical molecular diagnostics and the translation of complex genomic molecular assays from the research environment to the clinic. He received his M.D. from the Universidad Nacional Autonoma de Mexico and did subsequent postdoctoral fellowships at the University of Pennsylvania. While at the University of Pittsburgh Medical Center, he was the director of the clinical genomics facility of the University of Pittsburgh Cancer Center and the medical director of special laboratory informatics for the department of pathology. He currently served as the medical director of molecular diagnostics at The Methodist Hospital in Houston, TX. Monzon had been performing research with SNP array virtual karyotyping of cancer since 2006 and has more than 40 publications in peer-reviewed journals.

The organization and management team dilemmas facing iKaryos were two-fold: production capacity and business skills. While there was extensive scientific and laboratory expertise among the founders, none was ideally qualified from an experience perspective to be CEO or to handle the financial and sales and marketing tasks. Also, among them, Hagenkord and Monzon were the only two physicians qualified to write and sign off on the analytical lab reports, and both were limited by their universities to one day per week in outside consulting activities. Yet neither Hagenkord, as a single parent, nor Monzon, working full-time at Methodist Hospital in Houston, were willing or able to give up their full time faculty positions to work full time in the company. In addition, their travel schedules involving paper presentations, searching for additional financing, and marketing iKaryos's services also limited time in the lab.

iKARYOS' OPERATIONAL "TO-DO" LIST

LEGAL

The first step required to become a real company was to create a legal business entity. This process would involve choosing a law firm to represent them, along with drawing up documents to form a legal partnership or corporation. Choosing a law firm seemed to be the most straightforward decision. Klein was concerned that qualified expertise was not available locally, nor would a local firm give them access to VCs and other resources available on one of the coasts. Forms of business choices seemed a bit more complicated: either creating a Limited Liability Corporation or a regular C corporation. The LLC form was simpler administratively, operating more like a partnership from a financial accounting and tax standpoint but carried two significant liabilities. The first was the downside of having to file incorporation papers in multiple states, should the company expand and multiple locations be required. The second was the inability of the LLC to offer multiple classes of stock, which precluded many potential funders' preferences for preferred convertible stock. A C corporation, while more cumbersome to form initially, provided for multiple stock classes and a one-time incorporation that was recognized across states. However, the C Corp was administratively more demanding reporting-wise and had the disadvantage of possible double taxation on both corporate income as well as on any dividends paid out.

OWNERSHIP AND EQUITY

The team of Klein, Kash, and Hagenkord had developed fruitful professional relationships and personal friendships while they were all employed at Deltagen. As years passed and each moved on to other positions, they kept in contact and grew to appreciate the rare professional chemistry they once shared. Hagenkord left industry to complete her pathology residency and then met Monzon during her fellowship training at the University of Pittsburgh. Within a few months, she recognized she had the same personal and professional chemistry with Monzon that she had 10 years earlier with Klein and Kash.

Klein launched his biotech career as a Genentech scientist, but as the first employee at Deltagen, he was centrally involved in all aspects of financing, legal, and acquisitions as the company grew to over 500 employees, had several rounds of VC financing, and a successful IPO in August of 2000. Klein then went on to Rinat Neuroscience, again providing both scientific and business leadership as a key player in the largest all-cash biotechnology acquisition to date by Pfizer. Dr. Klein held several patents and was a respected scientist, but his start-up experience, business experience, and connections in Silicon Valley made him a valuable CEO who actually understands the science. Dr. Kash joined Deltagen as employee number 25 and remained at the company for 10 years, ultimately becoming vice president of operations and being one of three remaining employees before coming to Creighton.

Creighton University technology transfer office's initial position and written policy was that the university owned 58% of any technology developed at Creighton. The terms of Hagenkord coming to Creighton involved a three-year start-up grant, which paid for her equipment and provided research support, and the ability to do consulting work 20% of the time. Like other faculty, Hagenkord was still expected to teach residents and medical students, contribute to clinical service work, and assume administrative responsibilities. Kash's salary was paid from Hagenkord's research funds.

INTELLECTUAL PROPERTY AND REGULATION

As academics, Monzon and Hagenkord had not given much thought to intellectual property as they were developing the technology and clinical applications. Their focus was on publications, on grants, on the development and maintenance of a virtual karyotyping wiki, and on ultimately being able to provide better patient care. Thus, many of their ideas had been disclosed for over a year in a variety of public domain outlets. The team wondered if they could patent the process used to develop and perform the analysis, and recognized that a business method patent would require disclosing trade secrets that could be improved upon quickly by labs having greater resources. Hagenkord had successfully validated SNP array karyotyping as a laboratory developed test (LDT) per CAP requirements in Creighton's CLIA-certified lab. She also had, with co-authors, recently published several papers on the SNP virtual karyotyping diagnostic process. Without clear patent protection, Hagenkord turned to the possibility of trademarking the iKaryos brand name. However, here too were potential barriers: Apple's protection of the "i" prefix in a wide variety of categories, and with the existence of a mark for a Karyogenomics, a firm specializing in computer software for annotating genes, genomes, gene expressions, and gene and genome sequences.

MARKETING

Oncologists who had heard of and used Hagenkord's diagnostics were impressed and had become confirmed customers. An example was Ralph Hruban, director of the Maryland-based pancreatic cancer research center at Johns Hopkins University, who recently sent Hagenkord tissue from a cancer patient. "It's a field that is moving forward in some very exciting ways," Hruban said of Hagenkord's area of expertise. "It helped guide the patient's therapy."

Despite the early successes, market penetration was proceeding at glacial speed, due to the lack of time, money, and expertise. With the expected entry of better-funded, more-established competitors such as Lab Corp and Signature Genomics, Hagenkord knew that the only way for the new company to succeed was to be first to get the product and iKaryos's name known in various oncology and pathology channels. If they had time, Hagenkord, Kash, and Monzon could conduct additional research and publish more scholarly journal articles about the clinical utility of SNP array karyotyping of more types of cancer. Over the last few years, nearly every type of human cancer had been studied in research laboratories using SNP arrays, and the publications around clinical utility had been pouring out—like drinking from a fire hose. As research labs, these investigators were happy to get publications and grants with the data, but they could not offer SNP array karyotyping as a clinical test because their laboratories were not CLIA-certified. However, their publications kept adding credibility and, indirectly, free marketing for iKaryos, since they were the only ones so far who were offering it as a clinical test.

A main marketing question facing iKaryos included how and to whom they should market their diagnostic. It was not like Oncotype Dx, which was targeted only to breast oncologists. iKaryos's technology worked for all cancers. In addition, there were strong arguments on both sides as to whether to focus on oncologists or pathologists as their target audience. The best answer was probably "both", but it is a daunting task for a small company to target *all* pathologists and *all* oncologists for *all* tumors. Some even suggested that they should try direct to consumer marketing and advertise directly to the patients to try to get the patients to encourage the doctors to request testing.

None of the founding team had expertise in branding, marketing or web development, which they knew were crucial in establishing a first to market position. Not only did they need to develop marketing materials and a website, they realized that the adoption of new technologies typically resulted from getting a firm's name in the marketplace through personal selling and by providing detailed information to those attending oncology meetings such as the Association for Molecular Pathologists (AMP), the American Society of Hematology (ASH), the U.S. and Canadian Academy of Pathology (USCAP), the American Society for Clinical Oncology (ASCO), and the American Association for Cancer Research (AACR). Attending these conferences would consume Hagenkord's time, requiring trips to New Orleans, Washington, D.C., and Chicago among other venues, at the rate of about three days every two months.

FINANCING AND VALUATION

Given their bootstrapping approach, their business had been profitable from the beginning. The lab was operational and functional, and it had an inherent value. However, several troubling issues needed to be solved. First, Hagenkord had only one day per week to interpret and sign off on tests being run by Kash. Her time had not been factored into the overall cost structure, and she was limited in the number of tests that she could interpret each week. She felt that if she had better systems and assistance in the lab, she could increase her productivity. As yet, her laboratory was processing about 20 samples per month with only word of mouth advertising. With the new marketing materials and campaign underway, she needed to articulate how the operation was scalable, and where the thresholds were for new FTEs, new capital equipment, and additional pathologists. She wasn't sure how much capital would be needed to grow rapidly. How much money should the team take in an angel round? What should the pre-money valuation be? How would the capitalization table affected by different investment scenarios?

HELP ON THE WAY

With decisions looming large and time constraints demanding that these decisions be made quickly, Hagenkord used her cell phone to return the Creighton PSM program director's call, agreeing to speak to her class. Upon hearing more about Hagenkord's dreams for her company and the key decisions that she was facing, the director suggested that an objective but knowledgeable and free mentor would be ideal. She put Hagenkord in touch with an M.D./Ph.D. who taught in Creighton's PSM program and had been a manager at Genentech and Amgen, ultimately founding and selling his oncology-focused biotechnology firm, Hana Biosciences. She felt that he would be an objective person to assist iKaryos with valuation, equity negotiations and recommendations for IP and legal counsel.

Hagenkord reflected back on why she wanted to start her company in the first place, often getting by with four or five hours' sleep and feeling like "her hair was on fire." "Sometimes it is devastating information that I am giving," she thought. "But one of my philosophies is that knowledge is power against cancer. If it were me, I would want to know if I should quit my job, hug my kids a lot, and live like I am dying. Likewise, I'd want to know if I have a slow-growing cancer and that I shouldn't quit my job and that I can still yell at my kids if I need to," she joked to herself as she headed home to feed her children and put them to bed (Clark, 2009).

APPENDICES

APPENDIX 1: HOME PAGE OF IKARYOSDIAGNOSTICS.COM

Source: www.iKaryosDiagnostics.com

APPENDIX 2A: TRADITIONAL VS. VIRTUAL SNPS ARRAY-BASED KARYOTYPING

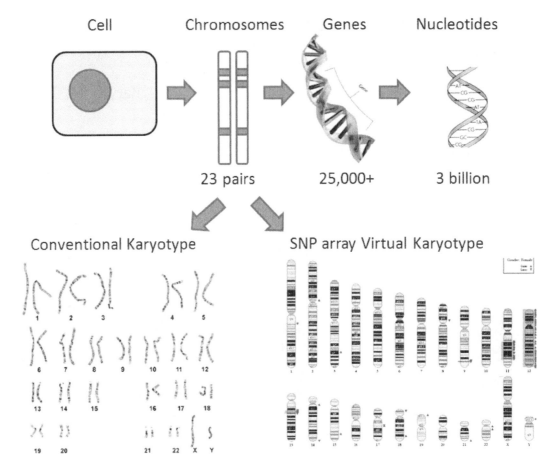

Source: www.iKaryosDiagnostics.com

APPENDIX 2B: TRADITIONAL VS. VIRTUAL SNPS ARRAY-BASED KARYOTYPING

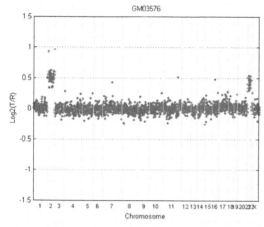

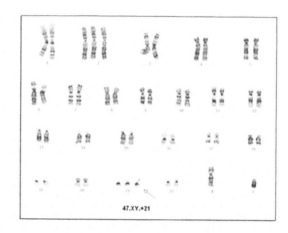

Virtual karyotype, low resolution array

Conventional karyotype

- **Copy number arrays**
  - Assess relative DNA copy number and generate virtual karyotypes.
  - Can be done on formalin fixed paraffin embedded (FFPE) tumors.
- **Conventional karyotypes**
  - Low resolution and are technically laborious
  - Can see balanced translocations, Inversions, and tetraploidy.
  - Requires culture (cannot be performed on FFPE tumors)

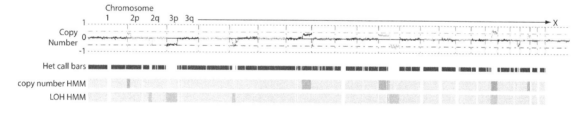

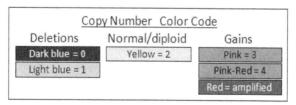

Source: www.iKaryosDiagnostics.com

APPENDIX 3A: VIRTUAL SNP-BASED ARRAY KARYOTYPING

Loss of Heterozygosity

Each allele can be either A or B
- AA (homozygous A)
- BB (homozygous B)
- AB (heterozygous)

Knudson Two Hit Hypothesis of Tumorigenesis

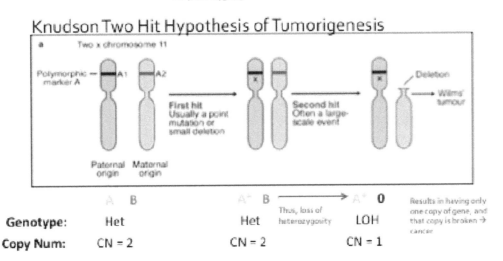

	A B	A* B	A* 0	Results in having only one copy of gene, and that copy is broken → cancer
Genotype:	Het	Het	LOH	
Copy Num:	CN = 2	CN = 2	CN = 1	

Source: www.iKaryosDiagnostics.com

APPENDIX 3B: VIRTUAL SNP-BASED ARRAY KARYOTYPING

Cancer Genome = All of the DNA in a cancer cell

 DNA  Digitalize the DNA

Once the DNA is digitalized, computational tools can be used to manipulate and interrogate the cancer genome.
1. Software to rebuild the cancer genome *in silico*.
2. Genome browsers to 'surf' the cancer genome, much like people use Google to surf the internet.
3. Identify DNA changes that are specific to a patient's cancer cells and tailor treatment accordingly.
4. Powerful research tool for biomarker discovery.

Sources: Left photo permission granted, Dr. Jill Hagenkord; middle and right photos, www.affymetrix.com.

APPENDIX 3C: VIRTUAL SNP-BASED ARRAY KARYOTYPING

 iKaryos Diagnostics | # Acquired UPD of p53 in CLL Creighton MEDICAL LABORATORIES

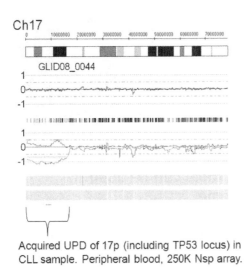

Acquired UPD of 17p (including TP53 locus) in CLL sample. Peripheral blood, 250K Nsp array.

Sequence analysis of key exons in *TP53* revealed a homozygous c.14070G>A (R248Q) mutation, which is one of the most common somatic mutations in *TP53*.

Loss of p53 in chronic lymphocytic leukemia (CLL) is a bad prognostic indicator and *directly impacts patient management*.

This copy neutral LOH was missed by conventional cytogenetics and FISH, and it would have been missed by arrayCGH.

SNP array karyotyping readily detects copy neutral LOH.

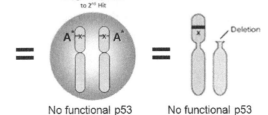

Hagenkord JM, et al. Array-based Karyotyping for Prognostic Assessment in Chronic Lymphocytic Leukemia: Performance Comparison of [16] Affymetrix™ 10K2.0, 250K Nsp, and SNP6.0 Arrays. *J Mol Diagn*, in press.

Source: Reprinted from J Mol Diagn. 2010, 12:184-196 with permission from the American Society for Investigative Pathology and The Association for Molecular Pathology

APPENDIX 3D: VIRTUAL SNP-BASED ARRAY KARYOTYPING

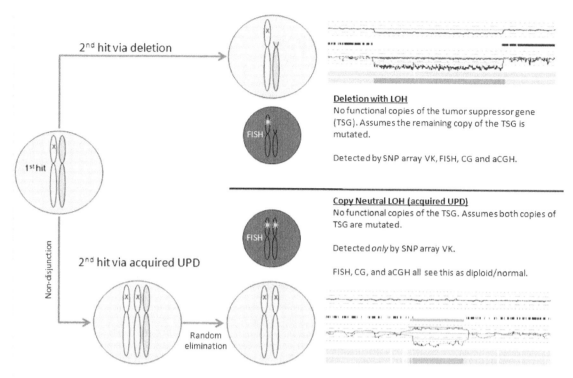

2nd hit via deletion

1st hit

Non-disjunction

2nd hit via acquired UPD

Random elimination

FISH

FISH

Deletion with LOH
No functional copies of the tumor suppressor gene (TSG). Assumes the remaining copy of the TSG is mutated.

Detected by SNP array VK, FISH, CG and aCGH.

Copy Neutral LOH (acquired UPD)
No functional copies of the TSG. Assumes both copies of TSG are mutated.

Detected *only* by SNP array VK.

FISH, CG, and aCGH all see this as diploid/normal.

VK = Virtual Karyotype   FISH = Fluorescent in situ hybridization   CG = Conventional Cytogenetics   aCGH = Array Comparative Genomic Hybridization

Source: www.iKaryosDiagnostics.com

APPENDIX 4: SAMPLE IKARYOS LAB REPORT

Creighton University
Department of Pathology
c/o iKaryos Diagnostics
601 N. 30th St., Suite 2469
Omaha, NE 68131-2197

iKaryos Diagnostics

Jill Hagenkord, M.D.,
Medical Director and CMO
jhagenkord@ikaryos.com
402.280.3963 [T]
402.280.3042 [F]

Patient Name: XXXXXXXXXX
Gender: X
Date of Birth: XX/XX/XXXX
Specimen Type: OCT-embedded
Submitters Name: Dr. XXXXXXXXX
Submitters Institution: XXXXX

iK Accession Number: iK09-GDXXXXX
Date specimen obtained: XX/XX/XXXX
Date specimen received: XX/XX/XXXX
Report date: XX/XX/XXXX
Test: SNP array karyotype
Test Indication: Renal mass, 4.5 cm

Result:
Tumor, right kidney, SNP oligonucleotide microarray karyotype (SP09-XXXX):
    Loss of chromosomes 3p, loss of 14q, and additional cytogenetic abnormalities, see comment.

Comment: Loss of 3p supports the morphologic diagnosis of conventional clear cell renal carcinoma. Loss of 14q in renal clear cell carcinoma has been associated with adverse prognosis. Additional chromosomal changes are present in this tumor, but the clinical relevance of these additional changes, if any, is not known. They are listed in the table in the interpretive component of this report.

Interpretation: Nearly 100% of conventional clear cell renal carcinomas have loss of 3p.[1] They may also have additional cytogenetic abnormalities with prognostic significance, such as loss of 9p or 14q.[2] Loss of 9p is an independent predictor of poor survival in patients with conventional clear cell RCC and should be integrated into prognostic models.[2,3] It has been consistently reported that loss of chromosome 14q is associated with higher grade and stage.[4-6] Furthermore, these have been linked with adverse prognosis and diminished disease-specific survival.[5] However, loss of 14q was not an independent predictor of survival.[3,5]

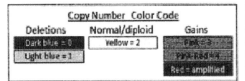

Copy Number Color Code		
Deletions	**Normal/diploid**	**Gains**
Dark blue = 0	Yellow = 2	Pink = 3
Light blue = 1		Pink-Red = 4
		Red = amplified

Whole genome view SNP array karyogram for this sample. Chromosomes are plotted in numeric order from left (chromosome 1) to right (X chromosome). A) Log2ratio, zero = copy number of 2. B) Allele-specific analysis of copy number. C) The copy number Hidden Markov Model (HMM) is color-coded as indicated to the left.

Creighton University
Department of Pathology
c/o iKaryos Diagnostics
601 N. 30th St., Suite 2469
Omaha, NE 68131-2197

Jill Hagenkord, M.D.,
Medical Director and CMO
jhagenkord@ikaryos.com
402.280.3963 [T]
402.280.3042 [F]

Copy Number	Chromosome	StartPos	EndPos	Length [Mb]	StartCytoband	EndCytoband
1	1	825852	119736169	118.910	p36.33	p12
1	3	48603	142238276	142.190	p26.3	q23
3	5	81949	180625439	180.543	p15.33	q35.3
3	6	119769	170791203	170.671	q25.3	q27
1	8	180568	36219874	36.039	p23.3	p12
2*	10	148946	135295604	135.147	p15.3	q26.3
1	11	79477795	117685151	38.207	q14.1	q23.3
3	12	50446	132377151	132.327	p13.33	q24.33
1	14	51160067	106355525	55.195	q22.1	q32.33
1	18	210071	76115293	75.905	p11.32	q23
3	20	17408	62374986	62.358	p13	q13.33
1	23	142664	154353200	154.211	p22.33	q28

Genomic position and cytogenetic break points for CML09-GDXXXXX. Asterisk indicates copy neutral loss of heterozygosity (acquired uniparental disomy).

Summary of Cytogenetic Abnormalities in Renal Epithelial Tumors

Several renal tumors have characteristic cytogenetic abnormalities that can aid in the classification of morphologically challenging cases. The cytogenetic abnormalities can be identified by SNP array virtual karyotyping of fresh or formalin-fixed paraffin embedded tumor samples. [7, 8] Nearly 100% of conventional clear cell renal carcinoma have loss of 3p. They may also have additional cytogenetic abnormalities with prognostic significance, such as loss of 9p or 14q. [2] Papillary renal cell carcinomas classically show trisomies of chromosomes 7 and 17. [9] Chromophobe renal cell carcinomas are hypodiploid with losses of multiple chromosomes, most commonly chromosomes 1, 2, 6, 10, 13, 17 and 21. [9] Oncocytomas are benign renal epithelial tumors with no or few cytogenetic abnormalities. When cytogenetic abnormalities are present in oncocytomas, the most common is complete or partial loss of chromosome 1 (40% of cases), but monosomy 14 (15%), and trisomy 7 (5%) can also be seen. [7]

Methods: DNA was extracted from OCT-embedded tumor following pathologist review of the H&E stained slide. Whole genome comparative genomic hybridization was done using Affymetrix 250K Nsp SNP array which can detect uniparental disomy and copy number changes as small as 500kb when performed on fresh DNA. Genetic changes present in less than 20% of the cells may not be detected by array-based karyotyping. The assay was performed according to the manufacturer's protocol. Analysis was performed using Affymetrix™ GTYPE 2.0.1, CNAGv3.0 [10], and oneClickCGH (InfoQuant LTD, London, UK) software programs. The normal reference DNA used for analysis was chosen by the CNAG software from a library of data files obtained from normal specimens. All controls performed as expected.

References:

1. Kim H, Shen S, Ayala A, et al. Virtual-Karyotyping with SNP microarrays in morphologically challenging renal cell neoplasms: a practical and useful diagnostic modality. *American Journal of Surgical Pathology.* In press 2009.
2. Brunelli M, Eccher A, Gobbo S, et al. Loss of chromosome 9p is an independent prognostic factor in patients with clear cell renal cell carcinoma. *Modern Pathology.* 2008;21(1):1-6.
3. Klatte T, Rao PN, de Martino M, et al. Cytogenetic Profile Predicts Prognosis of Patients With Clear Cell Renal Cell Carcinoma. *Journal of clinical oncology.* Jan 5 2009.

Creighton University
Department of Pathology
c/o iKaryos Diagnostics
601 N. 30th St., Suite 2469
Omaha, NE 68131-2197

Jill Hagenkord, M.D.,
Medical Director and CMO
jhagenkord@ikaryos.com
402.280.3963 [T]
402.280.3042 [F]

4. Alimov A, Sundelin B, Wang N, Larsson C, Bergerheim U. Loss of 14q31-q32. 2 in renal cell carcinoma is associated with high malignancy grade and poor survival. *Int J Oncol.* 2004;25(1):179-185.
5. Mitsumori K, Kittleson JM, Itoh N, et al. Chromosome 14q LOH in localized clear cell renal cell carcinoma. *The Journal of pathology.* Sep 2002;198(1):110-114.
6. Herbers J, Schullerus D, Muller H, et al. Significance of chromosome arm 14q loss in nonpapillary renal cell carcinomas. *Genes, Chromosomes & Cancer.* May 1997;19(1):29-35.
7. Kim H-J, Monzon FA. Kidney: Renal Oncocytoma. *Atlas of Genetics and Cytogenetics in Oncology and Haematology.* April 2008.
8. Hagenkord JM, Parwani AV, Lyons-Weiler MA, et al. Virtual karyotyping with SNP microarrays reduces uncertainty in the diagnosis of renal epithelial tumors. *Diagn Pathol.* 2008;3:44.
9. van den Berg E, Storkelvan S. Kidney: Renal Cell Carcinoma: Atlas Genet Cytogenet Oncol Haematol; 2003
10. Yamamoto G, Nannya Y, Kato M, et al. Highly sensitive method for genomewide detection of allelic composition in nonpaired, primary tumor specimens by use of affymetrix single-nucleotide-polymorphism genotyping microarrays. *Am J Hum Genet.* Jul 2007;81(1):114-126.

REFERENCES

American Cancer Society. Statistics for 2009. http://www.cancer.org/docroot/STT/STT_0.asp

Clark, C. 2009. Creighton pathologist launches next-generation cancer diagnostics company. www.irm.org.

Frost and Sullivan. 2002. U.S. SNP *Detection Technology Markets.*

Frost and Sullivan, 2008. U.S. Cancer Diagnostic Markets: Clinical Diagnostics Healthcare Group.

Hagenkord JM, Chang CC. The rewards and challenges of array-based karyotyping for clinical oncology applications. *Leukemia.* 2009 May 23(5):829-33. No abstract available. PMID: 19436309

Hagenkord JM, Parwani AV, Lyons-Weiler MA, Alvarez K, Amato R, Gatalica Z, Gonzalez-Berjon JM, Peterson L, Dhir R, Monzon FA. Virtual karyotyping with SNP microarrays reduces uncertainty in the diagnosis of renal epithelial tumors. *Diagn. Pathol.* 2008 Nov 6;3:44. PubMed PMID: 18990225; PubMed Central PMCID: 2588560

Kim HJ, Shen SS, Ayala AG, Ro JY, Truong LD, Alvarez K, Bridge JA, Gatalica Z, Hagenkord JM, Gonzalez-Berjon JM, Monzon FA. Virtual-karyotyping with SNP microarrays in morphologically challenging renal cell neoplasms: a practical and useful diagnostic modality. *Am J Surg Pathol.* 2009 Sept 33(9):1276-86. PubMed PMID: 19461508

Lyons-Weiler M, Hagenkord J, Sciulli C, Dhir R, Monzon FA. Optimization of the Affymetrix GeneChip Mapping 10K 2.0 Assay for routine clinical use on formalin-fixed paraffin-embedded tissues. *Diagn Mol Pathol.* 2008 Mar;17(1):3-13. PubMed PMID: 18303412

Lusky, K. April, 2008. SNP to it – moving arrays into clinical use. *CAP today.*

Monzon FA, Hagenkord JM, Lyons-Weiler MA, Balani JP, Parwani AV, Sciulli CM, Li J, Chandran UR, Bastacky SI, Dhir R. Whole genome SNP arrays as a potential diagnostic tool for the detection of characteristic chromosomal aberrations in renal epithelial tumors. *Mod Pathol.* 2008 May;21(5):599-608.

Epub 2008 Feb 8. PubMed PMID: 18246049

Ruggles, R. 2009. Creighton doc offers cancer breakthrough. *Omaha World Herald*, Metro section. May 2009.

Biocon: From Local to Global

ASHISH HAJELA, SHAD SHAHID, & MOHAMMAD AKBAR
Indian Institute of Management, Lucknow, India

- **Key issue/decisions**: The key issue is Biocon's strategy to emerge as a global biotechnology company. In its effort to find a global footprint, the case analyzes Biocon's balancing the strategic decisions between generics business versus innovative drug development, short term cash flows versus long term growth, expansion versus becoming a takeover target, and retaining a leadership position in India versus entering the overseas market.

- **Company:** Biocon started as a joint venture in 1978 and became one of India's largest, fully integrated biopharmaceutical enterprises, focused in healthcare.

- **Technology:** Biocon patented its fermentation technology called Plafractor, used for producing enzymes and pharmaceuticals for Indian and foreign markets. Biocon has sought to use its capability as India's biggest producer of insulin and other cell based formulations for establishing alliances with multinational companies, as well as entering overseas markets.

- **Geography**: India, global
- **Keywords**: Biotechnology, globalization, strategy, alliances

---

* *This case was prepared as a basis for class discussion rather than to illustrate either effective or ineffective handling of an administrative situation.*

INTRODUCTION

In the autumn of 2005, Kiran Mazumdar–Shaw, founder and CEO of Biocon, a Bangalore, India-based biopharmaceutical company, was scheduled to chair a board meeting in a few hours which would decide the future course of the company. Mazumdar-Shaw had dreamt of making Biocon a major global biopharmaceutical company. The competitive pressures and environmental threats, however, had raised significant risks in managing growth and profitability. Strategic choices had to be made about which product lines to pursue and where to expand globally. As she thought about the board meeting, she reflected on the challenges of starting Biocon and nurturing it to this day—and wondered whether these same strengths would be the basis for future success. As Mazumdar–Shaw noted:

> *"When I set up Biocon a little over 25 years ago I guess I was just focused on building a good company and I guess over the years it has evolved into a great potential for me and the way I look at it today is with a very different sense of purpose. I certainly want to build a globally recognized biotech company and that's the path I am on today. I think Biocon to me stands for this particular potential, and doing something like this out of India means a lot more to me as well" (Hahn, 2004).*

BIOCON COMPANY BACKGROUND

Kiran Mazumdar-Shaw started Biocon Ltd., India's largest biotech company. She had not set out to become a biopharmaceutical entrepreneur. She aspired to be a brewer, and graduated from Melbourne University in 1974 with degrees in malting and brewing. Her professor mentioned her talent and potential to Les Auchinloss, the owner of Biocon Biochemicals in Ireland, who sourced raw materials from India to manufacture enzymes in Ireland. After returning from Australia, Kiran spent a frustrating three years trying to find a job as master brewer in the male dominated beer industry. Kiran was about to leave for Scotland after finding a job with a malting unit there when Auchinloss contacted her and convinced her to start producing enzymes for his company. Kiran did not know much about manufacturing enzymes (Appendix A) and took a six month training course in enzyme manufacturing in Ireland (Sekhar, 2000). Mazumdar Shaw remarked:

> *"When this offer came my way from my Irish partner, in fact I did say to him, are you quite sure you want me as your partner because a) I am a woman and that too in a country like India - I'm not even a business woman so I don't have any formal business training, and most importantly I'm not even a rich woman. So I guess I tried to convince him that maybe I wasn't really the right choice, but I guess he had more confidence in me than I had in myself at that time" (Hahn, 2004)*

In November 1978, Kiran started Biocon India in a shed in Bangalore with an investment of Indian Rupees (Rs) 150,000 (US$3,440.00) with equity holding of Kiran (70%) and Biocon, Ireland (30%). She started producing two enzymes used in brewing: papain and isinglass. It took her weeks to rent space because landlords thought she wouldn't be good for the rent. She couldn't find anyone willing to work for her, so she begged a female friend to fill in as a secretary. Her first accountant, a man, left as soon as another job opened. By far the worst treatment came from raw-materials vendors who insisted she hire a male manager if she wanted their wares (Egan, 2004). Banks refused to lend to her, asking for a male guarantor, and were wary of the biotech industry. After being turned down several times, her big breakthrough came when she met a local banker at a national bank who became interested in Biocon. The bank eventually offered Ms. Mazumdar Shaw a credit line of Rs 500,000 (US$11,470) (Yee, 2008).

Figure 1: Kiran Mazumdar–Shaw, founder and chairperson, Biocon
Source: Wikimedia Commons; Jackbandjack.

In 1988, Biocon decided to scale up technology and create more sophisticated products which required additional capital. Mazumdar encountered another roadblock when banks turned away her requests for loans. In 1989, she met the chairman of what is today ICICI, India's largest private sector bank, who had just established a venture fund. When he heard Mazumdar Shaw's plan he offered Rs 750,000 (US$17,200) for a 20 percent stake in Biocon's new subsidiary—an injection of capital that prepared the company for its eventual move into biopharmaceuticals (Yee,2008). In 1989, Unilever acquired Biocon Ireland and merged it with Quest International (Sekhar, 2000).

In 1990, Biocon scaled up its in-house research program, based on a proprietary solid substrate fermentation technology, from pilot to plant level. In 1993, Biocon's commitment to build world class quality manufacturing resulted in receiving an ISO 9001 certification from RWTUV, Germany (originally a so-called "TÜV" technical inspection organization-a body in Germany largely responsible for statutory inspections). In 1994, Biocon established Syngene International Pvt. Ltd. as a Custom Research Company (CRC) to address the growing need for outsourced R&D in the pharmaceutical sector. In 1996, the commercial success of Biocon's proprietary fermentation plant led to a 3-fold expansion. Biocon leveraged its technology platform to enter biopharmaceuticals and begin production of statins (HMG-CoA reductase inhibitors are a class of drugs that lower cholesterol levels in people) (Biocon, 2009). By then Biocon had started moving toward more advanced work, and Syngene contracted work from multinational pharmaceutical companies such as Bristol Myers Squibb, Novartis, and Merck (Yee, 2008).

Unilever tried and failed to buyout Mazumdar-Shaw's stake. In 1997, the food giant sold its specialty chemicals division, which included Biocon, to Imperial Chemical Industries (ICI). ICI had no interest in drug research and a year later was willing to sell its ownership of Biocon for $2 million, on the condition that it be supplied with enzymes for two years. Mazumdar-Shaw came up with the money by persuading her husband, John Shaw, a former textile executive, to sell his London home and join Biocon. In 1998, the newly liberated business required additional capital and this time Mazumdar-Shaw had enough of a track record and name recognition to raise the cash less painfully (Egan, 2004).

In 2000, Biocon decided to expand and installed the PlaFractor™, a novel fermentation technology , which was completely automated to increase manufacturing quality and reliability. Financing for the expansion came from the ICICI Venture Fund company (IVC), which had decided to make an equity investment of Rs 175million (US$4 million) in Biocon India for a minority stake in the company (Businessline, 2000). Biocon also established Clinigene, India's first clinical research organisation to pursue clinical research and development. In 2001, Biocon became the first Indian company to be approved by US FDA for the manufacture of lovastatin, a cholesterol-lowering drug (Biocon, 2009).

In 2003, Biocon also entered into the biotechnology market with recombinant human insulin. Human insulin, which replaced animal insulin for the treatment of diabetes since Genentech's development of Humulin in 1982, had a global market of over US$ 1 billion per annum in 2003. To capture the growth in the Indian insulin market, Biocon India established a 50:50 joint venture with Shantha Biotechnics (Emerging Markets Economy, 2003). Biocon's aspiration was to become a major player in the insulin market, however, there were many challenges for product development and delivery technologies. For example, Eli Lilly worked with Alkermes to develop inhalable formulations; Novo Nordisk developed the pulmonary version of insulin in collaboration with Aradigm Corporation; and Pfizer developed its inhalable Exubera® and partnered with Aventis to source insulin crystals.

Noting these competitive trends, Biocon partnered with the Hannah Research Institute of Scotland, and its venture, FFAST Solutions for oral insulin. Biocon's subsidiary, Clinigene was responsible for conducting the human clinical trials (Madhumathi, 2003). Further in 2004, Biocon partnered with Nobex to develop oral insulin (PR Newswire, 2004) and became the second player after Wockhardt in the oral insulin segment. Its Insugen® was the world's first insulin made from pichia (yeast) (Businessline, 2004). Further, Biocon signed a product licensing agreement with Bentley Pharmaceuticals, Inc. for its intranasal spray formulation of insulin. The license agreement covers 85 countries, granting Biocon exclusive as well as co-exclusive rights to develop and market throughout Asia, Africa, and the Middle East.

As in the case of Nobex and Bentley, strategic alliances had developed into a core competency for Biocon. In 2003, Biocon established Biocon Biopharmaceuticals Pvt. Ltd. (BBPL), a joint venture with the Cuban Institute CIMAB to develop and market a range of MAbs and Cancer Vaccines. BBPL will benefit from CIMAB expertise in developing and manufacturing immunotherapy products (which today is largely imported into India) and Biocon's R&D and state-of-the-art manufacturing capabilities. In 2004, Biocon and Vaccinex, Inc. embarked on a broad, strategic partnership to discover and co-develop fully humanized antibodies focused on cancer, inflammation and autoimmune diseases. The collaboration combines Vaccinex's unique capabilities to discover fully human monoclonal antibodies using its proprietary antibody discovery technology and Biocon's proven expertise in clinical research and biologics manufacturing. In 2005, Biocon signed a Memorandum of Understanding (MoU) with the Karolinska Institute, Sweden to collaborate in research and research-education initiatives.

By 2005, Biocon had leveraged its technology platform, manufacturing facilities and skilled work force to come out with a range of products for diverse uses. Biocon's product range comprised APIs (active pharmaceutical ingredients) for contract manufacturing, biologicals and multiple dosage forms. Apart from products, Biocon also offered various services in custom and clinical research. Syngene, the custom research organization, and Clinigene, the clinical research company, worked with pharmaceutical and biotechnology companies. Such partnerships were expected to benefit the discovery portfolio (Biocon, 2009) and Biocon was already developing advanced biopharmaceuticals such as:

- **IN-105:** an insulin molecule which was orally deliverable for treating diabetes (Biocon, 2009).
- **Anti-CD6 (T1h):** a monoclonal antibody which was targeted to nullify the impact of antigen named CD6, found on the surface of T cells found in the immune system. T1h was expected to cause reduction in the inflammatory conditions in the joints and to make life easier for the patient (Biocon, 2009).
- **BVX-20:** a monoclonal antibody for the treatment of Non-Hodgkin's Lymphoma (NHL); cancer in the lymphoid organs. The antibody was expected first to bind CD20-a protein present on B-cells and then kill B-cells using body's immune system. BVX-20 was being developed in partnership with Vaccinex (Biocon, 2009).

BIOCON'S PEOPLE AND CULTURE

True to her entrepreneurial roots, Mazumdar-Shaw approached business with a sense of purpose and spirit of challenge. She believed in working hard and working differently to achieve success. For her, wealth was not about material wealth but it was about intellectual capital (Biocon, 2009) and people as she noted:

"I have a great team who has helped me build Biocon, I was very fortunate to be able to share my vision with a group of people who really were as excited about challenges as I was. And I think it is this team of people who have actually helped me build Biocon and I think success in any enterprise is about a team effort, and I think that's what it has been" (Hahn, 2004).

The desire to make Biocon a great place to work didn't mean the company escaped going through the growth pains most start-ups suffer. In the early 1980s, labour unrest prevented the company from servicing its loan from the Karnataka State Finance Corporation (KSFC) and tested her resolve. Her company was declared an offender and the government-firm released advertisements in newspapers stating that its assets would be auctioned. Mazumdar-Shaw was not intimidated, convinced KSFC to wait, and sacked the troublemakers.

Biocon continued to believe that human resources was of central strategic significance to its business. It strove to hire, nurture and retain the best brains in the biotechnology sector. Biocon believes that each member of it 1,800 work force is unique and endeavours to provide an enabling environment, learning opportunities, challenging career prospects, recognition and rewards for meaningful contribution within the overall aim of promoting organizational, professional and personal growth (Biocon, 2009) (Appendix B for management team and board of directors). As she noted:

"I think in terms of corporate philosophy I've always believed that you've got to treat people in a very very egalitarian manner in the sense I like to treat people on a one-to-one basis. And I like people to take on a lot of responsibilities because I think with a sense of responsibility also comes a sense of purpose. To me that's a very important part of the corporate philosophy: we have a very flat structure, I encourage a lot of informality in our workplace. Everyone at work calls me Kiran, which is a very different kind of a culture to have especially in a country like India where people are very reverential about the heads of companies" (Hahn, 2004).

FINANCING GROWTH

To finance Biocon's aggressive growth in products and services, the company completed a share offering in 2004 which was oversubscribed 32 times and made Biocon the first company in India worth more than US$1 billion. With a 65 percent ownership of Biocon, the 52-year-old "biotech baroness" became India's richest woman. While the recognition of Biocon's success was exciting, Mazumdar–Shaw knew that further gains would only come from navigating the challenges ahead (Appendix C for company financials).

BIOPHARMACEUTICAL INDUSTRY DYNAMICS

India was emerging fast as a significant biotechnology centre in Asia Pacific and was achieving critical mass in order to become a global player. The biotechnology sector was growing sizeably through emergence of new companies, more products developments, more patent filings, and new product launches. India was becoming a popular location for vaccines, bioinformatics, genetically modified (GM) products, clinical trials and clinical research organizations (Asia Pacific Biotech News, 2004). There were around 800 companies—mostly generic manufacturers, support and service providers—which employed about 20,000 people and were involved in some way with biotechnology. The small and medium enterprises were critical to industrial growth.

The top 20 companies earned over US$ 290 million in revenues or 59 percent of industry revenues (Asia Pacific Biotech News, 2004). In 2003, market leaders in terms of revenue were Biocon (US$ 54.3

million), Panacea Biotec (US$ 36.1 million) and Wipro Health Science (US$ 21 million). Wockhardt, Haffkine Bio-pharmaceutical, Eli Lilly & Co., and Nicholas Piramal were the other big companies (Asia Pacific Biotech News, 2004). In addition, there were several well financed companies establishing biopharmaceutical companies in India including:

- Reliance Life Sciences, owned by Mumbai-based Reliance Industries one of India's largest industrial groups. It planned to invest US$ 25 million in cell biology and to create one of the largest cord-blood repositories in the world (Business India Intelligence, 2002).
- Shantha Biotech, based in Hyderabad, was known for development of a Hepatitis B vaccine. Shantha planned to develop monoclonal antibodies, human insulin, and marine biotechnology. Shantha was undertaking collaborative research with seven Indian organisations and one in Korea (Business India Intelligence, 2002).
- Bharat Biotech, based in Hyderabad, was the second company to develop an indigenous Hepatitis B vaccine. It worked with the Centers for Disease Control and Prevention, National Institutes of Health, and Stanford University in the US, for the development and production of a rotavirus vaccine (Business India Intelligence, 2002).
- Avesthagena Technologies, based in Bangalore, focused on expressing plant genes. It had a partnership with Genetic ID of the US, making it the first Indian company to offer globally approved testing and certification for genetically modified organisms (Business India Intelligence, 2002).
- Bangalore Genei, based in Bangalore, provided products and services for the biotech community in India. It was one of around 25 companies producing restriction enzymes in the world (Business India Intelligence, 2002).

Another key industry driver was India's implementation of the Trade-Related Intellectual Property Rights (TRIPs) agreement for biopharmaceutical products. TRIPs required India to provide IP protection for innovators regardless of the country of origin. As a result, the establishment of an international depository in India will facilitate the patenting of microorganisms that are new and are created using inventive steps (Asia Pacific Biotech News, 2004).

Recognizing the potential for the biotechnology industry to be transformational for India's economy, policy formulation for developing biotechnology capabilities was initiated in 1980 with the implementation of India's Sixth Five year Plan (1980-1985), which proposed efforts in immunology, genetics, and communicable diseases. The Government established the National Biotechnology Board (NBTB) in 1982 as an apex body to identify priority areas and to evolve a long-term plan for the development of the biotechnology industry. Later in 1986, NBTB graduated to a full-fledged government department called the Department of Biotechnology (DBT) responsible for regulatory approvals for investment and development in the sector. The Working Group on Science and Technology for the Tenth Five Year Plan (2002-2007) proposed an outlay of US $460 million for the 2002-2007 period—or a sharp in increase of 234% from the prior plan—focusing on human genome sequencing, proteomics, structural biology, and bioinformatics (Asia Pacific Biotech News, 2004).

In addition, many state governments were developing infrastructure projects to enable biotechnology industry development. For example, Tamil Nadu planned Ticel, a collaborative effort between TIDCO and Cornell Business & Technology Parks to train bio-processing engineers. Andhra Pradesh was developing Shahpoorji Pallonji Biotech Park, a joint venture between the state government and property developer Shahpoorji Pallonji at Turkapalli, near Hyderabad. Karnataka planned Dharwad Marine Biotech Park with R&D centers, biotech companies, and incubation facilities. Parks were also coming up in Maharashtra, Madhya Pradesh, Orissa and Uttar Pradesh; and centers of excellence in biotechnology/bio-informatics were being established at Allahabad University, IITM, Gwalior and in regional engineering colleges (Business India Intelligence, 2002).

EMERGING SCENARIOS AND DETERMINING STRATEGIC DIRECTION

The strategic landscape facing Biocon in 2005 was rapidly evolving and Mazumdar-Shaw was keenly aware that their past success did not ensure future competitiveness and growth. On the one hand, multinational pharmaceutical companies in the West continued to increasingly seek growth in Asia Pacific countries due to the high cost of drug discovery and development, competitive intensity, and pricing pressures. This development put countries like India generally, as well as Biocon specifically, at an advantage for strategic alliances (Asiabiotech, 2007).

On the other hand, Biocon did not have the luxury of time. Biocon's core business of statins had come under price pressure in Europe, with 35-40 percent price reduction in the segment driven by aggressive pricing from Chinese generics. Beyond the immediate threat, the generic market was full of many players and profit margins were low. To be competitive in this segment Biocon would need to ramp up significantly to reduce unit costs.

Instead, investors were putting a premium on companies that developed novel molecules (Asiabiotech, 2007). Thus, Biocon could also focus on launching branded formulations. Insugen®, an insulin drug for diabetics, was already in the pipeline. The diversification strategy from being a bulk drug supplier to generic manufacturer and then to a branded player had its own challenges. Advertising and marketing expenses were huge and upfront, while returns took years to realize. Also, there was price competition in this segment as well as multinationals, such as Novo Nordisk and Eli Lilly's slashing prices by one-third in the insulin market (Babu,2005). In India, Ranbaxy and Dr. Reddy's were also moving from generic manufacturing to innovative drug discovery and development (Mazumdar, 2004).

Another option of moving overseas and obtaining novel technology was through acquisitions. However, Biocon had Rs 4.3 billion (US$100 million) in reserves which was orders of magnitude smaller than prices being paid for biopharmaceutical companies on the world market. Thus, a merger and acquisition strategy would require significant sources of new capital. As such, another variation would be to sell Biocon to a major pharmaceutical seeking to establish or grow operations in India (Mazumdar, 2004)

As she got up to leave for the meeting of board of directors, Mazumdar-Shaw found herself at a crossroads. She pondered which option to choose as she reflected on her role and the challenges ahead.

APPENDICES

APPENDIX A: ENZYME MANUFACTURING PROCESS

The starting point for enzyme production is a vial of a selected strain of microorganisms. They will be nurtured and fed until they multiply many thousand times. Then the desired end-product is recovered from the fermentation broth and sold as a standardised product.

A single bacteria or fungus is able to produce only a very small portion of the enzyme, but billions microorganisms can produce large amounts of enzyme. The process of multiplying microorganisms by millions is called fermentation. Fermentation to produce industrial enzymes starts with a vial of dried or frozen microorganisms called a production strain.

One very important aspect of fermentation is sterilisation. In order to cultivate a particular production strain, it is first necessary to eliminate all the native microorganisms present in the raw materials and equipment. If proper sterilisation is not done, other wild organisms will quickly outnumber the production strain and no production will occur.

The production strain is first cultivated in a small flask containing nutrients. The flask is placed in an incubator, which provides the optimal temperature for the microorganism cells to germinate. Once the flask is ready, the cells are transferred to a seed fermenter, which is a large tank containing previously sterilised raw materials and water known as the medium. Seed fermentation allows the cells to reproduce and adapt to the environment and nutrients that will be encountered later on.

After the seed fermentation, the cells are transferred to a larger tank, the main fermenter, where fermentation time, temperature, pH and air are controlled to optimise growth. When this fermentation is complete, the mixture of cells, nutrients and enzymes, called the broth, is ready for filtration and purification.

Filtration and purification termed as downstream processing is done after enzyme fermentation. The enzymes are extracted from the fermentation broth by various chemical treatments to ensure efficient extraction, followed by removal of the broth using either centrifugation or filtration. Followed by a series of other filtration processes, the enzymes are finally separated from the water using an evaporation process. After this the enzymes are formulated and standardised in form of powder, liquid or granules.

Source: http://www.mapsenzymes.com/Making_of_Enzymes.asp

APPENDIX B: BOARD OF DIRECTORS & KEY MANAGEMENT

BOARD OF DIRECTORS

- **Dr. Neville Bain:** Chairman, Institute of Directors, UK
- **Prof. Charles L. Cooney:** Professor, Chemical & Biochemical Engineering, MIT, USA
- **Dr. Bala S. Manian:** Chairman and Founder, Reametrix Inc.
- **Mr. Suresh Talwar:** Partner, Talwar Thakore & Associates
- **Ms. Kiran Mazumdar-Shaw:** Chairman & Managing Director, Biocon
- **Mr. John Shaw:** Vice Chairman, Biocon
- **Prof. Ravi Mazumdar:** University Research Chair Professor, University Of Waterloo, Canada
- **Prof. Catherine Rosenberg:** Alternate Director, Biocon

KEY MANAGEMENT TEAM

- **Ms. Kiran Mazumdar-Shaw:** CMD, Founder Entrepreneur, 1978
- **Mr. John Shaw:** Vice Chairman with Biocon since 1999
- **Dr. Arun Chandavarkar:** Chief Operating Officer, Biocon, with Biocon since 1990
- **Dr. Goutam Das:** Chief Operating Officer, Syngene with Biocon since 1994
- **Dr. A.S. Arvind:** Chief Operating Officer, Clinigene, with Biocon since 2000
- **Mr. Murali Krishnan:** President, Finance with Biocon since 1981
- **Mr. M.B. Chinappa:** Chief Financial Officer, Syngene with Biocon since 1999
- **Dr. Harish Iyer:** General Manager, R&D with Biocon since 2001
- **Mr. Rakesh Bamzai:** President, Marketing with Biocon since 1995
- **Mr. Ravi C.Dasgupta:** Group Head, HR with Biocon since 2007

Source: www.biocon.com

APPENDIX C: BIOCON FINANCIALS (RUPEES)

Income Statement	2001	2002	2003	2004	2005
Income					
Sales Turnover	122.31	160.52	276.46	530.91	688.28
Excise Duty	0	0	20.82	29.03	37.84
Net Sales	122.31	160.52	255.64	501.88	650.44
Other Income	1.28	4.95	1.78	10.18	18.16
Stock Adjustments	0	1.05	10.12	14.2	3.04
Total Income	123.59	166.52	267.54	526.26	671.64
Expenditure					
Raw Materials	0	83.69	131.79	269.57	351.27
Power & Fuel Cost	0	9.11	12.8	17.11	24.6
Employee Cost	12.62	19.65	27.31	35.27	42.04
Other Manufacturing Expenses	69.89	2.03	7.87	5.89	7.32
Selling and Admin Expenses	0	10.05	15.32	25.95	29.52
Miscellaneous Expenses	8.59	2.15	7.82	8.69	8.91
Preoperative Exp Capitalised	0	0	-0.17	-0.71	0
Total Expenses	91.1	126.68	202.74	361.77	463.66
Operating Profit	31.21	34.89	63.02	154.31	189.82
PBDIT	32.49	39.84	64.8	164.49	207.98
Interest	4.56	4.67	5.06	3.17	2.69
PBDT	27.93	35.17	59.74	161.32	205.29
Depreciation	6.2	7.78	12.02	13.85	18.09
Other Written Off	0	0	0	0	0
Profit Before Tax	21.73	27.39	47.72	147.47	187.2
Extra-ordinary items	0	0	0	0	1.32
PBT (Post Extra-ord Items)	21.73	27.39	47.72	147.47	188.52
Tax	7.21	7.09	11.85	22.81	14.12
Reported Net Profit	14.52	20.31	35.87	124.67	174.39

Source: Company reports

BIOCON FINANCIALS CONT.

Balance Sheet	2001	2002	2003	2004	2005
Sources Of Funds					
Total Share Capital	1.5	1.82	1.84	50	50
Equity Share Capital	1.5	1.82	1.84	50	50
Share Application Money	0	0	0	0	0
Preference Share Capital	0	0	0	0	0.00
Reserves	54.51	83.7	122.96	490.04	644.51
Revaluation Reserves	2	2.18	1.91	1.59	1.43
Networth	58.01	87.7	126.71	541.63	695.94
Secured Loans	34.3	61.47	68.57	64.69	76.34
Unsecured Loans	3.89	5	0	0	0
Total Debt	38.19	66.47	68.57	64.69	76.34
Total Liabilities	96.2	154.17	195.28	606.32	772.28
Application Of Funds					
Gross Block	78.67	125.64	155.55	191.23	270.2
Less: Accum. Depreciation	16.41	21.37	33.56	47.12	65.37
Net Block	62.26	104.27	121.99	144.11	204.83
Capital Work in Progress	2.52	3.71	7.98	54.31	310
Investments	0.06	8.48	8.48	8.93	223.73
Inventories	21.76	23.38	46.7	83.95	71.29
Sundry Debtors	39.17	62.35	73.75	115.96	172.88
Cash and Bank Balance	0.01	0.03	0.02	1.51	1.99
Total Current Assets	60.94	85.76	120.47	201.42	246.16
Loans and Advances	5.99	10	15.68	28.73	38.1
Fixed Deposits	0	0.06	1	316	1.43
Total CA, Loans & Advances	66.93	95.82	137.15	546.15	285.69
Deffered Credit	0	0	0	0	0
Current Liabilities	35.57	52.48	77.76	132.91	221.7
Provisions	0	5.65	2.57	14.28	30.29
Total CL & Provisions	35.57	58.13	80.33	147.19	251.99
Net Current Assets	31.36	37.69	56.82	398.96	33.7
Miscellaneous Expenses	0	0	0	0	0.00
Total Assets	96.2	154.15	195.27	606.31	772.26
Contingent Liabilities	0	12.56	34.77	131.91	125.88
Book Value (Rs)	186.7	469.44	679.11	54	69.45

Source: Company reports

APPENDIX D: TOP 15 GLOBAL BIOPHARMACEUTICAL MERGERS, 2005.

Biopharmaceutical Mergers & Acquisitions, Top 15 in 2005	US$ Millions
Chiron refuses $4.5bn Novartis offer, later accepts $5.1bn	$5,424.0
Pfizer buys Vicuron for $1.9bn in cash	$1,900.0
Shire buys Transkaryotic Therapies for $1.57bn	$1,571.6
Genzyme acquires Bone Care International for $600mm	$600.0
Genencor sold to Danisco for $592mm in cash	$591.5
Crucell plans to acquire Swiss vaccines company Berna Biotech	$451.4
Invitrogen buys Dynal Biotech for $386mm	$386.4
Sigma-Aldrich buys JRH Biosciences for $377.7mm in cash	$377.7
Pfizer to acquire Idun Pharmaceuticals in cash transaction	$298.0
Valeant buys Xcel for $280mm	$280.0
Applied Biosystems buys Ambion unit for $273mm in cash	$273.0
Takeda acquires Syrrx	$270.0
Matrix Laboratories acquires Docpharma	$252.1
J&J's Ortho-McNeil to buy Peninsula for $245mm	$245.0
J&J acquires TransForm Pharmaceuticals for $230mm in cash	$230.0

Source: Levin & Associates, 2009

REFERENCES

Asiabiotech. 2007. Company report, 11(14): 1011-1015.

Asia Pacific Biotech News. 2004. Overview of the Indian Biotech Sector, 8(17): 927-939.

Babu, V. 2005. Stress Test; Biocon's first quarter profits are down 20 per cent and the stock is trading below its list price. Can India's most hi-profile biotech company get hot again? *Business Today*, Aug 14, p62.

Biocon. 2009. Company website. http://www.biocon.com, accessed on September 17, 2009.

Businessline. 1999. India: Biocon looking at private placement, December 26:p1.

Businessline. 2000. ICICI Venture to invest in Biocon India group, January 25:p1.

Businessline. 2004. Biocon launches bio insulin, November 11: p1.

Business India Intelligence 2002. Biotechnology: The next big thing?, August pp 9-10.

Egan, M.E. 2004. Big Shot in Bangalore. *Forbes*, 174(8): 88.

Emerging Markets Economy. 2003. Biocon India & Shantha Biotechnics Float Human Insulin Joint Venture. Apr, 8:1.

Hahn, L. 2004. Chairman of Biocon, Kiran Mazumdar-Shaw Talk Asia Interview Transcript. *CNN*, http://www.cnn.com/2004/WORLD/asiapcf/09/13/talkasia.mazumdar-shaw.script/index.html, accessed on January 30, 2010.

Madhumathi, D.S. 2003. Pharma majors in race for painless insulin, *Businessline*, Oct 15:1

Mazumdar, S. 2004. First Lady. *Newsweek*,144(16): E40.

PR Newswire. 2004. NOBEX, Biocon to Develop Oral Insulin Product, New York, Oct 20: 1.

Sekhar, A. 2000. Brewing up a business, *Asian Business*, 36(2): 40.

Yee, A. 2008. Persistence amid prejudice, *FT.com*. London: Oct 14.

Adnexus: Strategic and Resource Considerations When Developing Novel Biotechnology Medicines

SUSAN SIELOFF, TUCKER MARION, JOHN FRIAR, AND RAYMOND KINNUNEN
College of Business Administration, Northeastern University

- **Key issues:** Adnexus (www.Adnexustx.com) was a small biotechnology company that was acquired by the Bristol-Myers Squibb Company in late 2007. The company uses its proprietary protein engineering technology, PROfusion™, to design and identify drug candidates from a novel class of proteins called Adnectins™. The company's lead product, CT-322, was in clinical development for the treatment of cancer. CT-322 has made considerable progress in clinical trials while the early-stage Adnectin pipeline has also been growing rapidly. To help prepare for a significant increase in new Adnectin protein therapies entering the clinical phase in a resource constrained environment, Bristol-Myers Squibb requested that Adnexus create an innovative process for more efficient product development. This is certainly not an easy task, as change is often difficult in large, established organizations. Adnexus was being asked to lead this change while simultaneously pioneering two novel technologies and managing a pipeline that was doubling in size each year.

- **Companies/institutions:** Adnexus Therapeutics Inc., Bristol-Myers Squibb Company

- **Technology:** targeted biologics, proteins

- **Stage of development:** early discovery through clinical proof of concept

- **Indication/therapeutic area:** oncology

- **Geography:** US

- **Keywords:** biotechnology, resources, FDA approval process, targeted biologics

---

* *This case was prepared as a basis for class discussion rather than to illustrate either effective or ineffective handling of an administrative situation.*

INTRODUCTION

In January of 2008, Adnexus was a company in transition. Just a few months ago, it had been acquired for $430 million by Bristol-Myers Squibb, which provided significant infrastructure and resources to expand research and clinical development of its technologies and products. John Edwards, President of Adnexus, had supported the acquisition, in part because he was aware that a new biotechnology product often takes over 10 years and costs approximately $1.2 billion to bring to market (Tufts, 2009). His growing pipeline would therefore need considerable later-stage resources for successful commercialization. Further, he wanted to ensure that Adnexus had enough capital to invest in further developing the Adnectin platform. Finally, Edwards also understood the importance of his company's lead program, CT-322, in building confidence in the novel Adnectin drug platform. Continuing to get favorable clinical results from this first program was critically important in assuring the regulatory, scientific and medical communities that Adnectins were indeed a viable new class of drugs. If this lead program was to falter, it could significantly impact his ability to obtain the necessary resources to continue to build his company.

Edwards was confident that the acquisition had been the best thing for Adnexus, but as he walked up to the front entrance of the Waltham, Massachusetts facility, the question of "where to best focus his team in developing their pioneering technology and the growing drug pipeline, without spreading his limited resources too thin" lingered in his mind. Any new drug must pass through an extensive review process (see Appendix B) before it can be approved for sale. This typically includes at least three clinical phases, often lasting more than 10 years at a cost of hundreds of millions of dollars. Given this expensive process, Adnexus clearly needed additional funds and resources to continue development of CT-322, as well as its growing early-stage pipeline. In addition, Edwards believed that their novel technology could be further developed in ways that would give Bristol-Myers Squibb an even greater competitive advantage.

While Adnexus had significant financial and infrastructure advantages over other independent biotechnology firms, Edwards knew that continued funding from Bristol-Myers Squibb was contingent on results. As he entered into his office, he was weighed down with several questions: How could he and his colleagues consistently hit their drug discovery and development milestones? What organizational structure was optimal to ensure this would happen? What additional investment in the technology was required, and how could he convince Bristol-Myers Squibb to make that investment?

MARKET BACKGROUND

Adnexus was in the business of making protein-based drugs. In 2009, the global targeted protein therapeutics market was approximately $42B (Adneuxs internal research), with an expected CAGR of 12% from 2010-2010 (RNCOS, Global Protein Therapeutics Market Analysis Report, August 2009). The fastest growing segment of this market was targeted protein therapeutics. These were protein drugs that were designed to specifically block, stimulate, or otherwise modulate a protein "target" in the body that was the cause of an underlying disease process. There were only two drug classes in this market segment: monoclonal antibodies and soluble receptors. Monoclonal antibodies had the largest share, and had enjoyed tremendous commercial success and growth. However, there remained unmet needs from traditional biologics such as antibodies "in terms of efficacy, tolerability, safety, convenience and cost" (Adnexus website, (www.adnexustx.com) 2009). The Adnexus team continued to believe, as they did when company was formed, that these unmet needs could potentially be addressed with Adnectins.

COMPANY BACKGROUND

At Adnexus, we are opening new therapeutic horizons by unleashing the power of our science and our people to improve life (Adnexus, 2009).

Adnexus Therapeutics, Inc. was founded in 2002 as Compound Therapeutics, Inc. and focused on developing a novel class of next-generation targeted biologics[1] called Adnectins.

Adnectins are a novel, proprietary class of targeted biologics derived from human fibronectin, an abundant extracellular protein that naturally binds to other proteins[1]. Adnectins offer numerous potential advantages compared to traditional targeted biologics including: speed of discovery, ease of manufacturing, as well as the ability to create multi-domain molecules that bind and can modulate a broad range of therapeutic disease processes (targets).

Adnexus uses Adnectins to design drugs that specifically attacked a disease target while minimizing interactions with unrelated targets. Adnectins are engineered to have many of the desired properties needed for a successful drug: "high potency, specificity, stability, favorable half life, and high yield *E. coli* production" (FierceBiotech, 2009). Adnexus's lead drug, CT-322 was being developed to treat cancer Not only was CT-322 being used to demonstrate the clinical potential of the Adnectin drug class, its mechanism of action offered important potential differentiation from related FDA approved drugs.

Adnectins are designed and identified using PROfusion, a proprietary protein engineering system, which allowed Adnexus to create and then sort through over a trillion unique Adnectins that target a specific disease of interest. Utilizing PROfusion allowed Adnexus to speed drug discovery and development, as well as design novel drugs with desired pharmaceutical properties.

Adnexus is headquartered in Waltham, Massachusetts, with approximately 100 employees. The management team has considerable life sciences experience, which includes involvement in the development or commercialization of over 50 biotech products.

Adnexus was initially funded by well-known venture capital firms such as Atlas Venture, Flagship Ventures, Polaris Venture Partners, Venrock and HBM BioVentures. Edwards talked about the approach that Adnexus took as a company in its early-stage development:

> *Given that we were developing a novel therapeutic technology—essentially, a new class of drug—it was critical that we work as efficiently as possible to prove our technology in the eyes of existing and future investors. One way to do this was to focus on a number of programs to showcase different ways that the technology could be used, for instance, in different therapeutic areas, in different formats, with different delivery mechanisms etc. This strategy would allow us to diversify our risk by putting our eggs in multiple baskets. But, given the resource constraints of a venture funded small biotech company, we also knew that this strategy would not allow us to take any one program very far in development. Another way to prove value was to largely focus on a single program and take it through to clinical development. The risk was that if the program failed, it could kill the company. But, we were willing to take this calculated risk because it would enable us to address the key questions that the scientific, medical and regulatory community would ask about any new class of drug: does this new class of drug really work in humans? This is the strategy we successfully followed—as we took our lead drug, CT-322, into Phase I clinical development. The strategy paid off—we caught the industry's attention with our early favorable clinical data, including that of Bristol-Myers Squibb, leading to the success we have experienced to this very day.*

---

1 *A biologic is defined as any substance derived from a biological source, for instance the human body, and can be used to treat or prevent disease.*

The approach Adnexus took is not inexpensive. For the five year period of 2002 to 2006, Adnexus spent $33.9 million on research and development, and for the first six months of 2007, an additional $10.4 million was spent, most of it related to early clinical trials with its lead program, CT-322.

BRISTOL-MYERS SQUIBB ACQUISITION

In February 2007, Bristol-Myers Squibb and Adnexus announced a major strategic alliance "to discover, develop and commercialize Adnectin-based therapeutics for important oncology-related targets." (Businesswire, 2009) The alliance provided Adnexus with approximately $30 million in committed research funds over three years. It also meant Adnexus could receive development milestone payments of up to $210 million per product, along with sales-related royalties. John Mendlein, Ph.D., J.D., the CEO of Adnexus at the time, stated:

> *Bristol-Myers Squibb has world-class expertise in oncology, and we look forward to working together using our PROfusion and Adnectin combination to discover potential therapies for people with cancer.* (Businesswire, 2009)

The strategic alliance was so successful it lead to Bristol-Myers Squibb's decision to acquire Adnexus for $430 million just eight months later in October 2007. At the time, Bristol-Myers Squibb CEO Jim Cornelius stated:

> *Bringing Adnexus into the Bristol-Myers Squibb family builds upon a successful and productive collaboration between the two companies in oncology and is an important step in accelerating the strategic transformation of our pharmaceutical business to a biopharma business model ... Biologics are one cornerstone of our growth strategy. This investment in biologics discovery complements our continued investment in a growing biologics pipeline and portfolio...* (Adnexus Therapeutics, Inc and Bristol-Myers Squibb joint press release, September, 2007)

Adnexus' management viewed this acquisition as a confirmation that the company had a very promising biologics technology and drug platform. Mendlein remarked:

> *This is an exciting milestone for our scientists, investors, and company and is a unique opportunity to further accelerate advancement of Adnectin-based medicines...We are proud to bring the strength of our science, team, and intellectual property to Bristol-Myers Squibb. We have enjoyed a highly productive and collaborative relationship to date, and look forward to helping Bristol-Myers Squibb advance its innovative pipeline.* (Adnexus Therapeutics, Inc and Bristol-Myers Squibb joint press release, September, 2007)

Adnexus' management also knew that this would provide a level of financial stability that had not previously existed. In 2006, the company burned through approximately $14 million in cash, and that amount would only increase as the company's clinical pipeline evolved.

From both a business and scientific perspective, Adnexus' management knew that as a subsidiary of Bristol-Myers Squibb it would be better positioned to reach its goal to generate "a unique pipeline of best-in-class targeted biologics across multiple therapeutic areas" (Adnexus website: www.adnexustx. com). Edwards added:

> *Bristol-Myers Squibb not only offered Adnexus greater financial security, but the in-*

credible opportunity to tap into a broad range of expertise and infrastructure to enable our technology and programs to be successful. This included extensive therapeutic area experience to help speed products to the clinic and through clinical trials, as well as support to increasing the efficiency of our PROfusion and Adnectin technologies.

As a recent acquisition for Bristol-Myers Squibb, Adnexus could have been fully absorbed into the larger pharmaceutical company. Bristol-Myers Squibb took a very different approach. According to Edwards:

The full integration approach works well when there is a single asset or technology of interest to the pharma company, but when the value also lies in the scientists, professionals, and company culture, full integration often fails as the top talent walks out the door. (Edwards, 2008)

Small biotech startups tended to be entrepreneurial and risk tolerant, often leading to greater innovation in discovery and higher R & D productivity. For example, of 103 FDA approvals from January 2006 to December 2007, 65% originated from the biotech industry (Czerepak and Ryser, 2008). Larger pharmaceutical companies offer the advantage of greater knowledge in specific therapeutic areas that can guide the discovery process and help identify successful drug candidates. In addition, larger firms often have more experience and a well-established infrastructure in later-stage drug development, as well as a 25% higher success rate in later stage clinical development. Edwards noted:

[in the last 10 years], the success rate of a Phase 3 trial for the average pharmaceutical product is 65%-75%, whereas that for a biotechnology product is ... 56%.

Edwards continued:

...as products move from discovery through development and then on to approval and commercialization, the preferred environment for optimal success also changes, moving from the need for innovation and less process in the discovery phase to greater levels of process and procedure as the complexity of taking a product to market grows.

Typically, the small biotech model is more naturally suited for discovery, whereas pharma is better positioned for late-stage development work and commercialization, and early development would work best if it can combine the culture aspects of the biotech company, with some of the established infrastructure of pharma. (Edwards, 2008)

Adnexus has maintained its name, location, and people, post-merger (see Appendix A for organizational chart), allowing it to continue to function as a successful biotech company without the need to adapt another organization's processes and procedures. The intent has been to focus Adnexus on the early drug development process, so that it could successfully deliver new candidates into Bristol-Myers Squibb's drug-development pipeline. As drug candidates move through the approval process, more of the BMS process and procedural aspects will be leveraged. Edwards noted:

Adnexus was acquired in large part for our discovery capabilities, which leverage our PROfusion discovery engine to create Adnectin drug products, our ability to move programs rapidly through clinical proof of concept, as well as for our lead compound, CT-322, which was nearing Phase 2 clinical trials.

THE DRUG DISCOVERY AND DEVELOPMENT PROCESS:

The discovery of a new drug proceeds in three major steps. The first step requires understanding of a disease process and identifying an appropriate biological target, that when modulated reduces the disease severity. The second step involves identifying a drug that stimulates or blocks the target of interest and has desired disease-altering activity in laboratory settings, such as in vitro (in "test tubes") or animal models. The third step involves demonstrating the safety and efficacy of the drug in human studies and developing methods to manufacture the drug.

To begin the process, researchers study the science behind the disease to better understand the genes or proteins involved with the cause or progression of the disease. In the area of cancer research, cellular pathways that are not properly controlled can lead to rapid and unchecked growth. Using a variety of laboratory and animal models, researchers will study alternative approaches to regulating the expressed gene product or other means to impact the disease symptoms.

Once these biological targets are identified, the second step in the discovery and development pathway begins. Typically, large compilations of chemicals, sometimes more than a million, are screened for those chemicals that can best modulate the biological target. This activity often yields a lead candidate that requires chemical optimization to create the final drug candidate. Chemical libraries can come from natural products or designed compounds. Sometimes smaller libraries of compounds are screened when much is known about the class of biological target. With targeted biologics, often known as protein therapeutics, much larger diversity can be screened. For example, in the case of monoclonal antibodies, animals are immunized and vast antibody biological diversity is screened to identify the best antibodies with activity that modulates the target of interest. With Adnectins, such screening can be done without the need for animals.

After a biologically active lead molecule is identified (i.e., the drug candidate), the discovery process next requires the use of animal models before human testing can begin. The aims of these studies include the demonstration of safety, efficacy, and appropriate pharmacokinetics (how long the drug will last in the body). With an appropriate demonstration of safety and efficacy in animal models, researchers then develop processes to manufacture the substance in larger quantities for human clinical testing and if clinical studies are successful, ultimately for commercial production.

THE REGULATORY DEVELOPMENT AND REVIEW PROCESS

The US and EU have a highly regulated process for oversight of products seeking approval for marketing. The process typically consists of three phases of development that typically take six to eight years for most products and consists of three major disciplines: (1) non-clinical which encompasses all the in-vitro (cell-based) and in-vivo (animal toxicology, pharmacokinetics, etc.) studies, (2) drug product manufacturing and testing, and (3) conducting human clinical studies to demonstrate safety and efficacy of the drug. Development in each of these areas is intentionally staged to support the clinical development phase of the product. A typical pharmaceutical company will invest in multiple parallel activities that increase costs, but lowers the overall development risk.

The drug product manufacturing and testing activities are seldom developed to a commercial level while the product is in early clinical studies. As the development program progresses, manufacturing process/scale changes are implemented and the final product is optimized to support commercial production and distribution upon approval. For Adnexus, investments in manufacturing and comprehensive testing/characterization studies are strategically implemented using a "phase-appropriate" strategy.

Prior to initiating human clinical trials, toxicology, pharmacokinetics, and other applicable animal studies are required. If a drug shows signs of promise in clinical trials, the completion of more comprehensive animal studies must be carefully planned to meet the requirements for a marketing application.

Human clinical studies represent the largest investment as well as the largest uncertainty of any drug development program. They are sequentially staged to first establish safety and tolerability of the drug, then to determine a therapeutic dose level and schedule, and lastly to demonstrate efficacy. Once safety and tolerability are established, decisions regarding clinical study designs, target patient populations, and product registration strategies are made.

THE SCIENCE BEHIND ADNEXUS' LEAD PROGRAM

CT-322 was Adnexus' first drug candidate. Specific to oncology, CT-322 inhibits tumor growth by blocking a process called angiogenesis or new blood vessel growth.[2] Tumors are dependent on angiogenesis because the tumor's rapid growth rate creates a higher requirement for oxygen and nutrients provided by the blood. CT-322 targets the blocking of vascular endothelial growth factor receptor-2 (VEGFR-2), a specific signaling mechanism that stimulates angiogenesis.

Another method of blocking this VEGFR-2 pathway is to block the vascular endothelial growth factor (VEGF), a small protein produced by tumors that stimulates VEGFR-2 and thus endothelial cell proliferation and the resultant angiogenesis. Genentech, a biotechnology firm in business for over 30 years and now fully owned by F. Hoffmann-La Roche Ltd., was the first to prove that inhibiting VEGF and thus inhibiting angiogenesis could inhibit tumor growth. Genentech created a monoclonal antibody called bevacizumab that specifically bound and blocked VEGF. Bevacizumab is approved for the treatment of numerous cancers, including colorectal cancer, NSCLC, and breast cancer. According to the company:

> *In 1989, Napoleone Ferrara, M.D., and a team of scientists at Genentech first isolated human vascular endothelial growth factor (VEGF), a protein now believed to be one of the most potent sources of angiogenesis. The need for oxygen and nutrients triggers tumor cells to produce and release the VEGF protein, which leads to the formation of new blood vessels to feed the tumor. In addition to supporting tumor growth, these new vessels provide a "highway" along which tumor cells can travel through the bloodstream to other parts of the body. This may lead to the formation of new tumors and spread of cancer (metastasis). Sustained angiogenesis is a hallmark of most, if not all cancers. Without angiogenesis, a tumor would not likely grow beyond a few millimeters, the size of an average pencil eraser. (Gene.com, 2009)*

Subsequent to the demonstration that bevacizumab had anti-tumor activity in humans, two small molecule drugs, sorafenib and sunitinib, that in-part block VEGFR-2, showed efficacy in renal cell carcinoma. Combined, these products validated the VEGF signaling pathway generally in oncology. Developing a new technology such as Adnectins, has a high degree of risk. By applying a new technology to validated or proven targets, helps to limit the overall investment risk. Therefore, the VEGF/VEGFR-2 pathway was an attractive target for the first Adnectin product due to the biological validation of the pathway[3] (Appendix C illustrates the role of VEGF in forming new blood vessels that support tumor growth) (Gene.com, 2009).

Adnexus had filed its IND on CT-322 in early 2006, with the Phase I clinical trial being initiated

2   *Angiogenesis refers to the formation of new blood vessels, and is critical for the growth of tumors.*

3   *VEGFR-2 Pathway refers to the Vascular Endothelial Growth Factor Receptor Pathway. The pathway has a receptor and multiple activator proteins, which are in the VEGF family. According to the National Center for Biotechnology Information (NCBI), this protein is a glycosylated mitogen that specifically acts on endothelial cells (the cells that form the interior surface of blood vessels) and has various effects, including mediating increased vascular permeability, inducing angiogenesis, vasculogenesis and endothelial cell growth, promoting cell migration, and inhibiting apoptosis.*

in August 2006. This initial trial was completed in February 2009. Edwards described the importance of this initial clinical study:

> This study provided important data related to validating the potential of the Adnectin platform. There is still work to be done, but this study played a crucial role in significantly increasing the confidence related to the potential for the novel Adnectin class.

The trial was conducted on 40 oncology patients in the United States, patients with advanced solid tumors or non-Hodgkin's lymphoma. The rationale behind the Phase I clinical trial had been that "CT-322 may stop the growth of solid tumors or non-Hodgkin's lymphoma by stopping blood flow to the cancer." (Clinicaltrials.gov, 2009) Although final study results had yet to be posted, preliminary results were consistent with the company's strategic objective, which included:

- Favorable Adnectin safety and tolerability profile
- Favorable Adnectin immunogenicity profile[4]
- Favorable Adnectin pharmacokinetic profile[5]

Phase II trials were the initiated in the area of recurrent glioblastoma multiforme (GBM), the most common and aggressive primary brain tumor, colorectal cancer and lung cancer. The trials were in the process of recruiting patients, and the first study was due to be completed in September 2010, with subsequent studies finishing in mid-2011 (Clinicaltrials.gov, 2009).

COMPETITORS

Due to the tremendous growth in the therapeutic application of targeted biologics, a handful of small biotech companies set out to create next-generation protein therapeutics. However, due to the speed of discovery and development of the platform, Adnexus was one of the first to advance a next generation drug into the clinic. The company believed there were a number of advantages it had vis-à-vis other competitors. First and foremost, it believed it had selected the optimal starting material for Adnectin drugs—that is, human fibronectin. This created a number of advantages, namely, that the particular domain of human fibronectin used by Adnexus was: (1) inherently stable; (2) had low immunogenic potential due to its human extracellular origin; and (3) was cysteine-free, which made it suitable for expression in *E. coli*. These features made Adnectins more readily discoverable, easy to work with, and straightforward to manufacture. On top of that, Adnexus' PROfusion discovery engine capitalized on the inherent advantages of Adnectins through the rapid and efficient engineering, evaluation, and optimization of Adnectin product candidates. This had the potential to give Adnexus a competitive advantage by reducing the time and cost of identifying high quality drug candidates.

Three competitor companies, Avidia, Ablynx and Domantis, stood out in terms of advancing next-generation protein therapeutics. The first used a polymeric protein construct called avimers and the second two developed their drugs by making further alterations to the traditional antibody scaffold. In late 2006, an avimer entered the clinical trials. By 2009, both antibody-based companies had their respective drugs in the clinical trials. During this time, a number of pharmaceutical companies were also exploring next-generation protein therapeutics via partnerships and the acquisition of fundamental intellectual property. For example, in late 2006, Amgen acquired Avidia and in 2007, GlaxoSmithKline acquired Domantis. Edwards knew that as time marched on, the success of Adnexus, and a handful of other companies, would invite more competition. He firmly believed continued investment in the

---

4   *Immunogenicity relates to a substance's ability to cause or produce an immune response.*
5   *Pharmacokinetics relates to a drug's process of absorption, distribution, metabolization, and elimination within the body.*

technology, to exploit unique ways that Adnectins could be applied to the treatment of diseases was critically important.

LOOKING AHEAD

Risk is inherent in the drug development process. Bringing a novel technology forward in a highly competitive and regulated industry with limited resources requires taking considerable risk. As Adnexus evolved from a small biotech company with an innovative technology, its strategy changed from a high-risk single target platform to one that was capable of broadening into new areas as the expertise and resources expanded following the acquisition by Bristol-Myers Squibb. Overall, Edwards was pleased with the company's progress:

> *The evolution of Adnexus didn't stop with the acquisition. In fact, the company has emerged stronger and more successful than ever. This has been made possible due the strategic decision to approach the acquisition of Adnexus with the idea of leveraging the best of biotech and the best of pharma, rather than focus on integrating the entire company or leaving it completely independent. This novel approach has resulted in one of the industry's most successful biotech acquisitions.*

However, as he reached his office, he thought about the challenges that remained, knowing it would take considerably more funding to realize the full potential of the Adnectin platform as the growing discovery pipeline enters the clinic.

APPENDICES

APPENDIX A: ORGANIZATIONAL STRUCTURE OF ADNEXUS WITHIN BRISTOL-MYERS SQUIBB

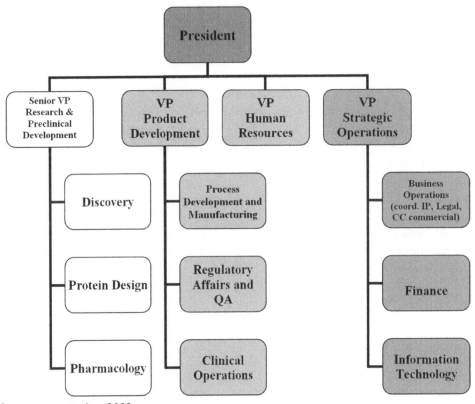

Source: Adnexus presentation, 2009

APPENDIX B : THE DRUG DISCOVERY, DEVELOPMENT, AND APPROVAL PROCESS

	Preclinical testing	File IND at FDA	**Clinical Trials**			File NDA at FDA	FDA		Phase IV
			Phase I	Phase II	Phase III		FDA		Phase IV
Years	3.5		1	2	3		2.5	12 Total	
Test popula-tion	Laboratory and animal studies		20 to 80 healthy volunteers	100 to 300 patient volun-teers	1000 to 3000 patient volunteers		Review process / Approval		Additional Post marketing testing required by FDA
Purpose	Assess safety and biologi-cal activity		Determine safety and dosage	Evaluate effective-ness, look for side effects	Verify effective-ness, moni-tor adverse reactions from long-term use				
Success rate	5,000 compounds evaluated		5 enter trials				1 approved		

Source: PhRMA, 2010 (www.phrma.org)

APPENDIX C: ROLE OF VEGF IN FORMING NEW BLOOD VESSELS THAT SUPPORT TUMOR GROWTH

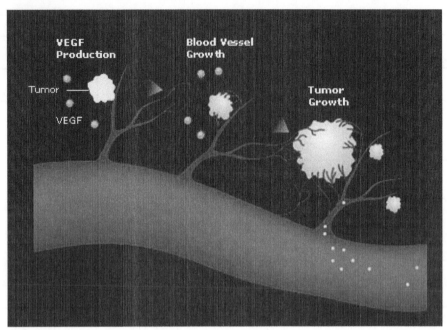

Source: http://www.gene.com/gene/research/focusareas/oncology/angiogenesis.html

REFERENCES

Accessibility.com. Online article. http://www.accessibility.com.au/news/emerging-8-set-to-join-big-5-targeted-cancer-therapy-blockbusters; The seven major markets are: US, Japan, France, Germany, Italy, Spain and UK, first accessed April, 2009.

Alacra.Pulse.com. http://pulse.alacra.com/analyst-comments/ONYX_Pharmaceuticals_Inc-C2000773, first accessed April 2009.

Adnexus Theraputics. Company Website. http://www.adnexustx.com/, first accessed January 2009.

Bristol Meyers Squibb, Company Website. http://www.bms.com/News/Pages/home.aspx, first accessed June 2009.

Businesswire.com, Industry news on Bristol Meyers Squibb. http://www.businesswire.com/portal/site/bms/, first accessed June 2009.

Centers for Disease Control and Prevention. Government Website. http://www.cdc.gov/nchs/FASTATS/lcod.htm, first accessed April 2009.

Clinicaltrials.gov. Government Website. http://www.clinicaltrials.gov/ct/show/NCT00374179?order=1Angiocept, first accessed January 2009.

Czerepak, E.A. and Ryser, S. March 2008. Drug approvals and failures: implications for alliances, *Nature Reviews Drug Discovery*, 7, pp. 197-198

Edwards, J. 2008. Corporate presentation. Newton, MA.

Federal Drug Administration. Government Website. http://fda.gov, first accessed January 2009.

FierceBiotech. Industry newsletter. http://www.fiercebiotech.com, first accessed June 2009.

Forbes.com. Online article. http://www.forbes.com/2009/03/13/pfizer-sutent-trial-markets-equities-pancreatic-cancer.html, first accessed April 2009.

Free-press-release.com. http://www.free-press-release.com/news/200801/1201002399.html, first accessed April, 2009.

Gene.com. http://www.gene.com/gene/products/information/oncology/avastin/vegf-angiogenesis-cancer.html, first accessed June 2009.

Redorbit.com. Online article. http://www.redorbit.com/news/health/1576873/targeted_therapy_cancer_brands_to_achieve_sales_of_over_42/, first accessed April, 2009.

Simon, M. and Friar, J.H. 1995. Contrasting the implications of R& D funding strategies of US biotechnology companies, *International Journal of Technology Management*, 106-126.

Tareeva. Company Website. http://www.tarceva.com/patient/considering/about.jsp, first accessed May, 2009.

Tufts Center For Drug Development, 2009. TCDD Home Page. http://csdd.tufts.edu/NewsEvents/NewsArticle.asp?newsid=69, first accessed June 2009.

Wall Street Journal. http://online.wsj.com/article/BT-CO-20090514-721367.html, first accessed May, 2009.

Wyeth Corporate Report. Q109 10Q, p.48.

Gardasil: From University Discovery to Global Blockbuster Drug

ALAN COLLIER[1], MARK J. AHN[2] AND BRENDAN GRAY[1]

[1] *University of Otago;* [2] *Atkinson Graduate School of Management, Willamette University*

- **Summary and key issue/decision:** This case presents two key decision points in technology transfer and commercialization: first, a typical dilemma faced by university-based inventors on whether to publish or commercialize their discovery; and, second, a decision faced by a patent licensee of whether to risk the costs associated with pursuing patent rights in the face of an early legal defeat. All human papillomavirus (HPV) variants have an 8 kb circular genome enclosed in a capsid shell composed of the major and minor capsid proteins L1 and L2, respectively. Purified L1 protein will self-assemble to form empty shells that resemble a virus, called virus-like particles (VLPs). Frazer and Zhou at the University of Queensland discovered and, in 1991, reduced to practice the production of VLPs for one HPV serotype that is a principal cause of cervical cancer. At the time of their discovery, Frazer and Zhou were young researchers intent on building-up their research profile and peer esteem. They did not know if their discovery would have any commercial potential, but they did know that other researchers in the United States were close to a similar breakthrough. They were faced with the decision of whether to seek early publication and gain kudos for the discovery, or possibly to delay publication and seek a patent. The early research had been supported by CSL when it was a government agency involved in manufacturing blood plasma and other products, but planning on broadening its product portfolio. CSL was floated as a publicly listed company on the ASX (Australian Stock Exchange) in 1994. The Frazer and Zhou patent, as well as patents by three separate research groups in the United States, were deemed by the United States Patent and Trade Mark Office to contain valid claims to a single patentable invention. The resulting interference proceedings lasted eight years and finally granted patent priority to Georgetown University over CSL. CSL, as the principal licensee, and still a young public company, had to make a decision at the commencement of this interference on whether to risk a significant portion of their capital to pursue its patent rights in United States courts.

- **Companies/institutions**: Merck, CSL, National Institutes of Health, University of Queensland, University of Rochester, Georgetown University

- **Technology:** Human papillomavirus (HPV) vaccine

- **Indication/therapeutic area:** Prophylactic vaccine for the prevention of human papillomavirus and anogenital warts caused by HPV serotypes 6, 11, 16 and 18.

- **Geography:** U.S., Australia

- **Keywords:** Gardasil, human papillomavirus, virus-like particles, cervical cancer, genital warts

* *This case was prepared as a basis for class discussion rather than to illustrate either effective or ineffective handling of an administrative situation.*

"I was able to give the very first vaccine, first licensed vaccine…in Australia to the first girl to receive it, and being involved in it right through from back at the beginning when we were just talking about papilloma viruses in 1982, right the way through to getting a vaccine delivered to women, that's really exciting. But it's also really exciting to see that if you put your mind collectively to a problem—a whole group of scientists working together on something—then you can come up with the answers." – Professor Ian Frazer

INTRODUCTION

The Russian proverb reminds us that "success has many fathers, but failure is an orphan." Gardasil[1] has many fathers. Indeed, paternity was claimed by no fewer than groups, and the ensuing actions to decide the issue of invention occupied many years in the United States Patent and Trademark Office (USPTO) and the United States Court of Appeals for the Federal Circuit.

The Gardasil story is about the invention of a key component of the product, the several fathers who claimed to have invented this key component and how their claims were resolved, the role of an Australian company in identifying and nurturing the early research and defending the patent, how the drug was proven and approved by the regulators, and how negotiations succeeded in ensuring that all the patent rights needed to make the drug were acquired. The story is of particular interest for two reasons: it involves the first blockbuster drug discovered in an Australian university; and it is the first prophylactic cancer vaccine approved by the United States Food and Drug Administration (FDA).

This story starts with a key discovery at the University of Queensland, Australia, in 1991 by researchers Zhou, Frazer and colleagues (Zhou, Sun, Stenzel, & Frazer, 1991) concerning virus-like particles (VLPs) that led to a patent application made in July 1991.[2] As Frazer noted:

"…the problem started with hepatitis B virus and the research work that I was doing in Melbourne when I first came out to Australia in 1981, where we were looking at the causes of chronic liver disease. And the group of men that I was studying at that time were men who had sex with men and this was before the HIV AIDS epidemic. Indeed, before there was even knowledge that there was a virus which we now call HIV AIDS. But one of the things that became clear was that the significant immune problems that these young men had led to them having a great deal of difficulty getting rid of infection with papilloma viruses that were causing genital warts, and a colleague of mine, Gabrielle Medley, in Melbourne, suggested that I should also be interested in whether they were having problems with any cancer as a result of the papilloma infections that they couldn't get rid of, because at that time, or shortly before that time, Professor Harald zur Hausen in Germany had for the first time hypothesized that in fact human papilloma viruses might be responsible for cervical cancer.

---

1   *Gardasil is the trade name of the drug manufactured by Merck to prevent human pappilomavirus (HPV) infection serotypes 6, 11, 16, and 18. A comparable drug manufactured by GSK (for HPV serotypes 16 and 18 only) is called Cervarix. In terms of serotypes 16 and 18 there is little difference between the products, however there is a difference in the respective adjuvants used, with Merck using a purely aluminum adjuvant, while GSK uses a combination of aluminum and Monophosphoryl Lipid A (MPL).*

2   *Patents represent a set of exclusive rights granted by a national government to an inventor or their assignee for a limited period of time in exchange for a public disclosure of an invention. Typically, a patent application must include one or more claims defining the invention which must be new, non-obvious, and useful or industrially applicable. The exclusive right granted to a patentee in most countries is the right to prevent others from making, using, selling, or distributing the patented invention without permission.*

"It had been known for a very long time that cervical cancer had an epidemiology which suggested a sexually transmitted infection. Indeed the first observation on that was made by a pathologist in Verona, Rigoni Stern, in about 1840. He made a very simple observation from an epidemiological point of view that nuns never got cervical cancer and prostitutes very frequently did and quite correctly assumed that there was something transmitted sexually which caused the disease. But it was really only with zur Hausen's work that it was finally pinned down that it was papilloma virus. Indeed, it was about 10 years after he first hypothesized that before it became generally accepted that that was the case. And here we were sitting in 1984 with evidence of another cancer, anal cancer, now being caused by the same virus."

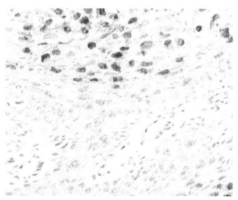

Figure 1: Immunohistochemistry of HPV type 16. Cells expressing the L1 protein of the HPV 16 virus can be identified by the intense dark labeling.
Source: www.bdbiosciences.com

An Australian company involved in the development and manufacture of vaccines and plasma protein biotherapies, CSL, became the university's key partner in commercializing the patent of the invention by Zhou and Frazer. In turn, CSL established a strategic alliance with Merck, a large multinational pharmaceutical company that provided the route to U.S. regulatory approval, as well as manufacturing and distribution capability needed to get the discovery to market.

While the discovery by Zhou and Frazer at the University of Queensland was essential to making a HPV vaccine, it was not sufficient. Gardasil comprises technology from a number of patents controlled and ultimately cross-licensed between Merck and GSK (GlaxoSmithKline). Merck and CSL supported the Frazer-Zhou patent, while GSK was associated with the other three research groups claiming priority to the intellectual property. So, as far as Merck and GSK were concerned, once the cross-license occurred, they didn't care which inventor was granted the patent on VLPs, as long as it was granted to someone, because they would have a license shared between them. But it made a great deal of difference to the inventors—only the registered patentee would garner the royalties, and CSL would only make money if Frazer and Zhou were the registered patentees.

In 2005 the United States Patent & Trademark Office (PTO) decided that another research group had priority over Frazer and Zhou and duly registered that group as patentee. By this time CSL was 11 years old as a listed company generating revenue exceeding $2 billion and a market capitalization exceeding $5 billion; Phase III clinical trials were essentially complete; and an appeal from the USPTO, if it proceeded, had to be run in the U.S. at considerable further expense. As the CSL chief executive said at the time: "Without being certain that the drug will pass clinical trials I am being asked to risk many millions of dollars to defeat the view of the U.S. Patent Office on who owns rights to this process? Let me think about that."

PUBLISH OR PATENT?

THE INVENTORS

Dr. Ian Frazer and Dr. Jian Zhou filed a provisional patent claiming to be the first to make the relevant discovery that made Gardasil possible, namely demonstrating that human papillomavirus L1 and L2 proteins together, but not L1 alone, result in the formation of small VLPs described as "incorrectly assembled arrays" of subunits (Zhou, *et al.*, 1991).

Frazer trained as a renal physician and clinical immunologist in Edinburgh, Scotland before emigrating in 1981 to Melbourne, Australia, to pursue post-doctoral studies in viral immunology and autoimmunity at the Walter and Eliza Hall Institute of Medical Research. In 1985, he moved to Brisbane, Australia to take up a teaching post at the University of Queensland, and has remained there since.

Researchers are constantly seeking sources of funding from outside their university, and Frazer had financial support from several private and government foundations and granting agencies. This financial support eventually enabled Frazer to fund Zhou's position, which he did out of profits obtained from research contracts performed by his team at the university and other research grants he obtained. During 1988, CSL contacted Frazer because of his work in vaccines. This led to CSL providing support in-kind, such as materials, for Frazer's research but, importantly for this story, CSL was also a potential partner for any future commercialization.

Frazer now holds a chair as head of the Diamantina Institute Centre for Immunology and Cancer Research (CICR) and teaches immunology to undergraduate and graduate students at the University of Queensland. He is also a director of a biotechnology start-up company, Coridon, president of Cancer Council Australia, and has been an adviser to the World Health Organization and the Bill and Melinda Gates Foundation on papillomavirus vaccines. He won the 2005 CSIRO Eureka Prize for Leadership in Science, was made the Australian of the Year in 2006, and won the Prime Minister's Prize for Science in 2008.

Although raised and trained as a medical researcher specializing in vaccinia in China, Zhou left China in 1989 to work at in the laboratory of Dr. Lionel Crawford at the University of Cambridge in the UK. Frazer met Zhou for the first time in 1989 while on sabbatical leave at Cambridge. Frazer believed that the work being done by Zhou was important in advancing his research in Queensland and encouraged Zhou to come to Queensland, which he did in 1990 as a post-doctoral researcher at CICR. Zhou had demonstrated considerable skill in the production of recombinant viral vectors and had a reputation as one of the world's leading researchers in recombinant viruses. Frazer had the view that Zhou's experience expressing L proteins of the viral capsid of the papillomavirus could prove valuable in creating a synthetic papillomavirus.

As well as his knowledge and experience, Zhou brought experimental L1 gene constructs to the CICR, which formed the basis of the VLPs that he and Frazer would develop into an effective vaccine. It was during the period between July 1990 and May 1991 that Frazer and Zhou made two key technical breakthroughs that enabled the production of VLPs from these L1 genes and hence enabled the production of Gardasil. In 1996, Zhou left Australia to work at Loyola Medical School in Chicago, ultimately returning to Australia in 1998 and adopting it as home because it offered his family permanent residence, rejoining Frazer and the CICR in Queensland as a senior research fellow.

While many researchers throughout the world were working on the same goal, it was Frazer and Zhou who, in 1991, were first able to demonstrate that the L1 protein could self-assemble to form empty shells that resemble the HPV virus. This achievement is more remarkable because they had not previously published on this topic, a fact that later raised suspicion among researchers elsewhere about the accuracy and fidelity of their findings.

Once they had created the VLPs, Frazer and Zhou had to make an election whether to patent their HPV vaccine discoveries or publish them. Frazer and Zhou did not wish to delay publication of their results because, as academic faculty members, their peer esteem and academic promotion were predicated principally on their publication record, and they were both at a stage of their careers where they were still establishing their reputations. To delay publication could allow researchers in other laboratories around the world to publish similar results and claim the credit before Frazer and Zhou. As Frazer explained:

> "...we realised we would have to build the coat of the virus—the shell of the virus—by expressing the viral capsid proteins, and that's where Jian's skills came into the picture, because he was a very good molecular virologist who had skills in gene cloning, and he cloned by polymerase chain reaction the major capsid proteins, L1 and L2, of the virus at a time when cloning 1600 base pair genes by PCR was really quite an achievement. And it was good that he did it that way, because basically he cloned from clinical material the genes that were responsible, and then we worked out expression systems using vaccinia virus to express those.

> Basically we said, 'This is going to be the basis of a vaccine, let's go and tell everybody.' So that's what we did. We went off to a meeting in Vancouver in the northern hemisphere autumn of 1991 and the international papilloma virus workshop and said, 'Look, if you do this, you get virus-like particles.'

> Well, from our point of view the next step was to go and talk to CSL, because we'd already been talking with them about the therapeutic vaccine, indeed for the two years prior to that, and they kind of knew what we were doing and we knew that they were interested in papilloma virus and vaccines in that area. So we went and talked to them about it. But also at the same time virtually every pharmaceutical company worth a mention that was working in the vaccine field came to talk to us, because once we'd talked about it at the meeting in Seattle and once we had published the paper in the Journal of Virology that was published in 1991, we kind of laid out the road map, and the last paragraph of the paper said, 'This will be the potential for a vaccine against cervical cancer.' So it wasn't exactly difficult to work out that we'd get a lot of visits, and indeed we did, from every company from Merck through to MedImmune and a whole range of other companies in between. So there was a lot of interest in the area."

Some researchers had a long record of research in this area. For example, Harald zur Hausen of the German Cancer Research Centre in Heidelberg, Germany discovered HPV to be a heterogeneous family of viruses, some of which cause cancer, which, in turn, led to an understanding of mechanisms of HPV-induced carcinogenesis. He was a co-recipient of the 2008 Nobel Prize in Physiology or Medicine for his innovative work in the 1970s and 1980s where he postulated that oncogenic HPV caused cervical cancer. Other research groups at the National Institutes for Health (NIH), the University of Rochester and Georgetown University were also engaged in developing a way to make HPV VLPs.

In addition, it was far from clear that a patent was likely to be of great significance in 1991. Indeed, Frazer and Zhou doubted that the discovery would lead to a vaccine for at least two reasons which relate largely to the nature of the disease rather than the technology. The first of these was the uncertainty, at the time, of the extent to which cancer was caused by the papillomavirus, and the second was uncertainty over the number of people that got infected by the virus. The researchers could not be sure that cervical cancer was the rare consequence of a common infection (HPV), rather than the common consequence of a rare infection. If it was the latter, there was little likelihood of a commercial market for a vaccine. The potential for their discovery to make a great deal of money was, at that stage, unthinkable.

"... we learned that colleagues overseas had tried the same thing and failed. So we locked out in one way in that we did of what in those days would have been regarded as very primitive comparative genomics and wrote out the sequences of the papilloma virus capsid proteins on bits of paper and shuffled them around and realised that it was quite likely that to get the correct authentic L1 protein we would need to express the gene not from the first initiation code on but rather from the second one in the gene. And, indeed, we'd been trying from the first and it hadn't worked. But when we tried to from the second, it eventually worked. "

Nonetheless, Frazer and Zhou sought a provisional patent in 1991 because it was clear to them that without a patent they would not get a commercial company interested in developing a vaccine if their discovery should prove efficacious. At the same time they could not neglect their academic careers, so Frazer had prepared a paper on their discovery and was scheduled to speak at a conference on papillomavirus in Seattle in July 1991.

Frazer and Zhou filed their provisional patent in Australia on 20 July 1991 to ensure it was protected prior to Frazer's disclosure of their work at the Seattle conference. The conference paper was made publicly available on 21 July 1991, so destruction of the right to patent (by virtue of prior publication of the process) had been avoided by the finest of margins. Thus the University of Queensland became the patentee of Frazer and Zhou's discovery in 1991 and continues to own the patent jointly with CSL. For convenience, the patent is referred to in this chapter as the Frazer and Zhou patent because of their inventorship. Tragically, Dr Zhou died suddenly in March 1999 at the young age of 42 as a result of complications arising from hepatitis contracted while he was a child in China.

THE DECISION TO RISK MILLIONS IN A PATENT FIGHT

THE UNIVERSITY

The University of Queensland is located in the sub-tropical city of Brisbane in the State of Queensland, Australia. The city itself has a population of about one million, with about two million in its conurbation. For many years the State Government encouraged the development of a life sciences industry as part of its *Smart State* strategy (UQ, 2010), and a cluster of research and clinical entities developed over a period of time (CICR, 2010).

The University of Queensland has more than 37,000 students, including about 10,000 engaged in post-graduate studies. The university employs more than 2,000 academic faculty members and 3,000 general staff. Of note, it is host to the state's oldest medical school with an enviable research reputation. Among its many research centers and institutes was the CICR where Frazer and Zhou did their work on VLPs.

To assist in the commercialization of research the university has a Technology Transfer Office (TTO), UniQuest Limited, which is a wholly-owned subsidiary company of the university (CICR, 2010). UniQuest started in 1984 and was still quite a small office when Frazer and Zhou applied for their patent in 1991. Since then it has grown to become the largest university TTO in Australia, a profitable enterprise, and would rank as one of the ten largest TTOs in the United States if it were located there.

The relationship between the university and CSL, initiated by Stirling Edwards of CSL with Frazer, was managed by UniQuest. Because of the interest shown by CSL, its in-kind support of the research, and its potential to assist in commercialization, the university granted CSL the first right of refusal to license any patents that interested the company.

This contract between the university and CSL was broad and included all the papillomavirus research in the hope that a vaccine may result. When the researchers identified the potential of VLPs to offer prophylactic protection, this was within the broad field contemplated by the agreement. As the agent of the university, UniQuest applied for a provisional patent on behalf of the university and CSL at its own cost.

Before expiration of the twelve-month provisional patent term UniQuest and CSL took the decision to apply for a full patent in 1992. This decision, wise in retrospect, was not an easy one for the university given the cost involved and the uncertainty about the prospects of the technology. When the university owns the intellectual property, it licenses it solely to UniQuest in the first instance, effectively giving UniQuest complete power to negotiate with external parties. This arrangement gives UniQuest the ability to operate as a separate and independent business, while also providing an incentive to strike the best commercial bargain possible. Largely because of its inexperience in commercializing new drugs, the UniQuest engaged a U.S. consultant to assist in its negotiations with CSL. As a result of these negotiations, UniQuest granted CSL a right of first refusal to commercialize the discovery.

Australia has no legislation such as the U.S. *Bayh-Dole Act* (1980) which, subject to certain requirements, grants all right, title and interest in patents derived from federally-funded research to the university in which a discovery is made. However, all Australian universities are required to comply with a national intellectual property code endorsed by the principal granting agencies (ARC, *et al.*, 2001) to be eligible for government funding. Within the scope of this code, each university publishes its own intellectual property policy and rules.

The intellectual property policy and rules published and used now by the University of Queensland were not in place in the early 1990s when the key events were happening at the university. In fact, the rules that were previously in place had expired and no replacement published. This created a potentially dangerous hiatus that required goodwill and negotiating on the part of the university and inventors to resolve. In fact the discovery by Frazer and Zhou provided the stimulus to update the university's intellectual property rules, although negotiations about royalties derived from Gardasil were not fully resolved until 2005.

In general terms, the university's present intellectual property policy covers two particular issues that were raised as a result of the Gardasil negotiations: ownership of intellectual property; and sharing the profits from commercialization of discoveries.

On the first issue, there is a common assumption that discoveries by researchers employed by a university become the property of that university as a result of the employer-employee relationship. This assumption is restated in the present intellectual property rules of the University of Queensland, which say that "subject to any specific agreement to the contrary, the University owns all IP created by staff in the course of their employment (UQ)." While all but one Australian university claims ownership of intellectual property generated by academic faculty members, there are many different expressions of this rule. The assumption that universities own intellectual property generated by academic faculty members has been challenged in Australia by a ruling in the Federal Court of Australia which, subject to the terms of the employment contract, will usually grant the actual inventors ownership of intellectual property ahead of the university.[3] In the case of Gardasil, Frazer and Zhou assumed all along that the university owned the relevant intellectual property rights arising from their discovery.

On the issue of profit-sharing, university policy states that profits accruing to the university from commercialization are shared one-third to each of UniQuest, the faculty or institute involved, and the inventors. This policy, which is common among universities in the United States and Australia, evolved largely from requirements contained in the *Bayh-Dole Act* that the benefits of university commercialization must be shared with the inventor. As well as the lack of a formal university policy on profit share, verbal agreements on this issue between the inventors and the TTO Director between 1991 and

---

3   *University of Western Australia -v- Gray and others (No 20) [2008] FCA 498; upheld in all material particulars on appeal to the Full Federal Court in University of Western Australia v Gray [2009] FCAFC 116; leave to appeal to the final appellate court has been refused.*

1994 were not documented, so serious negotiations had to take place between the researchers and the university to resolve the latent uncertainty. This highlights two of the challenges that occur when there is a long gap between disclosure of an invention and its commercial realization—many of the agreements may have been lost or forgotten; and university rules change.

The agreement between the university and CSL, as licensee, is and remains confidential. However, the university receives a royalty from CSL which it then shares with the inventors, the institute and UniQuest. In broad terms CSL receives from Merck a royalty of 7% of sales plus milestone payments and, from this, the university receives a royalty of 2% of sales plus milestone payments.

THE COMPANY - CSL

CSL as a listed company emerged from the former Australian Government agency, the Commonwealth Serum Laboratories (CSL). CSL started its existence in 1911 as the Commonwealth Vaccines Depot, and became the Commonwealth Serum Laboratories in 1916, with principal responsibility for supplying Australia with vaccines and other bacteriological products. In 1961 it became the Commonwealth Serum Laboratories Commission, and in 1991 CSL Limited, all the time remaining in Australian Government ownership. In 1994 the Australian Government sold CSL Limited to the public through an Initial Public Offering (IPO), which saw the company listed on the Australian Stock Exchange with a capitalization of A$299 million.

Since its listing CSL has been an ambitious company prepared to take a risk and stamp its presence in world markets. It is driven by a competitive and talented chief executive, Brian McNamee, himself a physician, who was employed by the Commonwealth Serum Laboratories Commission (1990) and has been chief executive since the company was corporatized in 1991 and eventually floated in 1994, making him a chief executive of exceptionally long-standing. Employees of the company are similarly imbued with a sense of ambition for the company and pride in its achievements which have seen it grow from a stock market value of A$299 (US$269) million on floating in 1994 to a stock market value approaching A$20 (US$18) billion in 2010, delivering a total shareholder return of about 18% per annum for the previous ten years.

With its head office in Melbourne, Australia, CSL employs over 10,000 people and has presence in twenty-seven countries with major operations in Australia, Germany, Switzerland and the United States. In 2009, the company achieved revenue exceeding $4.5 billion and a net profit over $1 billion.

In the late 1980s, prior to its public listing, CSL made a decision to identify programs outside the company from which it may be possible for CSL to manufacture vaccines in order to expand its business. One research scientist, who had been working on the herpes virus was also interested in the papillomavirus when its role in cervical cancer was postulated. CSL identified the leading researcher in this field as Frazer at the University of Queensland and formed its association with him in 1988.

After floating, CSL concentrated on becoming a world leader in blood plasma protein manufacturing. In 1994, it did not have the intention of becoming a vaccine manufacturer so that, in one sense, the relationship with Frazer and Zhou was a legacy the company retained after it became a listed company as a result of its long association with the University of Queensland.

As part of its contribution to his research, CSL provided Frazer with materials such as vaccinia virus as well as support for patenting. Frazer's obligation was to keep CSL informed about his research activities. CSL saw Frazer as a world leader in papillomavirus and used him as a source of the best information about developments in this field with the added prospect that he may discover something that CSL could commercialize. CSL's investment in the early stage was, essentially, blue sky discovery research; the development of VLPs to prevent papillomavirus was only proven by Frazer and Zhou two years later, in 1991.

Frazer, Zhou and CSL had their belief that VLPs could form the basis of a vaccine reinforced when, during 1991 and 1992, Frazer was invited to speak to at least twenty different companies in the

United States, at their invitation, to discuss the prospect of a license. Largely as a result of his scholarly publications, many prospective licensees contacted Frazer. Merck also had a longstanding commercial relationship with CSL and expressed interest in licensing the VLP technology shortly after the patent application had been published. With CSL as joint patentee (with the university) and possessing first right of refusal over the technology, CSL ultimately exercised its rights and became exclusive licensee for the life of the patent and entered into a contract with Merck.

Between filing the provisional patent in 1991 and the full specification in 1992, important work was done on expanding the HPV serotypes investigated. Frazer had done his original work with the cancer-inducing HPV serotype 16, but was encouraged by the Australian-based patent attorneys to expand this to include other serotypes. This led to a full specification that included the two principal anogenital wart-inducing HPV serotypes 6 and 11 and bovine papillomavirus (BPV), which four types were included in the full patent specification. This additional investigative work increased the value of the patent in terms of therapeutic benefit and financial value. At this stage CSL invested several tens of thousands of dollars in the research leading to the full specification.

While there have been challenges in the relationship, it has worked because of the goodwill and flexibility of all the parties. A comment by the university makes this clear:

> *"CSL has been an excellent partner, you couldn't ask for a better partner... they understand the way universities operate... we've always managed to overcome any differences or difficulties and keep the thing moving forward without any big fallouts. If only all of our partners were like CSL." This praise reflects the approach of the executives of CSL who are "very fair, very reasonable... and they keep to their word."*

BIG PHARMA - MERCK

Merck approached Frazer about his research into VLPs as a result of presentations he had made in the United States, after which discussions began between Merck and CSL in mid-1993. As the relationship developed, Frazer participated in advising Merck research laboratories about the potential for developing prophylactic vaccines. Despite the growing relationship and confidence between the parties, Merck kept its powder dry by making no commitment to the production of Gardasil, and retained its research into this field, until Gardasil had been proven after the successful completion of Phase 2 clinical trials in 2002.[4]

The fact that Merck initiated discussions with Frazer as early as 1993 suggests that Merck was well aware of the developments in the field, such as VLPs, and the potential for an HPV vaccine to be developed, giving it some advantage over competitors in this area.

During and following the Phase II clinical trials Frazer and the Australian research team had to take more of a back seat and allow Merck to take the lead because of the potential conflict of interest that could arise from their involvement in the proof of efficacy and safety of the vaccine. Equally importantly, Merck possessed the knowledge, skills and deep pockets needed to take the vaccine through clinical trials, regulatory approval and to a global market.

The discovery by Frazer and Zhou was a necessary but not complete component of Gardasil and Cervarix. While there are numerous patents involved in manufacture of the vaccine,[5] four patents in particular had to be negotiated. The first of the four key patents needed—that of Frazer and Zhou—was exclusive to Merck. A patent from the NIH was non-exclusive to Merck and GSK through MedImmune (subsequently acquired by Astra Zeneca), while the remaining two came one each from the University

---

4   *Phase III trials were conducted January 2002 through March 2003.*

5   *For example, GSK had an exclusive license on technology to disassemble and re-assemble in a more usable form VLPs which was essential to manufacture of the drug, as well as holding an exclusive license over a Genentech patent concerning the use of a yeast culture in which Merck was producing the vaccine.*

of Rochester and Georgetown University, which had licensed these to GSK through MedImmune.

Negotiations between Merck and GSK in 2004 to discuss cross-licensing were held in Belgium where GSK has a major manufacturing facility. CSL worked with Merck in the negotiations and had to convince GSK that the Frazer-Zhou patent was going to be upheld in the United States because Merck was relying on the validity of this patent as its principal bargaining chip in these negotiations. Prior to this time GSK was unconvinced that it needed Merck because Merck had only the Frazer-Zhou patent to bring to the table, while Merck needed several patents over which GSK had rights. It could be seen that, in this negotiation, Merck did not hold a strong hand, but the assistance provided by CSL was important in convincing GSK to complete a cross-licensing agreement.

Once the cross-licensing deal was completed, Merck and GSK had only a limited commercial interest in the outcome of the Frazer-Zhou patent interference because one or other of them would end up with rights to this patent which they had already agreed to cross-license. However, between the four participants in the interference, CSL, NIH, Georgetown University and University of Rochester, the result had significant implications in terms of professional esteem and financial benefits.

THE PATENT BATTLES

Australia is signatory to the major patent treaties and grants a patent priority date from the date of filing a provisional patent. A provisional patent-holder then has twelve months in which to file a full patent specification lest the application lapses. In common with most pharmaceutical patents of any significance, key patents are sought in the major market, the United States. And if a patentee outside the United States follows proper formalities, the priority date for U.S. patent protection is the date of the original filing in the home jurisdiction.

Frazer and Zhou filed their provisional patent in Australia on 20 July 1991 to ensure it was protected prior to Frazer's disclosure of their work at a papillomavirus conference for which he had prepared a paper. A dispute arose in the United States concerning who had priority over the process described in the patent that had been lodged in Australia, with claims by three other parties that they had priority over the relevant discovery: Georgetown University, the University of Rochester, and the National Institutes of Health (NIH).

In one sense, Frazer and Zhou were not especially concerned about the outcome of the interference at this time because they had got the recognition they needed for their careers—but the university and CSL were vitally interested as the U.S. market was commercially the most important and their ability to profit from the drug depended entirely on the outcome of the interference. Merck was less concerned as it had already cross-licensed with GSK, although it still needed a patent to be granted to someone in order to ensure a monopoly could be obtained; there was always a risk that a patent judge could find the discovery was not patentable for reasons such as obviousness or prior publication, which would have left Merck with nothing to offer GSK through the cross-license.

In addition, Georgetown University filed its provisional patent in June 1992. It had expressed the L1 protein of a different serotype to that investigated by Frazer and concluded, and stated in its patent, that it did not form VLPs. Also, the NIH filed its provisional patent in September 1992 and a second provisional in March 1993. The second filing included a claim to wild-type HPV serotype 16 and the admission that they had used the prototype in their earlier disclosure. They also argued that the prototype was used by the Australian team, a position all three parties were to adopt during the interference. The fourth party, the University of Rochester, filed its first provisional patent in March 1993.

According to these filings, Georgetown University would be the first to succeed if it could prove that Frazer-Zhou had failed to reduce the discovery to practice at the provisional filing date. However, it faced the additional hurdle of arguing against itself by claiming that use of the L1 protein did not result in a VLP.

To defeat Frazer-Zhou and each other, the NIH and the University of Rochester had to show a priority date according to United States law. This required each of them to satisfy an adjudicator that they had reduced the process to practice on a date before Frazer-Zhou had done so as they claimed in their 1991 provisional patent. This could occur because at that time, United States law gave priority according to the date of discovery, not the date of provisional patent filing, while the rest of the world granted priority generally according to filing date.

With three other eminent research groups claiming priority over the Frazer-Zhou patent, the decision is made, in the first instance, by the USPTO. Very few universities would have the money and the will to pursue patent litigation, so it fell to CSL, as the sole licensee, to decide whether and how to pursue its claim. With Phase II trials not yet commenced, CSL took the risky decision to commit nearly a significant amount of capital at the time to pursue the Frazer-Zhou patent before the USPTO.

The USPTO held that there were four parties with competing claims over the same invention, and sorting this out involved what is called an 'interference'—in this case a four-way interference.[6] The four-way interference was terminated after about 18 months and the USPTO initiated six two-way interferences to be run concurrently. The first step in this process was to establish what the invention actually was—initially it had to be resolved whether the invention concerned a wildtype or a prototype. The NIH alleged that the Frazer-Zhou patent described a prototype and that a prototype was useless because it had not been reduced to practice and lacked utility. The university had lodged material the subject of the patent developed by Frazer and Zhou with the American Type Culture Collection (ATCC), and conducted tests on this material to establish its efficacy. They established that it was a wildtype and not a prototype, which meant that Frazer and Zhou had made the process work. Unfortunately, this fact did not alter the further stages of the interference. While the USPTO agreed that Frazer and Zhou had developed the wildtype, not the prototype, it refused to admit it into evidence, holding that this should have been placed in issue at an earlier time.

In fact Frazer and Zhou had the best priority date, but they could not introduce it into evidence because, under U.S. law, they were not permitted to use their laboratory records nor rely on the fact that they had used the wild-type protein. Prior to 1995, the use of laboratory notebooks to establish date of invention only applied to work actually performed in the U.S. The only option left open to them was to attempt to destroy the evidence of the other parties to the proceedings.

There was concern about Georgetown University being granted the patent because it was the only applicant for a patent, which did not describe the method of production of VLPs, which was the basis of the vaccine.

There was concern, too, about NIHs priority because the NIH claimed to have discovered VLPs and reduced their production to practice, but had to withdraw its claim because it had, in fact, not generated a virus like particle but used a clone of the virus that did not produce VLPs.

One move recommended to Frazer and Zhou by their patent attorney in 1991 was depositing material with ATCC, a global bioresource center, a sample of the original clones that Frazer and Zhou had used as part of their patent application. This move turned out to be very important in the patent dispute because the NIH had obtained and used a faulty L1 clone[7] from Germany which could not be used to derive VLPs, and hence the NIH and another party to the interference concluded that Frazer and Zhou could not have produced VLPs as described from the same faulty L1 clone. However, by having deposited the organic material that they had used, Frazer and Zhou could prove conclusively in the patent dispute that they had not relied in their experiments on the same faulty material that had been supplied to the NIH. In fact, the L1 and L2 clones used by Frazer and Zhou had been brought by Zhou from a separate source in Europe.

The opposing parties also demanded that Frazer and Zhou demonstrate that the chain of events prove that the material deposited with ATCC was that which was used. Frazer and Zhou could prove

---

6   *An interference proceeding, also known as priority contest, is an inter partes proceeding to determine the priority issues of multiple patent applications.*

7   *This defect arose because the material had been derived from a cancer rather than from a patient lesion.*

an unbroken chain because the technician at the university who was given the disputed material could testify that what Zhou had delivered had been sent to ATCC and used in the experiments.

In its decision the USPTO declared Georgetown University to have priority on the invention by six days because it concluded that Frazer and Zhou had not adequately disclosed their invention in the United States. This decision created the perverse result that the party which, according to its patent application had not created VLPs, was held to be the holder of the patent to manufacture VLPs. Frazer and Zhou were held to have the next priority date over the remaining two parties.

Now, a crucial decision faced CSL. It had lost its case before the USPTO, yet the next step involved further substantial costs with no guarantee of success. A number of factors weighed on CSL in its decision whether to take the matter to appeal. First, it had only just lost before the USPTO; despite the fact that Frazer and Zhou had made and registered their discovery before the other claimants, CSL was refused permission to call its best evidence. CSL was advised and believed that a case properly presented before a court would permit it to introduce all relevant evidence.

Second, if CSL did not appeal it would get no value in the U.S. post-2012 from the discovery by Frazer and Zhou, which it had supported for so long. This arose from U.S. law dealing with subsequent patent applications that rely on material in an earlier claim: so-called divisional applications. In this case, divisional applications made in the U.S. covered an aspect of VLP manufacture, which guaranteed CSL royalties at least from Merck. But, according to U.S. patent law, granted under a divisional application has a term of 20 years from filing rather than the 17 years from grant, which would apply if the divisional application was linked to the interference application. 20 years from filing would have expired in 2011 (the original patent application having been made in 1991), while the 17 years would not commence until a valid U.S. patent had been issued (which would be a much later date). So, of course, CSL sought to have the divisional application linked to the interference to benefit from the longer patent term.

Third, while the Phase II trials had been positive so far, the Phase III trials were still underway. As yet there was no certainty about the efficacy of the vaccine.

Fourth, the cost of the appeal would amount to many millions of dollars [about A\$20 million (US\$18 million)] by the time a court gave a decision, money which would be spent win or lose. This represented over one-third of profits the company generated in 2001.

While the prospect of receiving royalties of around 7% of revenue from sales of Gardasil was a great incentive, pursuing and losing the matter before the courts would cost the company dearly in money, credibility, and a huge amount of management time.

The chief executive, Dr. Brian McNamee told his staff about advice given to him by a trusted mentor: 'if you believe in your intellectual property, support it to the hilt.' McNamee knew that pursuing the Frazer-Zhou patent would have a great influence on the future of the company: failure would cost it dearly in financial and reputational terms; success could repay the investment handsomely. He summoned his staff, most of whom had worked with him for many years, and whose advice he trusted: "Is the risk of litigating in the United States worth the potential reward?"

APPENDICES

APPENDIX 1: HPV AND GARDASIL RESEARCH CHRONOLOGY

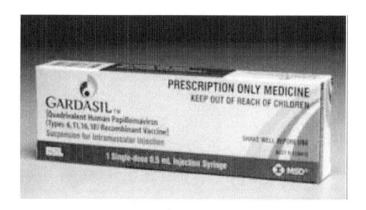

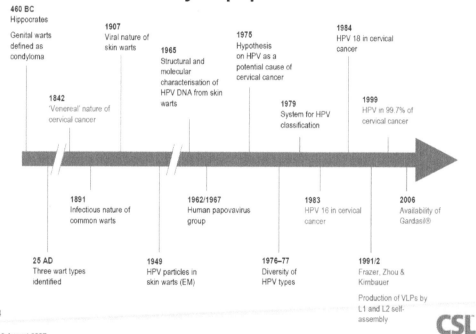

Source: CSL

APPENDIX 2: HPV MOLECULAR STRUCTURE

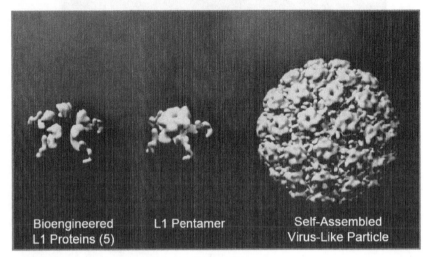

Source: CSL

APPENDIX 3: GARDASIL MECHANISM OF ACTION

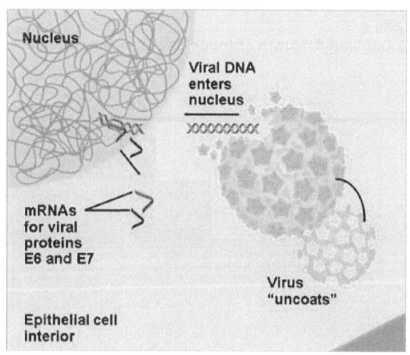

Source: Queensland University of Technology

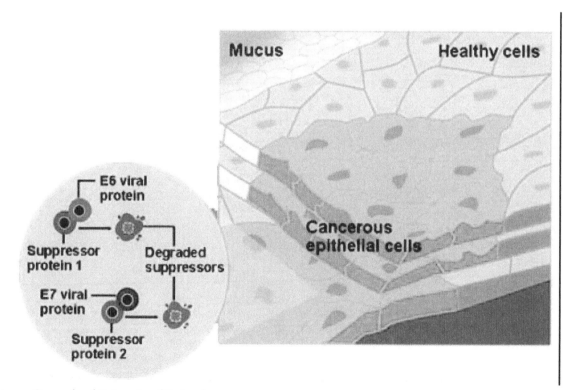

Source: Queensland University of Technology

APPENDIX 4: GARDASIL LIFE CYCLE MANAGEMENT

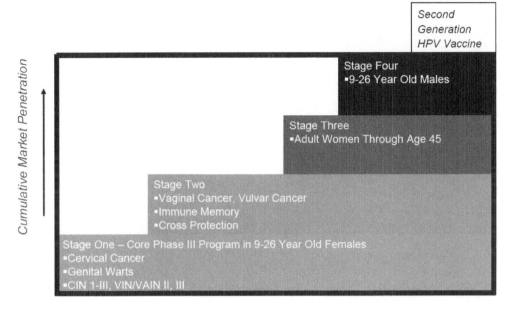

Source: CSL

REFERENCES

Australian Government/Research Council (ARC). 2001. *The National Principles of Intellectual Property Management for Publicly Funded Research*. Accessed March 10, 2010 http://www.arc.gov.au/about_arc/national_ip.htm.

Bayh-Dole Act (1980) Pub. L. 96-517, §6(a), 94 Stat. 3018 (35 U.S.C. 200 *et seq.*) Also known as the *University and Small Business Patent Procedure Act* of 1980.

CSL Limited, *Annual Report 2009.*

Diamantina Institute Centre for Immunology and Cancer Research (CICR). Accessed March 10, 2010 http://www.uniquest.com.au.

National Health and Medical Research Council. 2008. *Ian Frazer interview.* Accessed March 28, 2010, http://www.nhmrc.gov.au.

State of Queensland. 2010. *Smart State Strategy.* Accessed March 10, 2010 http://www.smartstate.qld.gov.au/.

University of Queensland (UQ). 2010. *Intellectual Property Policy for Staff, Students and Visitors.*

Zhou, J., Sun, X.Y., Stenzel, D.J., and Frazer, I.H. 1991. , Expression of vaccinia recombinant HPV 16 L1 and L2 ORF proteins in epithelial cells is sufficient for assembly of HPV virion-like particles, *Virology*, 185: 251-257.

Guru Instruments: Bootstrapping a Medical Device Startup

ANNE S. YORK

College of Business, Creighton University

- **Summary and key issue/decision:** Guru Instruments is the brainchild of Dr. Sameer Bhatia, a physician in the process of studying for an MBA and a participant in Creighton's Bioscience Entrepreneurship Program. During the course of his work in the Creighton University pathology lab, Dr. Bhatia found his colleagues constantly improvising, often with spotty success, work-arounds to create safer and more useful instruments and accessories. He began designing prototypes of products to solve these problems, beginning with multi-sized rubber and foam reusable grips for scalpels and vinyl rib protectors. He quickly realized that some of these products were useful in live surgery, which created the need for sterilizable versions of the same products. During the course of product development, interesting IP, product manufacturing, and sourcing issues arose, along with decisions regarding the type of financing that was most appropriate for the venture and the best channels for distribution. Essentially, Dr. Bhatia hoped to bootstrap his new company without obtaining significant outside debt or equity financing.

- **Companies/institutions:** Guru Instruments, Creighton University

- **Technology:** Guru Instruments designs, produces, and sources a variety of autopsy and surgery-related devices.

- **Stage of development at time of issue/decision:** The products fall into the FDA classification of Class I and II devices, that is, either not regulated or required to show substantial equivalence. Surgical products, because of the sterilization required for live operations, fall into Class II, while autopsy products, because they are not used on live subjects, fall into Class I. Each device has about a six month development and time to market cycle, with a strategy of continually introducing new products to market.

- **Indication/therapeutic area:** Class I and II medical devices for pathologists and surgeons.

- **Geography:** US-based, with international market potential

- **Keywords:** IP strategies, bootstrapping, new product development, distribution channels, medical devices, financing options

Research assistance provided by Creighton University Bioscience Entrepreneurship Program students: Sameer Bhatia, Chris DeGrande, Shannon Neubauer, Erik Olsen, and Mark Pepin

* *This case was prepared as a basis for class discussion rather than to illustrate either effective or ineffective handling of an administrative situation.*

INTRODUCTION

Taking a few moments to savor his recent graduation from the MBA and bioscience entrepreneurship programs at Creighton University, Dr. Sameer Bhatia knew that he couldn't afford to take a break from his grueling 18 hour a day schedule. By day, he was a physician working at the Creighton University Medical Center. Most nights, he ate dinner with his wife and new baby and then headed for his office at the Halo Institute, a not-for-profit incubator that was providing his company, Guru Instruments, with free space and other resources. He knew that his time in the incubator would expire soon and was worried about taking his fledgling company from one that was just breaking even to one having a solid growth trajectory into the future.

So far, Bhatia was Guru's only employee and investor, having put about $25,000 of his own money into the firm for legal work, corporate formation, and prototyping. Despite his ability to bootstrap by getting distributors to pay for initial product inventory and his decision to start with a low cost sourcing strategy for non-patentable products used in pathology labs, he knew that it would take more money and the right management team to achieve his ultimate goal of creating a successful company. But what sources of funding and what sort of organizational structure made the most sense for a start-up at this early stage? He didn't want to give away equity or control, but he also didn't want to jeopardize his family's financial future through debt financing, which due to the economic downturn was even harder and most costly to obtain. And who should be his next key hire? He needed marketing and financial assistance, as well as someone to monitor and update the website, while he invented new product lines and sought out relationships with other inventors to add to the product portfolio. While he was working as many hours in his business as he could, he was not yet in a position to quit his day job. He thought about how fortunate he was to have a supportive wife whose family had been long-time entrepreneurs and who herself was completing an MBA degree. As a secondary school science teacher and primary caretaker for their new baby, however, she had no time to give to the new venture.

A SHORT HISTORY OF GURU INSTRUMENTS

Guru Instruments was the brainchild of Dr. Sameer Bhatia, a university physician with goals of becoming a successful entrepreneur and inventor. Despite a family history of working in the medical profession (his father was an MD psychiatrist at a Veteran's Administration Hospital), Sam followed in those footsteps reluctantly, always dreaming of new and better medical products and ultimately of starting and running his own company. Bhatia had traveled to India for medical school, but while there he became a partner in an IT-related startup. He brought the startup back to the U.S. but in the end, the venture failed and he took a full-time position in a university pathology lab. Yet, the experience in India had convinced Bhatia that his passion lay in the realm of business rather than medicine, and he enrolled in a local evening MBA program. During the course of his daily work in the pathology lab, Bhatia found his colleagues constantly improvising workarounds, often with spotty success, to create safer and more useful instruments and accessories used to perform autopsies. Always a tinkerer, he began designing prototypes of products to solve these problems, beginning with multi-sized rubber and foam reusable grips for scalpels and vinyl rib protectors. He quickly realized that some of these makeshift laboratory products could be useful in live surgery, which would require sterile versions of the same products, and that they might even have applications in the funeral and veterinary industries. However, despite his MBA studies, Sam still knew little about how to start a business or protect his ideas.

Bhatia recalls his path to becoming an entrepreneur: "I was born in Omaha and attended parochial schools and then enrolled in Creighton. However, I left after a year for medical school in India, which was an opportunity to save time, returning to Omaha every 6 months for holidays. In India, I became friends with industrialists (Hero Honda, Hero Cycles, Deesons, Avon Cycles, Rothman Cycles, to name a few) and saw how one man can create an empire, which was contrary to what I previously perceived. Additionally, I found that I loved doing service work. Very little money can help villages or even cities, so I started identifying the gaps between the industrialized world and the third world. I began to question these gaps and started creating solutions that modeled both value systems—low cost products that could be profitable. Innovation and trying not to reinvent the wheel, along with the ability to negotiate and allocate resources from my contacts, proved to be successful."

> *"I came back to Omaha with a startup (El Mundo De Papel), and it failed. I also was focusing on an eBay business (over $10k first year) and was steered away from that. I resumed studying for my board exams but was still contemplating a different pathway. Back in the lab at Creighton Medical Center and the Nebraska VA hospital, I again kept finding the gaps between corporate hospital and university/government pathology lab equipment. Lots of times, the gaps were simple solutions that were not currently available in the marketplace. This is how the grips and other products under development came about. Currently, I also am in a partnership for 3 iPhone apps (psych, anesthesia, and pathology) and a laptop for telemedicine that enhances patient-physician interaction. Let's see if it works."*

When Bhatia read about a Bioscience Entrepreneurship Program (BEP) at Creighton, he was intrigued. The two-course, summer internship program was open to MBA, law, and science and health science students, including those in medicine. The courses would count as electives toward finishing his MBA. The timing seemed perfect for him to use the program as an opportunity to develop his ideas by working with professors, other students, successful entrepreneurs, and others involved in commercializing new technologies, including the directors of Creighton's Technology Transfer Office, who shared his interest in this emerging field. Over the course of the year, he used his time in the program to develop a business plan, pitch and presentation, incorporate his company as an LLC, put together a website, and secure a place in the local not-for-profit Halo Institute incubator. He had designed and sourced prototypes for two inventions—ergonomic scalpel grips and ergonomic forceps grips - and lined up a distributor to not only sell his products through their catalogues and reps but also to fund the inventory cost for the first order of grips. Through the program, he developed a relationship with a local IP attorney, whose advice he sought several times on potential patent issues. Also, the BEP offered him an opportunity to participate in a national workshop where, along with other bioscience startups, he was able to further refine his concept and plans, and also obtain a bit of free publicity (see Appendix 1).

While investors at the final business plan competition were impressed with Bhatia's energy and creativity, to most of them, Guru looked like "small potatoes," a lifestyle business unlikely to attract angel investors, even if Bhatia were interested in trading equity for growth capital. But Bhatia had other ideas for proprietary products and had been talking with other inventors with prototypes and patents for devices, but little expertise or interest in starting companies. If he could manage his cash flow and get the right people in place, he knew that Guru could fulfill his dream of becoming a successful entrepreneur, combining his business and medical skills with his passion for inventing better medical devices and accessories.

THE MARKET FOR MEDICAL ACCESSORIES

Guru's initial three products were variations of disposable instrument grips: non-slip ergonomic plastic grips for scalpel and forceps handles and a comfortable foam grip for scalpel handles. He had identified four main markets of autopsy, mortuary, veterinary surgical and human surgical. The autopsy and

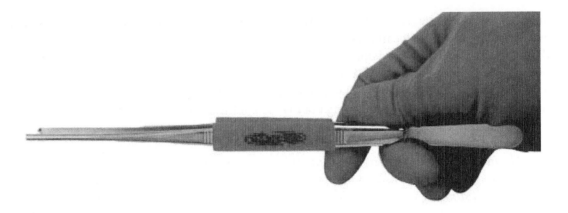

Figure 1: Guru's forcep ergo grip
Source: www.guruinstruments.com

mortuary markets do not require a sterile product, whereas the human and veterinary surgical markets are for sterile instruments only, as they are used on live subjects.

NON-STERILE: THE AUTOPSY MARKET

The initial non-sterile markets that Guru Instruments was targeting were autopsy and undertaking. He realized that non-sterile markets are the easiest to enter because there are no FDA requirements for instruments used on deceased patients.

The size of the autopsy market was a function of the instances in which autopsies were performed, which included answering questions about the prior health of the deceased, determining the exact cause of death, and resolving legal or medical concerns (Aurora Health Care 2009). Thus, not every death required an autopsy. According to the U.S. Census statistics, an analysis of the ten-year period ranging from 1996 to 2006 revealed a largely steady number of deaths annually, with a slight indication of growth. The average number of deaths was around 2.4 million a year (see Table 1). According to the Centers for Disease Control (CDC), of those 2.4 million deaths roughly 400,000 autopsies were performed each year.

Of the various instruments utilized during the course of an autopsy, ten have been identified which are nearly always used and could utilize one or more of Guru's disposable grip products.

- Osteotome (a chisel)
- Enterotome (a type of scissors)
- Hagedorn needle
- Scalpel
- Dissecting Forceps
- Straight mayo scissors
- Bone Saw
- Hammer with hook
- Breadknife
- Reciprocating saw (a vibrating saw)

NON-STERILE: THE MORTUARY MARKET

The mortuary market, on the other hand, targeted those who prepare the deceased for burial, which includes embalming the body. Nearly every body which is not cremated undergoes this procedure.

Year	Both sexes	Male	Female
2006	2,426,264	1,201,942	1,224,322
2005	2,448,017	1,207,675	1,240,342
2004	2,397,615	1,181,668	1,215,947
2003	2,448,288	1,201,964	1,246,324
2002	2,443,387	1,199,264	1,244,123
2001	2,416,425	1,183,421	1,233,004
2000	2,403,351	1,177,578	1,225,773
1999	2,391,399	1,175,460	1,215,939
1998	2,337,256	1,157,260	1,179,996
1997	2,314,245	1,154,039	1,160,206
1996	2,314,690	1,163,569	1,151,121

Table 1: Death trends for U.S. from 1996-2006 by total number.
Source: Centers for Disease Control

According to the Cremation Association of North America (CANA), the percentage of total deaths which were cremations in 2006 was 33.61% and preliminary figures for 2007 are 34.89%. Rates for 2010 are projected to be 39.03% and 58.89% for 2025. Therefore, approximately 66.3% of all deaths require burial. Thus 1.6 million bodies must be prepared for burial each year.

Guru had worked with its distributor and his costs to come up with an estimated price of $95 for a package of 10 grips. Given the above numbers of autopsies and an estimate that 25% of the market could be reached by Guru, revenues from the autopsy market were projected to be $9.5 million annually. Assuming the same average price of $95 per pack, that five of the above instruments would be used by morticians per body preparation, and again that 25% of the mortuary market could be captured by Guru, estimated revenue from the mortuary market was $19 million per year. Thus, the estimated total non-sterile autopsy and mortuary annual market revenue totaled $28.5M.

STERILE: VETERINARY SURGICAL MARKET

Veterinarians perform live surgeries on both companion (pets) and large (livestock) animals. The total number of operations performed is hard to calculate as this number is not as well documented as surgery numbers in humans. There were approximately 22,000-26,000 pet hospitals in the country in 2009 (AVMA, 2009) at which 62,196 veterinarians were employed (US Bureau of Labor and Statistics, 2009). The number of veterinarians was expected to rise to 83,956 by 2016. According to the American Veterinary Medical Association in 2009, there were a total of 153.8 million dogs and cats in the United States (AVMA 2009). Since the animals are alive during procedures, this market requires a sterile product. As such, it is a longer term target for Guru, given the need for FDA device approval and funds to collect the data required to gain approval.

The most common surgical procedure done for companion animals is to have them spayed or neutered. The number of these procedures provides a good lower bound for this market. The American Pet Products Manufacturers Association reported that 28% of dogs and 16% of cats are not spayed or neutered (Noyes & Creehan, 2007). Thus, of the total number of animals, 120.5 million undergo one of these procedures; the animals only undergo the operation once in their lifetime. Assuming that the average age for a companion animal is 12 years, then 1/12 of the spayed/neutered population must be replenished each year. An estimated 10 million spay or neuter procedures are performed each year. All of these operations use either scalpel or forceps. Assuming that 10% of the market was reached by Guru's grip products, the total annual revenue from spay and neuter operations would be $12 million.

The statistics for the number of other operations performed on companion animals and for those performed on large animals were not readily available, but a conservative estimate for the revenue from these other operations gives further revenue of $8 million. The total revenue from the veterinary surgical market thus was estimated at approximately $20 million per year.

STERILE: HUMAN SURGICAL MARKET

A second longer term market opportunity was to sterilize the grips and begin marketing them to surgeons performing live surgery on human patients. This market was difficult to define because of the large number of surgical operations performed each year. The most recent data from the National Hospital Discharge Survey in 2006 showed that there were roughly 45.9 million operations during that year. Figure 1 below shows the trends in the number of live surgeries performed over the most recent available 10 years of data.

Not every surgery performed would use either a scalpel or forceps. Also, due to the delicate nature of surgery, depending on the surgeon and the degree of finesse required by the operation, some procedures would require trading off safety the grips may provide for increased accuracy. An analysis into which types of surgeries might use a scalpel or forceps showed that roughly 62% of operations use a scalpel and 47% use forceps. Guru's estimated 25% market share in dollars at estimated prices of sterile scalpel grips and forceps grips was approximately $31.0 million for combined forceps and scalpel handles. Assuming that the market followed the upward trend illustrated by Figure 2, the sterile surgical market was large and growing, making it an attractive opportunity for Guru Instruments.

Number of Surgeries (In Thousands)

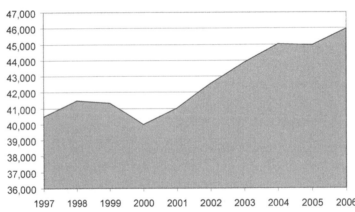

Figure 2: The total number of surgical operations performed over the span of 1997-2006. Note that there are a few instances of a decreasing total, but the overall trend is an increase in the number of surgeries.
Source: Guru Instruments

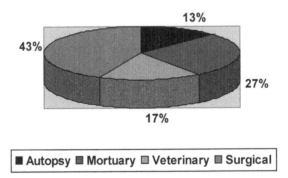

Figure 3: Breakdown of the total estimated revenue for the four main disposable grip markets
Source: Guru Instruments

Thus, in summary, Guru's estimated revenue from all sterile and non-sterile markets for scalpel and forceps grips was roughly $72M, broken down as Figure 3 illustrates.

COMPETITIVE ADVANTAGE

Table 2 shows several competitors in the instrument grip market. Bhatia believed that Guru's products provided advantages over other products in a number of ways. First and foremost, Guru's products were more economically priced than its competitors. Additionally, Guru's products were designed and tested by practitioners for practitioners, which provided an increased understanding of performance and comfort needs for the everyday user. Furthermore, Guru's products were one-time use grips that eliminated the need or expense of large and costly sterilization equipment, especially in pathology labs and mortuaries.

INTELLECTUAL PROPERTY

The idea of a comfort handle for various medical and surgical tools was not a new one. Bhatia initially hoped that his design might be patentable, and he knew that a patent search would be required not only to determine whether he might be able to file for a patent, but also to assure that his product would not infringe on existing patents. Guru Instruments' grips provided three potential features: their use in pathological and/or surgical areas, their slip-on disposable nature, and their eventual ability to be sold in sterile packages.

He had located numerous examples of patents on different types of handles for tools ranging from spoons, brooms, racquets, screwdrivers, medical tools and scalpels. Precedents existed in each area separately (or perhaps in two of the three), but a patent which incorporated all three of Guru's features in one product had not yet been issued. Thus, Bhatia felt that Guru's grips would not be infringing upon any device previously patented, and in the future they might be patentable under the novelty of new use and combination of the three points listed above.

Reversing the search from existing grip products back to patents, Bhatia's BEP team found that

Company	Multiple products	Distribution	Retail price	Applicable patents	Competitive advantage
Bradshaw Medical	Yes	Website Tradeshows Sales representatives	Unable to obtain quote	D543,433 S D557,584 S	Reusable Sterilizable
Minnesota Rubber & Plastics	Yes	Website Multiple American and international companies	$42.95 each plus $15,000 - $18,000 mold per tool.	No	Reusable Sterilizable
Teleflex OEM/ Beere Medical	Yes	Website Tradeshows Sales representatives	Unable to obtain quote	6,564,680 B1	Reusable Sterilizable
BD Medical	Yes	Catalog Multiple American and international companies Sales representatives	Scalpel Handle: $18.58 - $31.14 Disposable Scalpel: $27.55	Safety scalpel handle patent pending	Established company Reusable Sterilizable
Guru Instruments	Yes	Website MOPEC catalog MOPEC sales representatives	Foam grip: $7 Plastic vinyl scalpel grip: $15 Plastic vinyl forceps grip: $10	No	Low Cost Disposable

Table 2: Guru Instruments competitor comparison
Source: Guru Instruments

patents were currently held by Bradshaw Medical (D543,433 Rotatable Tool Handle, D557,584 Driver Handle For a Tool) and Teleflex Medical (D6,564,680 Hand Manipulated Torque Tool). However, it was determined that these patents did not have similar properties which Guru Instruments would be in danger of infringing. A reference list of related patents identified can be found in Appendix 3.

Because Bhatia was interested in developing an entire product portfolio, another intellectual property area that was important was trademark and branding. Guru Instruments, itself trademarked, planned to create a substantial market presence, a more complete product line, and a strong brand identity so that, should another competitor enter the market with a similar product, customer loyalty would help protect market share.

A review by Creighton's Technology Transfer Office had determined that Bhatia had invented the products on his own as opposed to on the job, which meant that Bhatia outright owned his inventions, as compared to being required to split all rights 42/58 with the university.

REGULATORY ANALYSIS

The FDA Center for Devices and Radiological Health is the division that is responsible for regulating firms who manufacture, re-package, re-label, and/or import medical devices sold in the United States. The first step required by the FDA is classifying the device by placing it into one of their three regulatory brackets. Once the device classification has been determined, device firms like Guru must evaluate the necessary forms and paperwork to ensure that their products do not violate any FDA regulations. Other requirements that must be taken into account are the labeling regulations outlined by the FDA, as well as who is required to register with the FDA with regard to inventors, manufacturers, and distributors.

DEVICE TYPE

Bhatia had determined that Guru's grips were a medical device, which the FDA defines as "an instrument, apparatus, implement, machine, contrivance, implant, *in vitro* reagent, or other similar or related article" in section 201(h) of the Federal Food Drug & Cosmetic Act. Medical devices encompass a wide variety of objects that range from simple tongue depressors to tools as complex as laser surgical devices.

DEVICE CLASSIFICATION

The FDA has three different classes of medical devices ranked on their level of danger to the patient when being used. These classifications are:

Class	510(k) Premarket Notification	Pre-Market Approval (PMA)	Additional Notes
Class 1	No	No	Usually less dangerous devices. Ex: Medical Gown
Class 2	Yes	No	General controls are usually not enough. Ex: Medical needles
Class 3	Yes	Yes	Most dangerous devices. Ex: Respirator system.

Source: United States Food and Drug Administration

The Federal Code of Regulations, specifically Code of Federal Regulations number 878.4800, indicates that the grips are a Class I Medical Device. 878.4800 shows that scalpel handles are considered a Class I device even when sterilized, meaning they are exempt from premarket notification.

DEVICE REGISTRATION

Registering the device requires completing an FDA-2891 form, which identifies the company as a specification developer, defined as one that develops specification for a device distributed under the establishment's own name but performs no manufacturing. This includes establishments that, in addition to developing specification, also arrange for the manufacturing of devices labeled with another establishment's name, such as a contract manufacturer. Along with the specification developer, the contract manufacturer who commercially distributes the device for the specification developer also must register, along with the contract sterilizer who distributes the device. A contract manager who does not commercially distribute the device for the specifications developer does not need to register. A contract manufacturer of subassembly or components as well as domestic distributors also does not need to register (US FDA, 2009).

MARKETING AND DISTRIBUTION

Guru Instruments decided to initially seek strategic partnerships to market, distribute, and manufacture its line of ergonomic grips for forceps and scalpel handles. One of the fundamental strengths of Guru Instruments was the strong relationship that Bhatia had formed with its marketing and distribution channel, MOPEC, Inc. MOPEC was an established distributor in the fields of pathology, histology, necropsy, and autopsy instruments and had the ability to market and deliver a product from manufacturer to purchaser. While not as large as other companies currently distributing similar products, Bhatia felt that MOPEC's values were a good fit with their strong emphasis on customer service and product satisfaction.

Guru's marketing strategy was to present the customer with a high level of customer service and delivery speed, while offering their products at a low price. The MOPEC partnership allowed Guru to make use of their partner's superior advertising and marketing techniques. For example, MOPEC could produce valuable advertising tools such as brochures and flyers, in addition to its on-line catalog, to educate prospective customers about Guru's products (see Appendix 4). In addition, MOPEC also had agreed to very favorable financing terms with Guru, including paying up-front costs for the manufacturing of initial inventory and the purchase of each lot once it was completed. MOPEC's website can be found at www.mopec.com.

While Bhatia felt that the arrangement with MOPEC was very helpful in getting the company up and running with a low outlay of cash, he also planned to offer sales of the custom ergonomic handle grips directly from Guru's own website: www.guruinstruments.com. His hope was that some day, once Guru had established itself as a well-known brand with multiple product lines, his company could become the primary distribution channel for its products so that he did not have to share such a heavy percentage of the profits with a distributor. Guru's interactive site allowed customers to see the mission of the company and read about Bhatia's background, as well as peruse a list of its products and equipment. Bhatia felt that the website increased the perception of legitimacy and allowed customers to connect with the company by making use of the contact information section of the page. There was also a section of the page that allowed for direct sales, which gave customers easy access to the products they desired (Appendix 5).

In the event that Guru should experience rapid growth and need other channels of distribution beyond those currently being utilized, Thermo Fisher was a larger company which Bhatia felt would be an excellent resource. Thermo Fisher's website can be found at www.thermofisher.com.

PRICING AND MANUFACTURING STRATEGIES

Guru's overall strategy, at least for its first, non-patent-protected, product lines, was to compete on price. Because Bhatia, through his connections, was able to manufacture and source these initial products from low cost manufacturers and because most medical products are sold through large distributors who often mark up product prices well above costs, Guru felt confident that it could take sales from competitors on the basis of price. The key was getting the word out to purchasing departments of hospitals and universities, which is why Guru had chosen to use the MOPEC distribution channel as well as its own website. MOPEC had also been very willing to negotiate with Guru Instruments. For example, MOPEC had contracted to purchase Guru's ergonomic custom scalpel handle grips for $3.50 each, with a suggested retail price of $7.00 and Guru's custom forceps grips for $2.50 each, with a suggested retail value of $5.00. MOPEC agreed to pay the initial total cost to manufacture the grip inventory and then remit 50% of their profit to Guru as the grips were sold.

Guru's initial manufacturing partners included Harman Corporation and Grab On Grips. Harman Corporation was established in 1963 and was a current supplier and in-house manufacturer of multiple plastic, vinyl, and foam products (including grips) for companies worldwide. Harman Corporation agreed to manufacture Guru Instruments' custom plastic vinyl grip product line at a rate of $0.758 per forceps grip (including cutting and printing), and $1.06 per ergonomic scalpel handle grip. More information on the Harman Corporation can be found on their website: www.harmancorp.com. Guru's foam scalpel handle grips were being manufactured by Grab On Grips in Walla Walla, WA. Grab On Grips began manufacturing foam grips in 1973 and currently made grips of all kinds, including custom products with custom branding for various inventors and distributors. For more information on Grab On Grips, see their website: www.grabongrips.com.

NEXT STEPS.....

Bhatia sat alone in his open space "office" at 9 p.m. in the Halo Institute, surrounded by several prototypes in progress and a white board with to-do lists and product development plans and sketches. He enjoyed the quiet and lack of distraction provided by working so late but also realized that his work hours did not allow much time for him to network with resources at the institute that could help him, nor did he have much interaction with the other entrepreneurs in the space. He was scheduled to meet in the next few days with an intern, a student in the bioscience entrepreneurship program who was tasked to research potential new markets and channels for one of Guru's new product lines. While this would help free up his time for other things, Bhatia realized that in the long run, he was going to have to solve his capital and human resource needs in order to succeed. While helpful, part-time, short-term interns were not going to take him where he needed to go.

Exactly who his first full-time hire should be was a critical question that he knew he must answer right away, along with how he should compensate this key employee. What would the ideal employee profile look like if he were to write a job description? Would equity, salary, commission, or some combination be the best compensation package? How many new products would he need to produce to reach his goal of being a company that could sustain his family comfortably on its profits? What amount of funding would be needed to achieve these goals, and what was the most appropriate source? Bhatia felt that Guru was at a strategic inflection point: it could become a learning experience or actually become a viable company. Which would it be?

APPENDICES

APPENDIX 1: OMAHA WORLD HERALD STORY ABOUT GURU INSTRUMENTS

Seeking a good prognosis for products being pitched

Published Thursday June 4, 2009
ROSS BOETTCHER
WORLD-HERALD STAFF WRITER

Omahan Sam Bhatia says he fully understands the constraints and stresses that a struggling economy places on inventors in the biomedical field.

Sam Bhatia, a MBA student at Creighton, presents a slideshow about one of his three low cost autopsy devices to a panel of judges during the Advanced Invention2Venture Conference in Creighton University's Harper Center on Saturday May 30, 2009.

He has learned firsthand that being an inventor is a 24-hour endeavor that requires copious sacrifices and a valiant work ethic.

Bhatia, who received his medical training in India after graduating from Creighton Prep, is now working toward an MBA at Creighton University. He also works full time at the Creighton University Medical Center conducting autopsies. On top of that, he is trying to introduce three autopsy devices into the biomedical market.

Bhatia said he hopes that each becomes available for $100 or less as low-tech, economically feasible tools for morgues and radiology and pathology departments.

Last weekend, Bhatia presented the business model for his inventions at an Advanced Invention2Venture Conference. The event drew seven teams from Creighton University, the University of Nebraska Medical Center and various other schools across the Midwest.

The teams vied not for tangible prizes but feedback from a panel of three judges: Tom Chapman, who represents insurance, bioscience and entrepreneurship at the Omaha Chamber of Commerce; Destynie Jenkins, a business professor at Creighton; and Todd Roseberry, an Omaha lawyer.

Seven teams made presentations at Creighton's Harper Center. Bhatia, who worked on the presentation with Mary Ann Wendland, associate director of Creighton's intellectual resources management office, was one of two groups to pitch low-cost inventions.

Bhatia and his wife, Kathryn, who also is pursuing her MBA at Creighton, said they have sacrificed over the last few months as Bhatia worked to make his inventions realities.

To participate in the weekend conference, for example, Sam had to skip a pregnancy workshop that Kathryn, who is expecting their first child in July, attended alone.

Despite the struggles, Bhatia said he was confident that his efforts will pay dividends.

"The autopsy tools are cost-effective, and they make your work more efficient," Bhatia said. "Compared to the other products we saw at the conference, I'll be the fastest to get to market.

"There are no major innovations in this field, and autopsies are pretty archaic."

One of the devices, called a manual cover slipper, applies a glue-type substance for binding cover slips to slides before they go under a microscopic lens. The device is an inexpensive alternative to the $40,000 machines typically used to do the job, Bhatia said.

"If you're a small lab, you can't afford the $40,000 cover slipper," Bhatia said. "My device is not groundbreaking and it's not going to change the world, but it's going to serve as a feasible, affordable alternative."

Other Omaha entrepreneurs who made pitches during the weekend conference included orthodontist Clarke Stevens with his Bio Bracket, an update to the 100-year-old bracket design used for braces; Joseph Norman, who introduced a meter able to detect oxygen levels of hospital patients; and Lora Frecks, an intellectual property manager at UNeMed at UNMC, who developed the A-Wrist-A-Trac system, a group of thin bracelets that helps people keep track of what they've eaten during a day.

Anne York, director of bioscience entrepreneurship at Creighton, said that of all the inventions, Bhatia's autopsy tools and Frecks' wristbands are likely to find the most success in the shortest period of time. They are inexpensive and can be quickly introduced on the market, York said.

UNMC released Frecks' A-Wrist-A-Trac device three months ago and so far has recorded $20,000 in sales, she said.

"It's surprising, but the little stuff like that is often the most successful," York said. "They've sold more than anybody else at the conference."

Frecks' bracelets, which sell for $20 per 30-unit pack, were the least technologically advanced invention presented, York said, but they have potential in the weight-loss market.

"We're afraid to fail, but we're also afraid of becoming a great success," Frecks said during her presentation. "The growth potential is huge."

York said the bracelets, if marketed correctly, could make a real splash.

"If I had to bet, it's probably going to be one of the most successful products to come out of UneMed," York said. "It's very hot."

APPENDIX 2: GURU INSTRUMENT PRODUCT DESCRIPTIONS

Ergonomic gross grip
Plastic vinyl hand-dipped in soft texture coating
Slides onto BD scalpel handle easily
Increases outer diameter of handle creating a more ergonomic grip
Non-slip surface
Disposable
Costs: Pack of 10 for $120, Single for $15

Foam ergonomic scalpel grip
Increases outer diameter of the most popular scalpel handle used worldwide
Disposable
Non-slip
Non-absorbing
Light-weight
Slides onto scalpel handle easily
Costs: Pack of 10 for $70

Ergonomic forceps grip
Plastic vinyl hand-dipped in soft texture coating
Slides onto long forceps easily
Increases feel and leverage of forceps grip
Less pressure needed to grip forceps, resulting in less fatigue
Disposable
Non-slip surface
Costs: Set of 2 for $5

APPENDIX 3: LIST OF REFERENCE PATENTS FOUND DURING INTELLECTUAL PROPERTY RESEARCH

Note: Patents are enforceable for 20 years from date of issue.

Patent D549327, Handle for Surgical Instruments. 2007.
Patent 5556092, Ergonomic Handle. 1996.
Patent 399722, Handle for Hand Tools. 1998.
Patent, D436801, Knife Handle. 2001.
Patent, 5860190, Expanded implement handle grip. 1999.
Patent, 701379, Balistreri. 1955.
Patent 3037783, Handle for Rapid Interchangeability of Various tools. 1962.
Patent, 6408524, Tableware Grip Structure with Comfortable Touch Feeling. 2002.
Patent, D536784, Handle for Medical Instrument. 2007.
Patent, D543433, Rotatable Tool Handle. 2007.
Patent, D557584, Driver for a Handle for a Tool. 2007.
Patent, 11/471,307, Handle and Method of making Thereof. 2007.
Patent 6500187, Scalpel with a Double Grind Blade Edge and Detachable Handle. 2002

APPENDIX 4: FLYER FROM GURU'S DISTRIBUTOR

Source: Guru Instruments

APPENDIX 5: GURU INSTRUMENTS WEBSITE

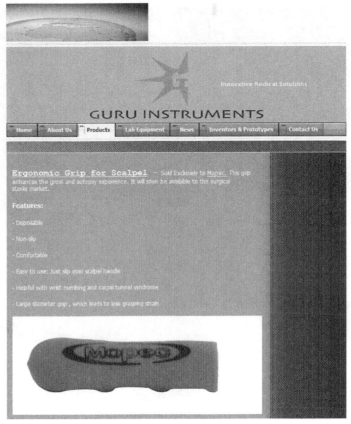

Source: Guru Instruments

REFERENCES

American Veterinary Medical Association, 2009. *Market Research Statistics.*http://www.avma.org/reference/marketstats, first accessed September 2009.

Aurora Health Care, 2009. *Autopsy.* http://www.aurorahealthcare.org/yourhealth/healthgate/getcontent. asp?URLhealthgate="14771.html", first accessed September 2009.

Centers for Disease Control and Prevention. 1997-2006. *National Hospital Discharge Survey.* http://www.cdc.gov/nchs/nhds/nhds_products.htm, first accessed September 2009.

Noyes, K and Creehan E. 2007. Spay or Neuter Your Pet. *Charityguide.org.*http://www.charityguide.org/volunteer/fewhours/spay-neuter.htm, 2007.

U.S. Bureau of Labor and Statistics, 2009. ftp://ftp/bls/gov/pub/special.requests/ep/ind-occ.matrix/occ_pdf/occ_29911-31.pdf, first accessed September 2009.

U.S. Food and Drug Administration Code of Federal Regulations 878.4800. 2009.

U.S. Food and Drug Administration. 2009. *Device Classification.* http://www.fda.gov/MedicalDevices/DeviceRegulationandGuidance/Overview/ClassifyYourDevice/ucm051512.htm, first accessed September 2009.

U.S. Food and Drug Administration. 2009. *How to Market Your Device.* http://www.fda.gov/MedicalDevices/DeviceRegulationandGuidance/HowtoMarketYourDevice/RegistrationandListing/ucm053165.htm#specification, first accessed September 2009.

U.S. Food and Drug Administration. 2009. Forms 2891 and 2892. www.fda.gov, first accessed September 2009.

Growing Pains at Camelot Biopharmaceuticals

LYNN JOHNSON LANGER
Johns Hopkins University

- **Summary of key issue:** This case study for Camelot Biopharmaceuticals (name disguised), a development stage biotechnology company, focuses on the board of directors' assessment of the CEO. Using management team interviews from a 360 degree performance assessment, the Board must assess the CEO's performance and potential for future growth. Although Camelot had advanced products into early stage clinical trials, the organization as a whole still maintained a highly scientific and research focus as the company was evolving towards a focus on product development and strategic alliances.

- **Key words:** leadership, decision making, management, scientist founders, leadership style

---

*   *This case was prepared as a basis for class discussion rather than to illustrate either effective or ineffective handling of an administrative situation.*

INTRODUCTION

In Fall 2008, Steve Barker, chairman of the board of directors convened a meeting with Julie Chan, board member and chair of the nominating & governance committee at Camelot Biopharmaceuticals.[1] Barker, a venture capitalist whose firm had invested in the first two rounds investment, had just met with a prominent healthcare investment banker, Lance Bryant, about next steps for the Boston, Massachusetts-based biotechnology company that was pursuing novel cancer stem cells therapies. Bryant had taken Arthur Morgan, president & CEO of Camelot Biopharmaceuticals, on a two week road show to gauge institutional investor interest to participate in a SPO (secondary public offering) to finance future growth. Bryant was candid and frustrated with the investor visits and reported to Barker: "Arthur is a an outstanding scientist, but investor feedback is very muted, particularly amidst this difficult financial market. Investor feedback indicated he struggled to articulate the company's future, as well as seemed uncomfortable in teaming up with the two other Camelot executives with whom he travelled. In short, investors are wondering if this company can grow and scale with Charles in charge and seemingly trying to do everyone's job."

Chan took in the seriousness of the discussion: "This is a good opportunity to review Arthur's performance and growth potential. The executive team just completed their 360 evaluations and I'm meeting tomorrow with the head of human resources, Gwen Hill, to review the results.[2] " Barker agreed, "Please be prepared to debrief the full board at our meeting next week. Let's be careful, Arthur has been critical to Camelot's success and we need to make sure we're not too hasty in responding to banker feedback."

COMPANY OVERVIEW

Camelot Biopharmaceuticals (company named disguised) was founded in the late 1990's with $3 million in venture capital funding. By 2001 the company had raised $70 million in private capital, one of the largest-ever financing deals for a U.S. biotechnology firm. For the first nine months of 2007, revenues were less than $1 million, and the overall net loss was more than $15 million. However, the company had more than $30 million in cash, cash equivalents, and marketable securities. By this measure, Camelot Biopharmaceuticals is considered a successful biopharmaceutical company even before it has products on the market.

The company's initial public offering was in 2005, and Camelot's growth strategy has focused on pursuing strategic alliance partnerships with large multinational pharmaceutical firms. The company's lead drug candidate is in phase II clinical trials. Of its 60 employees, approximately 20 are in business and 40 are scientists. Several executives were hired in the last few years to develop the company's first drug products. Arthur's natural style is collaborative and as a leader he emphasizes consensus building. As Camelot transitions from a science-focused company to one focused on the clinic and patients, the style of leadership that worked well in the past is not working as well with newer executives. Camelot

---

1   *The nominating & governance committee is responsible for assessing and nominating members of the board of directors; making recommendations regarding the board's composition, operations and performance; overseeing the company's CEO succession planning process; developing corporate governance principles; and helping shape the corporate governance of the company.*

2   *360-degree feedback is an evaluation method that incorporates feedback from the worker, his/her peers, superiors, subordinates, and customers. Results of these confidential surveys are tabulated and shared with the worker, usually by a manager. Interpretation of the results, trends and themes are discussed as part of the feedback. The primary reason to use this full circle of confidential reviews is to provide the worker with information about his/her performance from multiple perspectives. From this feedback, the worker is able to set goals for self-development which will advance their career and benefit the organization. 360-degree Feedback, or multi-rater feedback, was used by 90% of Fortune 500 companies last year.*

Biopharmaceuticals is in transition and experiencing growing pains.

THE EXECUTIVES

There were nine senior executives on the Camelot Biopharmaceuticals team. 6 of 9 executives were with the company less than two years, but all the newer hires were very experienced biopharmaceutical industry veterans with a diverse set of small and large firms. Key team members included:

- **Arthur Morgan, President & CEO:** co-founder and president and CEO since the company's inception in 1999. Prior to founding the company, he was senior scientist and group leader at a large biotechnology company. He earned his Ph.D. in human genetics and cell biology and published more than 50 scientific papers and holds several patents.
- **Lance Jones, Executive Vice President & Chief Business Officer:** joined the company in 2006 and is responsible for developing new commercial opportunities, directing all business development activities, and contributing to corporate strategy. Formerly, he was president and chief executive officer of a pharmaceutical organization, the president and chief operating officer at a therapeutics organization, and the senior vice president of business operations at another biopharmaceutical company. Prior to this, he worked more than 15 years with a major pharmaceutical company. He has an MBA degree in finance and pharmaceutical marketing.
- **Gil Knight, Executive Vice President and Chief Financial Officer:** past experience includes senior vice president and chief financial officer of a major biotechnology supply company, the chief financial officer, secretary and treasurer, corporate controller, accounting manager, and budget manager at other life sciences companies. He has an MBA.
- **Gavin Cornwall, Chief Medical Officer:** joined the company in 2006 and is responsible for all clinical, medical, regulatory affairs and is the key contact with the USFDA. In the past, he was group director of oncology for a major drug company. He is an MD and is a board-certified oncologist.
- **Tristan King, Senior Vice President, Product & Pharmaceutical Development:** joined the company in 2002 as a senior scientific director and has served as senior vice president of product and pharmaceutical development since 2006. Prior to joining the company, he worked at a major drug company as a principal scientist. He has a Ph.D. in molecular and cell biology.
- **Jeffrey Kay, General Counsel & Senior Vice President of Operations:** a co-founder who was a government attorney prior to his current position. He has a JD and MBA.
- **Luke Fisher, Vice President of Research:** joined the company in 2000 and became vice president of research in August 2007. Prior to that position he was an academic professor. He has a Ph.D. in biology.
- **Gwen Hill, Senior Director of Human Resources:** has worked for the company since 2001. Prior to 2001, she was director of human resources for several organizations and a private consultant. She has a M.S. degree.

CEO AS LEADER

Arthur's collaborative nature in decision-making worked well when the company was smaller and more oriented toward science. As the company grew, however, Arthur's decision-making skills did not evolve as quickly. He is viewed by many of the executives as indecisive. Arthur acknowledges that he will need to adjust his style if he is to continue to lead the company toward success. He noted that because he wants to be inclusive and collaborative, he has not always made difficult decisions quickly enough. Arthur concedes:

> *"It was really obvious to everyone, including me, that a certain situation needed be moved in a certain direction or a certain person needed to be moved out of a leadership position. But because I was so busy being inclusive and collaborative, I did not make the hard decision."*

Arthur admits that his difficulty with decision-making creates "confusion and resentment in the organization at the senior level." Some executives felt that "a decision for Arthur is putting off a decision." Other senior executives felt that Arthur should learn to be more decisive, or he should step down. But at least one executive, Gwen Hill, director of human resources felt that Arthur was making progress in changing his approach. *"Arthur used to be terrible at making a lot of decisions, but he has really grown a lot in making the big decisions; and he's also willing to delegate decisions better than he ever did before."* The consensus on the issue underpinning the lack of decisiveness was that Arthur has difficulty making decisions because he wants everyone to like him.

one executive recounted a recent incident in which both the chief medical officer and the executive vice president/chief business officer left Arthur's office after barging in shouting:

> *"The chief medical officer, Gavin Cornwall stormed to Arthur that he cannot make a change to a clinical trial that is already underway. Luke Fisher, VP of R&D countered that he wanted to tweak the process one more time to increase its efficacy. Earlier he had agreed with Luke with that changing the process may increase efficacy. The scientist in Arthur enjoyed engaging with Luke, like the old days, when they worked in the same lab. Of course, Gavin was right. Arthur could not change a process that had already been approved by the FDA, but he seemed to have a hard time making and sticking with a decision."*

Arthur engaged Harry Roman, an executive coach to help him come up with new skills for leading and managing. Comments from a 360 degree analysis instrument used by Harry reached the same general conclusion: Arthur is reluctant to make difficult decisions, is tentative in the face of a vocal minority, and then loses focus on the critical path. He needs to delegate decision making authority to others or garner insight through his people and make the decision himself. Rather than asking people what they think about a pending decision, he should ask them what they see as the consequences of making a particular decision. Arthur's greatest leadership challenge is his ability to make timely decisions.

Originally, decisions were made by all the executives reaching consensus. However, as the company grew and hired more executives, this approach became too cumbersome and the process changed. The addition of new executives caused some of the former executives to feel left out of certain decisions, and the new executives sometimes feel frustrated that they do not always have the autonomy they expect. Gil Knight, a long time employee and CFO, states:

> *"There are new folks in the mix and they have different definitions of roles and responsibilities and the way decisions should be made, versus some of the older folks. The challenge at the senior level is how we resolve that. The older folks are used to a very small group where everybody gets together and discusses a problem, and then actually it isn't really clear who has responsibility for an issue. In some cases, it wasn't clear who could actually make the decision and run with something, or who could veto it. Newer execs came in and have a more hierarchical orientation, and they say, 'This is my area, so it's my call. I'm happy to have your input, but at the end of the day, I decide.' In*

the leadership ranks, there's been confusion and testiness at times. But the guy at the top of any organization has to set the roles and responsibilities and the culture of how decisions are made. I would say that's an issue here. We started out with a very small entrepreneurial group and a lot of consensus. It's hard to move away from that as we grow bigger."

Gil noted that consensus worked in a small group "because it allows you to get buy in, get opinions, and make sure you're not making mistakes." Decision-making by consensus stopped working as the company size approached 20 employees. In a small company, everybody wears multiple hats by necessity; there are not enough people to cover all the administrative, executive and scientific tasks, so everyone chips in. When the number of employees reached 20, the administrative structure changed into teams with individual responsibility. When everybody is not involved in every task and each individual is not aware of everything that is happening, the consensus decision-making process slows down to accommodate the need to educate, negotiate, and reconcile the opinions of each individual. It no longer works efficiently.

On the company science, major decisions are no longer made with everybody, including the scientists, sitting around the table and involved in discussions. This has led to dissatisfaction in the research group, who now feel left out. Moreover, many of the scientists struggle with business decisions that need to be made, but which impact the science and research program. Compartmentalization of business and science sides of the organization occurred as the company grew and the potential products began transitioning into the clinic, and communication between these two essential sides of the organization became more formal and less inclusive. Management of the dynamic tension between the two groups is multifactoral and a challenge for all biotechnology companies. As Luke Fisher, VP for R&D pointed out, *"There is the managing of the personalities, which is much more of an organizational management; there is the managing of the science, which is much more of an expert domain. The people that are experts get to make the decisions."*

The need for a clear decision-making process is becoming critical as the company grows, but the process needs to be efficient. As one executive expressed:

> *"The counter side to the need for process is the need for speed. There's a dynamic tension between how much time you spend on getting tasks done and how much time you spend discussing and deciding how the task should be done."*

The entrepreneurs in the senior executive leadership felt that it is far easier to make decisions in a small organization. As one noted:

> *"In a small organization you can make the wrong decision and figure it out and change it much easier than you can in a large organization. But at least you start to move things when you make a decision."*

Senior executives articulated two key decision making and leadership traits that will be important for company success: ability to make difficult decisions on resource allocation and ability to adapt and change. Particularly in biotechnology companies that are financed by equity sales and not product sales, the volatility of the financial markets requires that cash outlay be carefully managed, since whatever cash is available may be the last the company can raise. One executive said:

> *"You can't run a large organization by treating everyone equally, since not every function is equal in an organization. There are some functions that are more important to company success than others. Resources should be applied to those tasks that are criti-*

cally important for an organization. And that requires making difficult decisions."

As the company grows and its products mature in development, the decisions of limited resource allocation will become more and more difficult. Arthur will also need to shift from a focus on science to "a corporate ambassador both of the board and to the investors." The company is at an inflection point and will need to adapt its management style to accommodate the internal change in focus from discovery to development and the external changes borne of the rigors of running a public company.

LEADERSHIP

The senior executives described Arthur's leadership role as someone who is the "synthesizer and the driver towards the vision," has "the ability to select and motivate the right people," and has "the ability to get the job done while adapting to changes in the internal and external environment." Communication skills and experience were also considered essential.

Several executives expressed concern that as the company grows, Arthur may not be as visible to the organization because his duties and responsibilities will bring him outside the company. At this stage of company growth, Arthur can still "walk the length of the building... seeing people and faces and recognizing that they aren't just a name on an org chart." As one executive worried:

> *"Arthur sets the culture. As long as he does that, people will continue to admire and work for him. If we get bigger and he... gets less visible, and travels a lot, you know, things permeate."*

All of the leaders at Camelot Biopharmaceuticals are white men. There is only one female in the executive suite and she is not a senior executive, but a director. She stated, "I don't like that we don't have enough female people in leadership positions. I am the lone female, and I am a fairly typical female leader. I am not confrontational. I don't like a lot of things that go with [being alone]."

Arthur used the words 'collaborative' and 'inclusive' to describe himself. These traits are evident in his use of the 360 degree assessment tool that he commissioned to gain more information about himself. Arthur feels that his collaborative nature and his ability to:

> *"...see a relatively clear vision and at the same time, be able to articulate and get buy in from various stakeholders and constituencies is one thing that I think has both seen [Camelot Biopharmaceuticals] through some very difficult times in the biotech world and some ups and downs in the last eight years in which many companies that were in our generation went out of business."*

Another example of Arthur's inclusive nature is his ability to work well with his board of directors.

> *"I was able to sit with all of those venture capitalists that were in my board room one by one and make them feel they are the part of very inclusive team which had total transparency to the management."*

Arthur has been extensively interviewed in trade journals as a young, successful entrepreneur. He described his views about leadership traits that can lead to failure in one interview (citation omitted to protect identity of CEO).

> *"Egomania is a destructive trait. Thinking that you are smarter than everybody else only means that you aren't. Thinking that your idea and way of doing things is always*

the best and should be adjusted only in the most dire circumstance is the kiss of death for any company, sooner or later."

The senior executives described Arthur as honest and non-confrontational. Another leadership attribute is "getting people excited and jazzed up, you know, empowering the heart." Several also said that Arthur admits failings, and that admission "makes you want to help fill in those bits and pieces. Those are the characteristics that make you want to follow him."

In the 360 assessment, respondents were asked to describe Arthur's greatest leadership asset: skill or talent. Words used to describe Arthur did not vary much from the comments made by the senior executives: humility, ability to seek and receive feedback, ability to communicate the vision both internally and externally, ability to inspire people to work hard, trustworthy, and putting people first with praise. When asked about his greatest leadership challenge or area for development, the same respondents also mentioned traits that the executives used: decision-making ability, emotional and overly optimistic.

CONFLICT

Tensions exist in Camelot Biopharmaceuticals between scientists and business people in part because the company still retains a strong science focus. The business executives ask:

> *"How do we take this cutting-edge science and turn it into something that will generate revenues, a return on investment for investors, and provide benefit to patients?*

Part of the tension results from the fact that many of the scientists have a strong academic focus, with little company experience. However, the research executives felt that the scientists were also very used to uncertainty and constant change because research is inherently uncertain, and that this lent them an edge over the business people, who have to adjust development and market plans regularly as the science, and its uncertainty, progresses. These changes, and the inability to follow a plan once it's articulated, drive the business side of the organization crazy:

> *"On the business side, it's best to have predictability. On the science side, you can't predict that a clinical trial is going to work or not work. You can't even tell whether it's going to take six months or a year. It just depends, and you can't say that on the business side."*

Part of the scientific process requires that scientists continue to work on a project until they have answered the questions they posed when doing the research. This process does not always work in a business environment, because projects often need to end by a certain date, or there are milestones that need to be met regardless of whether enough information has been obtained. This limitation may result in projects being discontinued if they cannot be resolved in a timely manner. The business execs ask "when do you quit something that isn't working out?" The scientist always says, "Well, I'll run another experiment and it might work this time," or "I have a new idea and I can try this new thing." And the business person says, "Well, you know, three strikes and you're out."

This difference between scientists and business people is a large source of conflict between the two. Most agreed that articulating milestones and goals and actually having to achieving them within a strict timeline, given the uncertainty in the scientific process, makes scientists uncomfortable. One research executive said:

> *"It's very difficult for a scientist to make that kind of organizational leap because, not only are you not trained to do it, you're actually trained to do the exact opposite. You*

work out a lot of things yourself. That's the one advantage I have on both the research
side and the business side: I actually like figuring out milestones, so it isn't a necessary
evil to articulate them; it's just a necessity."

As the organization changes to become more business focused, some executives felt conflicts would be inevitable and that it would play out mostly at the level of the executives, since they can't let the anxiety of running a business percolate all the way down to the rest of the organization. However, if an organization is not ready for change, then conflicts could extend to the Camelot Biopharmaceuticals as a whole. Most people want to maintain the status quo and are resistant to change, and this can play out as turf battles or anxiety over status or control.

Even though there is conflict within Camelot Biopharmaceuticals, most feel that change will be easier for them, a smaller organization, than larger organizations where a new idea or structure "goes through one layer of review to another layer of review…" seemingly ad infinitum.

POWER

Most executives in Camelot Biopharmaceuticals work in the business side of the company. Several people interviewed commented that there is conflict over power within the business area, struggles that may be inevitable as the company becomes more business focused. This power struggle appears to have permeated the support staff, and some support staff are becoming territorial.

Two of the executives described conflict with one particularly difficult executive, and there is not a mechanism in place in the company for dealing with difficult people. Each exec approached the conflict in a different way: one felt that "if you can frame things as clearly as possible in business terms, it goes a very long way toward resolution of tactical disagreements;" the other thinks that the way to avoid conflict is to not "overplay it and alienate everybody. Then it will never go anywhere. You've got to be smarter about the way you do it." Unfortunately, the second exec did not describe how to "be smarter" in difficult situations, and neither exec felt that Arthur was particularly supportive in rancorous discussions with the person in question.

CHANGE

Change is inevitable in a biopharmaceutical organization if it is to be successful. In particular, in companies headed by a scientist/founder, there is a critical transition from an early-stage science focus to that of a product-oriented business. The critical factor is an ability of Arthur, the management team, and the company to change and adapt as the business model evolves. In Camelot Biopharmaceuticals, Arthur has worked hard understand his strengths and weaknesses so that he could develop into a more decisive leader and alter his decision-making process from that which worked well when the company was small. Several executives commented on the transition from scientist to business person of Arthur:

> *"He has separated from the operational part of the Camelot Biopharmaceuticals and*
> *has become more of a CEO. He's lost some of his scientist persona because he had to;*
> *that was the only way he can survive, actually."*

This change is critical not only for the success of the company, but also for Arthur's long-term employment. As one of the top executives said:

> *"That is also a part of the cultural change of a startup biotech developing into to a com-*
> *mercial entity. The cultural change is something that's got to be driven by Arthur. And*

if he can't do it, there may have to be changes at the top."

Newer executives hired into the organization brought different ideas and way of doing things, and this has sometimes resulted in tensions among the group, including resistance to internal reorganization. Change is hard for many people even when they know it is inevitable and the reason for change is good.

> *"When we go to the next level, perhaps the whole senior executive group will have to make an important transition. Management will really have to become the business people running the company. When we were 30 people, if you had an issue, you walked next door and talked to the person who did it. There weren't that many layers. Then we got to 60 people, and it will be worse when we grow even more, and there are now operational groups with intermediate managers. The opportunities for informally solving problems are much fewer, and fewer problems get solved."*

Because growth in the company has been slow for the past few years (a result of a partnership opportunity that fell through), some feel they were given time to adjust to the coming changes. A few of the executives prepare for the change to a more business-focus by reading books or taking classes, while others do not think preparation is necessary. Arthur instituted training classes to instill team spirit, but many of the executives are too cynical or confident in their own experience to take them seriously. All want change to happen quickly and many are frustrated that the process is slow. The executives estimate that 10-15 percent of the company is very change-ready; another 10-15 percent is very change-resistant; and the balance are "figuring out who is going to win and then they will decide which team they are on."

SUCCESS AND FAILURE

All the executives indicated a strong desire to create better treatments for cancer rather than to simply create monetary value for the company, although all acknowledged that as a clearly necessary component of success. Even if they eventually fail as a company, if their science proves successful, the executives believe that constitutes a success.

Both management and scientists feel as if they are working towards a common goal. Even those who see the company with a strictly business focus acknowledge:

> *"You wouldn't launch a product that didn't provide patient benefit. As soon as the product starts going into humans, however, you have to start preparing for potential success, success meaning a product approval and commercial launch."*

Almost all the executives had suggestions for how the company can continue to be successful. They mentioned learning, being adaptable, having a strong science vision, having successful partnerships, planning for inevitable failures, perseverance, and good leadership as necessary for success. In an outside, published interview, Arthur advised,

> *"Seek out and listen to advice from experienced and successful entrepreneurs. Listen and learn from your employees. Listen and learn from your competitors and be flexible enough to incorporate what you learn."*

Several executives mentioned that successful companies plan for the failures that invariably happen by building in flexibility, either by redundancy or by the ability to change to a successful program.

And a strong vision of how the science can lead to drugs is critical. Given the business model of the company, strong partnerships are also critically important.

> *"Developing partnerships that will bring in funds during the development of your products and revenue from royalty on sales of your product when it's approved, even though you are not selling product yourself, will help you get to profitability, the holy grail of biotech."*

One executive is concerned that the company may not continue its former success with the senior executive team currently in place.

> *"It is my belief that the senior management team in place now will not, if the company is to be successful, be the same senior management team in the next two or three years. So that is my Draconian forecast. In order to be successful, I don't think the same team should be in place."*

Others believe that the path forward is now clear and the management is in place to drive success, although the uncertainty around the science is always a concern in the back of their minds.

All the executives offered reasons why most biotechnology companies fail: intellectual arrogance, unrealistic valuation expectations, inappropriate skill sets, and unprepared for change. Arthur stated:

> *"When entrepreneurs cannot accept the reality of the market for their idea, patent, company or services, they end up with nothing. Whether it's start-up companies seeking capital or later-stage companies trying to make deals, it's better to have a small piece of a big pie than a big piece of nothing."*

Some of the business executives felt that companies with scientist founders fail because many scientists lack the necessary skills to successfully run an organization.

> *"I think a lot of businesses fail, especially when a scientist from a university or the government starts a business. All the technology is there, or at least envisioned, but how to put it into practical application often is not. Scientists often aren't great businessmen."*

Two executives talked about failure as being inevitable in science and in building businesses. Planning for failure by redundancy in programs or multiple approaches is one strategy. In biotech, failure can be due to unanticipated scientific glitches in discovery or development, failure to meet clinical safety or efficacy endpoints, or economic conditions that preclude adequate financing to keep the products moving through their clinical paces. Or it could be that the employees or management are not able to adapt and change as the organization evolves.

VISION

Arthur has a clear vision of what Camelot Biopharmaceuticals can be:

> *"I think you've got to have a good vision of how the science can lead to drugs or solve some therapeutic issue or problem. In biotech, you've got to have the scientific vision. I guess that's a critical part of a success. Leadership is a good vision, in this case a scientific vision."*

Many in the company, including Arthur, center their vision on the science of the company rather than whether the science can be successfully developed and sold as a product. Although proof-of-principle for the science has not yet been demonstrated, no one interviewed said they should stop the effort.

> *"The vision of our founders and of our original investors was to do 'this' and we've stuck to it. The scientist CEO/founder articulates the vision. Just because it's taking us longer to get there, it's still working. That doesn't mean it's broken and that we should deviate from it. We should stick to it."*

GOALS

Most executives agreed that the company sets clearly defined goals and all are expected to meet them. Arthur stated,

> *"We set yearly goals as an organization and we formally review those goals in all of the departments. I communicate a lot around those goals, and I pay a lot of attention to whether or not the things that we are tactically and practically doing articulate them. The meanings of those goals are written with a very clear business focus and that helps the scientists learn how to frame their goals."*

The 2008 goals, in progress at the time of the interviews, are decided primarily by the scientific directors in concert with the VP of Research. Then the executives review the goals with the scientists and it becomes "socialized." The research scientists, however, are frequently uncomfortable with the timeframes associated with the goals.

> *"There are goals and there are milestones you're supposed to reach at a certain time. Sometimes it's earlier and sometimes it's later, and I think that makes putting the pro-totypical scientist into this framework uncomfortable."*

There is a staff meeting once a month, and once a quarter the executives review and adjust the goals. Arthur said that he has tried to create a corporate culture in which, if the goals are adjusted during the year, people are able to adapt. The goals are communicated to the employees through several channels, including face-to-face meetings. In addition, employees are given clocks that have the corporate goals printed on them. Each year, the goals are updated and new hires are given clocks with the goals. Executives are charged with making sure their own groups understand the goals and the executives are expected to talk with people about these goals in person. Although it is important for employees to hear how well the goals are being met, employees are not expected to worry about corporate goals, but to focus on their own job and let the executives do any worrying.

> *"We tell you what the goals are, but you need to think about the job you're doing, because for us to reach those goals we need you to do your job. We don't need you to be distracted and worried."*

The executives generate the goals for themselves and their own groups, and there is within the corporate culture a high threshold for vetting of a goal by other executives. Also within the corporate culture, the ultimate goal is to be proud of whatever it is that comes out of their efforts.

COMMUNICATION

When Camelot Biopharmaceuticals was smaller, people communicated regularly and informally. They also met and socialized with each other outside of work. Everybody knew everybody, and their wives, husbands, and children. As Camelot Biopharmaceuticals has grown, communication has become more of an issue. Communication is mainly done at staff meetings and follow-up meetings with senior executives. There is still cross-communication in the product research area, approximately 70% of the organization. The five research departments are collaborative and cooperative, and by necessity have to interact.

Because some in the organization complained that senior management had "lost touch with what was going on," a task force on communication was created. Arthur wrote mid-year goals for each of his direct reports that required that they had to, in two distinct ways, "every month move beyond their comfort zone to reach out to establish a deeper connection with the rank and file." That took on many different forms, depending on the executive, but all of them are accountable to Arthur on a monthly basis. This 'forced' communication resulted in mixed opinions from the senior executives. Some executives make a specific effort to talk with support staff and others in the organization. Some found a way to go into each of their subordinate's offices at some point during the month to chat. Some executives wonder if lack of communication is really an issue facing the company. Some believe Arthur has been "hijacked on this issue" and that Arthur may be overreacting. All think it is important that executives are available to answer questions, however, and set the vision and deliverables.

> *"It's their job to ask us what they don't understand or what they need to know. Part of our communication is just getting people excited about their jobs. And not only excited about their jobs, but they need to have confidence that the leaders know what they're doing and that they're worth following."*

CULTURE

Camelot Biopharmaceuticals began with five people. At that time, the culture was entrepreneurial and the there was no hierarchy. As Camelot Biopharmaceuticals has grown, so have its organizational requirements. Now there is a more hierarchical structure in place. Arthur noted:

> *"The culture that grew up out of the beginning was having everybody being more or less on the same level. No hierarchy. As we started to get bigger, I started to notice that the non-hierarchical approach was going to need to change as the company grew."*

The original core group of executives wanted to keep the flat structure as long as possible. It took time for the new hierarchy to be accepted.

> *"It's taken a while for us to build that into our culture. It was also a matter of our management team really showing the rest of the company that we could. By the time we went public, we had done it."*

The executives described the culture at Camelot Biopharmaceuticals as "respectful," "pleasant," "collegial," "family-like" and "pretty much everybody gets along." Some noted that this family-like atmosphere makes it difficult when an employee needs to be let go." However, some of the executives described two "subcultures" within the company, that of the business side and that of the scientists. Scientists make up about 70 percent of the employees at Camelot Biopharmaceuticals. They were portrayed by one executive as "people who really want to get the job done; people who are willing to put in long hours; people who like each other and don't tend to compete with each other. They see the greater good." This same executive described the business side of the company "a guarded competitive atmosphere" where there are tensions and frustrations among the employees.

Another aspect of the culture at Camelot Biopharmaceuticals is its public recognition of employees for work well done. Some think this is a good idea; others question whether it can sometimes be "patronizing." Arthur is proud of the way people are recognized at Camelot Biopharmaceuticals. He described the award for employees who are promoted and a program called Invention of the Year that is "similar to the Academy Awards." Arthur said that only once or twice has an employee refused to be publicly recognized, and one pleased employee told him that receiving the award was one of "the two most important things in his life."

PROCESS

The executives discussed the need for more processes as the company grows and increases its visibility to the industry and to the US FDA. A small company generally doesn't get a lot of regulatory attention. But it is critical to understand how to get a drug approved and what processes are required to comply with all the regulatory agencies that are involved in approval and commercialization. The balance for a small and growing organization is between adequate controls and processes and creation of an unwieldy bureaucracy.

Those who start companies are generally more focused on tasks rather than process because there is a large amount of work that needs to be done, time needs to be spent on the task at hand; however, processes need to be implemented to make sure there is consistency in the work. The executives described a tension between how much time can be spent on doing a task versus how much time should be spent designing the process for doing the task.

> "In a small company you don't have the level of resources that you typically would have in a larger company. You don't have folks who can really focus on process."

At least one exec described a need to have more training. He stated that some managers are not good at certain processes, such as evaluating people, but could improve with the proper training.

> "Here it's all very independent, so each manager fills in their form and you don't know how your people stack up versus other people. Those that do not improve could be moved to different positions. At a bigger organization, there's more infrastructure to train supervisors and bringing these skill sets to a standard. That's going to be a critical component of our growth."

RISK

Several executives described the risk they took in coming to a start-up biotechnology company. One executive described his previous job as being less risky and wondered, "what happens if the company falls on its face?" Others described being surprised at the high level of risk:

> "The reality in a biotech company is if you don't make your milestones, if you don't show progress on the money you've got, you're not going to be able to raise the next round of financing. In addition to the risk of the company failing financially and the uncertainty of the science, there is also the risk involved with so few key personnel. If one of the key staff were to leave or be unavailable for work, the perceived value of the company could drop."

Camelot Biopharmaceuticals mitigates some of this risk by developing some products that are more likely to succeed, but may not have as high a rate of return. Camelot Biopharmaceuticals also partners with other companies to bring in more revenues, while also developing some "higher-risk products that could have a big return for investors.

CONCLUSION

In line with best practice (Appendix A), Steve Barker, chairman of Camelot Biopharmaceuticals convened the board of directors meeting without the CEO present to review his performance evaluation. Barker reviewed the feedback from the latest banker-led investor meetings, as well as his own concerns about Arthur Morgan's performance. He then asked Julie Chan, board member and chair of the nominating & governance committee to summarize the results of the 360 evaluation which had been confidentially distributed to the board a few days earlier.

Chan began: "Gwen Hill, senior director of human resources, conducted a 360 evaluation and interviews she for Charles Miles. She walked me through an exhaustive review of the results, we distributed to the board, but two primary themes emerged. First, there is significant concern about decision-making and issues stemming from the overly collaborative style of the CEO. In turn, conflict and frustration that has evolved from the CEO's perceived ineffectiveness at decision-making. Although other themes emerged, they were much less prominent: change; success and failure; communication; culture; vision; goals; organization processes; and risk."

Barker addressed the other board members, and his mind wandered to the challenges of his beloved and beleaguered Washington Redskins who were on a losing streak, "Arthur has been the right leader to bring Camelot from founding to its current position as a publicly traded company. We're now rapidly progressing into new territory as we move towards advanced clinical development, partnerships with large multinationals, and commercialization. We can provide additional executive coaching[3] to see if Arthur can continue to grow to meet new challenges, change him to a position such as chief scientific officer which would leverage his core strengths, or replace him in favor of an experienced executive who has already led a larger organization. In short, we need to quickly decide if Charles is our quarterback of the future."

---

3   *Executive Coaching is a facilitative one-to-one, mutually designed relationship between a professional coach and a key contributor who has a powerful position in the organization... The coaching is contracted for the benefit of a client who is accountable for highly complex decisions with wide scope of impact on the organization and industry as a whole. The focus of the coaching is usually focused on organizational performance or development, but it may also serve a personal component as well.*

APPENDICES

APPENDIX A: CAMELOT BIOPHARMACEUTICALS, INC. NOMINATING AND GOVERNANCE COMMITTEE CHARTER

The primary responsibilities of the Board of Directors of Camelot Biopharmaceuticals, Inc. (the "Company") are to oversee the exercise of corporate powers and to ensure that the Company's business and affairs are managed to meet its stated goals and objectives. The Board recognizes its responsibility to engage, and provide for the continuity of, executive management that possesses the character, skills and experience required to attain the Company's goals and to ensure that nominees for the Board of Directors possess appropriate qualifications and reflect a reasonable diversity of backgrounds and perspectives.

In fulfilling their roles as directors, the Board will be guided by the following principles and objectives:

- Represent the collective interests of all stockholders of the Company;
- Discharge Board duties in good faith, with due care and in a manner he or she reasonably believes to be in the best interests of the Company;
- Possess independence, objectivity and the highest degree of integrity on an individual and collective basis;
- Be dedicated to understanding the business of the Company and issues presented to the Board;
- Be committed to active, objective, thoughtful, constructive and independent participation at meetings of the Board and its committees;
- Bring to the Board's deliberations their collective breadth of business, professional and personal experience to represent the interests of stockholders;
- Review fundamental operating, financial and other corporate plans, strategies and objectives;
- Evaluate on a regular and timely basis the qualitative and quantitative performance of the Company and its senior management;
- Review the process of providing appropriate financial and operating information, internally and externally;
- Assure adherence to proper policies of corporate conduct, including compliance with applicable laws, regulations, business and ethical standards;
- Assure maintenance of proper accounting, financial and other appropriate controls; and
- Evaluate and take steps to improve the overall effectiveness of the Board.

The Board has the responsibility to organize its functions and conduct its business in the manner it deems most effective and efficient, consistent with its duties of good faith, due care and loyalty. In that regard, the Board has adopted a set of flexible policies and practices to guide its governance practices in the future. These guidelines, set forth below, will be regularly re-evaluated by the Board's Nominating and Corporate Governance Committee (the "N&CG Committee") in light of changing circumstances in order to continue serving the best interests of the Company's stockholders. Accordingly, this summary of current practices is not a fixed policy or resolution of the Board, but merely a statement of current practices that is subject to continuing assessment and change.

A.   BOARD COMPOSITION.

1. **Size of the Board**. The Board of Directors currently has five members. The Board expects the size of the Board to increase, but the size of the Board will always be subject to change depending on circumstances existing from time to time. Accordingly, the Board will periodically review the appropriateness of the size of the Board.

2. **Role of Chair**. Because the Board believes it is important that the Board and the Stockholders be represented by a leading director, there shall be a non-executive Chair of the Board, who shall be selected by the entire Board on an annual basis. The role of Chair shall not be held by the Company's CEO or other member of senior management, but shall be a pre-existing, independent member of the Board. Generally, the Chair's role shall be to lead the Board and to provide a bridge between the outside directors and the executive management of the Company, including the CEO. In particular, the Chair shall (i) preside over all meetings of the Board, (ii) be available to discuss with any independent director his or her concerns about the Company and its performance, and relay those concerns, where appropriate, to the full Board, (iii) be available to consult and discuss with the CEO regarding the concerns of the directors, (iv) be available to discuss with the CEO or other senior executives of the Company any concerns such executive might have, and (v) maintain close contact with the chair of each standing committee of the Board.

3. **Majority of Independent Directors**. It is the policy of the Board that two-thirds of the directors will not be current employees of the Company and will otherwise meet appropriate standards of independence. In determining independence, the Board will consider the definition of "independent director" in the listing standards of the NASDAQ Stock Market ("NASDAQ") (Marketplace Rule 4200), as well as other factors that will contribute to effective oversight and decision-making by the Board. No relationship between any independent director and the Company should be of a nature that could interfere with the exercise of independent judgment in carrying out the responsibilities of a director. The determination of what constitutes independence for an independent director in any individual situation shall be made by the Board in light of the totality of the facts and circumstances relating to such situation and in compliance with the requirements of NASDAQ's applicable listing standards and other applicable rules and regulations.

4. **Selection of Board Nominees**. The Board is responsible for selecting its members, subject to stockholder approval on an annual basis. The N&CG Committee shall recommend candidates for election to the Board. The N&CG Committee considers nominees recommended by directors, officers, employees, stockholders and others using the same criteria to evaluate all candidates. The N&CG Committee reviews each candidate's qualifications, including whether a candidate possesses any of the specific qualities and skills desirable in certain members of the Board. Evaluations of candidates generally involve a review of background materials, internal discussions and interviews with selected candidates, as appropriate. The N&CG Committee may engage consultants or third-party search firms to assist in identifying and evaluating potential nominees. Upon selection of a qualified candidate, the N&CG Committee will recommend the candidate for consideration by the full Board.

5. **Board Member Criteria**. The N&CG Committee is responsible for reviewing the appropriate skills and characteristics required of directors in the context of prevail-

ing business conditions and composition of the Board. The primary qualifications to be considered in the selection of director nominees include the extent of experience in business, finance, science or management; the extent of knowledge of the biotechnology and biopharmaceutical industries; and personal attributes, including without limitation, integrity, loyalty and the overall judgment to advise and direct the Company in achieving its corporate objectives and in meeting its responsibilities to stockholders, customers, employees and the public. In addition to these primary qualifications, the highest consideration will be given to those candidates who bring a critical set of skills to the Board (e.g., biopharmaceutical or biotechnology development, marketing or operations). It is also preferable for the candidate to have prior experience on a board of directors of a widely-held company of at least similar size and complexity to the Company, and to have significant experience such as CEO, president, COO, or CFO of a significant operating business or business segment. In evaluating candidates for the board, the N&GC Committee will consider all of these qualifications and criteria in light of specific needs of the board and the company at that time. The objective is to have a board that brings to the company a variety of perspectives and skills derived from high quality business and professional experience.

6. **New Director Orientation; Continuing Education**. Orientation materials will be made available and appropriate meetings will be held to acquaint new directors with the business, history, current circumstances, key issues and top managers of the Company. Directors are encouraged to also participate in continuing education programs in order to maintain the necessary level of expertise to perform his or her responsibilities. The Company's Secretary shall work with the Chair of the N&CG Committee as necessary to periodically provide materials that would assist directors with their continuing education.

7. **Directors Who Change Job Responsibility**. The Board does not believe directors who retire or change their principal occupation or business association or who experience another material change in the circumstances surrounding his or her employment or business associations or interests should necessarily leave the Board. However, promptly following any such event, the director must notify the Chair of the N&CG Committee and the Board Chair, giving appropriate detail of the reasons for and other circumstances surrounding such change. The Board, through the N&CG Committee, shall then consider such change and review the continued appropriateness of Board membership under the new circumstances.

8. **Retirement; Term Limits**. Although the Board currently believes that neither a fixed retirement age nor term limits are necessary or appropriate, it shall periodically review those positions.

9. **Other Board Memberships**. Without specific approval from the Board, no director may serve on more than 4 public company boards (other than the Company's Board). Without specific approval from the Board, the Company's executive officers should not serve on any public company boards (other than the Company's).

10. **Board Evaluations**. The N&CG Committee will conduct an annual assessment of the overall effectiveness of the organization of the Board and the Board's performance of its governance responsibilities. The Committee will report its findings to the whole

Board for discussion and the Board shall conduct an annual performance review of itself based upon such report.

B. BOARD MEETINGS AND MATERIALS

1. **Frequency of Board Meetings; Attendance**. Currently, the Board has four regularly-scheduled, in-person meetings each year, with additional telephonic meetings as required from time to time. At least one regular meeting per year will be scheduled as an all-day meeting. The Board considers its current meeting schedule to be adequate, but the number of regularly-scheduled meetings may be adjusted as necessary to meet changing conditions and needs. A calendar of Board meetings will be developed and circulated as far in advance as practicable. The Board intends that the location of its 4 regular meetings will rotate between the San Francisco Bay Area and New York City. Members are expected to attend all meetings barring special circumstances.

2. **Agenda; Distribution of Meeting Materials**. The Chair, with assistance and input from the CEO, develops the agenda for Board meetings. The agenda is circulated in advance and Board members may suggest additional items for consideration. As much information and data as practical on the meeting agenda items and the Company's financial performance is sent to Board members as far in advance of the meeting as is reasonable and practical. In addition, in months in which the Board is not scheduled to hold a regular meeting, the Company's executive management will provide to the Board reports or other materials concerning the Company's operations, finances and other appropriate matters.

3. **Independent Director Discussions**. It is the policy of the Board that the independent directors meet separately without management directors at least twice per year to discuss such matters as the independent directors deem appropriate. The Company's independent auditors and counsel may be invited by the Board to attend all or a portion of these sessions.

4. **Access to Senior Management**. The Board encourages the presentation at meetings by managers who can provide additional insight into matters being discussed or who have potential that the CEO believes should be given exposure to the Board. Beyond Board meetings, all Board members shall have access to senior management, provided such contact is minimally disruptive to the business operation of the Company.

C. BOARD COMMITTEES

1. **Standing Committees**. The Board currently has four standing committees: Audit Committee, Compensation Committee, Finance and Transaction Committee and N&CG Committee. From time to time the Board may establish a new committee or disband a current committee if the circumstances so warrant. The Board has adopted a charter for each standing committee of the Board; each committee shall annually review their charter and recommend to the full Board any changes it believes are appropriate.

2. **Committee Member Selection.** The N&CG Committee is responsible for reviewing and recommending to the Board the assignment of directors to various committees. The N&CG Committee will also recommend to the Board, subject to applicable membership requirements and as practical, an appropriate rotation process, to ensure diversity of Board member experience and variety of exposure to the affairs of the Company. The members of the Audit, Compensation and N&CG Committees will consist solely of independent directors and will have all such other qualifications as are required by NASDAQ listing rules or applicable law. The Board will make committee assignments on an annual basis at the first regular meeting of the Board following the Company's annual meeting of stockholders.

3. **Committee Functions.** The number and content of committee meetings and other matters of committee governance will be determined by each committee in light of the authority delegated by the full Board to such committee, the committee's charter and applicable regulations or principles. The Company shall provide to each committee access to employees, counsel and other resources to enable committee members to carry out there responsibilities. The full authority and responsibilities of each committee is fixed by resolution of the full Board and/or the committee's charter, if any. Committee charters shall be posted on the Investor Relations page of the Company's web site.

D. MANAGEMENT RESPONSIBILITIES

1. **Management Succession Planning.** The CEO will annually review succession planning with the full Board as it relates to elected corporate officers, and make recommendations to the Board with respect to individuals to occupy these positions in the event of those officers' termination of employment, disability or death. The entire Board shall annually recommend and approve the succession plan relating to the CEO. The Board, at all times, will have plans for the immediate replacement of the Chair, CEO and CFO.

2. **Financial Reporting, Legal Compliance and Ethical Conduct.** The Board's governance and oversight functions do not relieve the Company's executive management of the primary responsibility for preparing financial statements that accurately and fairly present the Company's financial results and condition. Executive management shall maintain systems, procedures and a corporate culture that promote compliance with legal and regulatory requirements and the ethical conduct of the Company's business.

3. **Corporate Communications.** The Board believes the executive management has the primary responsibility to communicate with investors, the media, employees and other constituencies that are involved with the Company, and to set policies for those communications.

REFERENCES

Charan, R. 2005. *Boards that Deliver.* New York: John Wiley & Sons.

Terry A. Beehr, Lana Ivanitskaya, Curtiss P. Hansen, Dmitry Erofeev and David M. Gudanowski. 2001. Evaluation of 360 Degree Feedback Ratings: Relationships with Each Other and with Performance and Selection Predictors. *Journal of Organizational Behavior,* 22 (7), 775-788.

Tornow, W., London, M. 1998. *Maximizing the value of 360-degree feedback.* San Francisco: Jossey-Bass Inc.

Sandhill Scientific: Where to Manufacture?

STACI D. SANFORD AND ARLEN MEYERS
University of Colorado Denver

- **Key issue/decision**: Sandhill Scientific is the world's largest manufacturer of diagnostic esophageal monitoring equipment and other devices associated with gastrointestinal function. This case follows Rick Jory, CEO of Sandhill, as he faces the key strategic challenge of deciding where to relocate his overseas manufacturing operations. At issue was that Sandhill faced rising manufacturing costs in the face of globalization and competition.

- **Companies/institutions**: Sandhill Scientific, Inc., www.sandhillsci.com

- **Technology**: Sandhill designs, develops, and manufactures medical diagnostic equipment including products used to measure the motility of the esophagus and devices that measure acid or non-acid reflux from the stomach into the esophagus. This is part of the evaluation of patients with an array of swallowing disorders as well as those with signs and symptoms suggestive of gastroesophageal reflux disease (GERD).

- **Stage of development at time of issue/decision**: Sandhill has been selling its products since 1981 to customers throughout the world. Sales in 2008 approached $20 million. During the global downturn in 2009 Sandhill's revenues remained flat, but are expected to grow in the low double digits in 2010 and beyond.

- **Indication/therapeutic area**: Swallowing and esophageal motility disorders and gastroesophageal reflux disease.

- **Geography**: U.S., Europe, Asia

- **Keywords**: esophageal motility disorders, gastroesophageal reflux disease, GERD, global manufacturing, global supply chain

---

*   *This case was prepared as a basis for class discussion rather than to illustrate either effective or ineffective handling of an administrative situation.*

INTRODUCTION

Sandhill Scientific is the world's largest manufacturer of esophageal monitoring equipment and other devices associated with gastrointestinal function. Sandhill was founded in 1981, with the company's primary focus on the development of diagnostic instrumentation used by gastroenterologists. It offers reflux monitoring recorders and equipment, the associated reflux monitoring probes, impedance/manometry esophageal function diagnostic equipment, including high-resolution impedance manometry, and systems which diagnose liver stiffness (cirrhosis). A separate section of the company focuses on world-class clinical education offered to physicians and clinical practitioners regarding the company's products.

In 1994, the company established much of its current management team, including Rick Jory, CEO. Jory holds a degree in industrial engineering and an MBA in Finance. His past experience included work with both Cobe Laboratories and Valleylab. He is past board chair of the South Metro Denver Chamber of Commerce and former president of the Colorado Medical Device Association. At the time of Jory's appointment, Sandhill Scientific was using a U.S.-based contact manufacturer to produce a high-volume, single-use (disposable) catheter. However, due to relatively high U.S. labor costs, competitive pressures to reduce medical device pricing, and healthcare reform cost challenges, Sandhill could not afford to continue to manufacture their product in the U.S.

Around 2000, Sandhill made the decision to start manufacturing their catheter product line themselves, and set up their own manufacturing facility in the Czech Republic. However, several events moved the cost structure associated with the Czech Republic upward. These included the falling of the U.S. dollar, the impact of the Czech Republic joining the European Union, plus inflation in the Czech Republic. This prompted Sandhill to investigate whether relocation of their manufacturing operation would make sense—and to evaluate options for a new location. In evaluating their options, Rick Jory narrowed the choice down to three different locations: Costa Rica, India, and Vietnam.

COMPANY HISTORY AND FINANCIALS

Sandhill Scientific, a privately owned company, was founded in 1981 by two engineers and a retired executive from Atlantic Richfield Company (ARCO). Since Sandhill's founding, almost all of the company's focus has been toward the development of diagnostic instrumentation used by gastroenterologists. During the company's early years, sales were minimal and the company struggled financially. When the current management team arrived in 1994, sales were under $3 million, today they approached $20 million.

In the early days, the executives at Sandhill recognized a number of factors taking place: (1) smaller hospitals were buying their equipment, (2) hospitals were having staff turnover, and (3) equipment and medical procedures were getting more and more complicated. These three factors suggested the need for better clinical education to increase efficiency and reduce errors. Sandhill knew they could offer education without a large expenditure of money and/or investment, so they established what would become a world-class clinical education program. This program was impressive enough to attract the attention of the former president of the American Gastroenterological Association, Dr. Donald Castell. He joined Sandhill Scientific as the medical director in 1998, and continues the hold this position today. This clinical education program also serves as one of Sandhill's major marketing tools.

TECHNOLOGY, PRODUCTS, AND PIPELINE

GASTROESOPHAGEAL REFLUX DISEASE

Gastroesophageal reflux disease, commonly referred to as GERD, is a condition in which the gastric contents of the stomach regurgitate (reflux) into the esophagus (Ferri, 2009). GERD occurs when the lower esophageal sphincter (LES) fails to close properly (see Figure 1). The LES is a ring of muscle that is located at the bottom of the esophagus and acts as a valve between the esophagus and stomach (Ferri, 2009). When these gastric contents reflux back into the esophagus, they cause a burning sensation often called heartburn, and can inflame and damage the lining of the esophagus. The current definition for GERD is "a condition that develops when the reflux of stomach contents cause at least two heartburn episodes per week and/or complications" (Ferri, 2009).

The main symptoms of GERD are persistent heartburn and acid regurgitation; other symptoms can include pain in the chest, hoarseness in the morning, and trouble swallowing. GERD can also cause a dry cough and bad breath (Ferri, 2009). GERD does have the potential to cause serious complications. Inflammation of the esophagus can cause bleeding and/or ulcers. Scar tissue may also form in the esophagus and lead to difficulties in swallowing. In extreme cases, people develop Barrett's esophagus, a condition in which the cells in the esophageal lining take on abnormal shape and color, which over time can lead to cancer (Ferri, 2009).

GERD is typically treated, in escalating order of severity, by patient lifestyle modification (e.g., decreased fat intake, cessation of smoking); treatment with antacids and over-the-counter acid suppressants; and acid suppression with proton pump inhibitors (PPIs) are the mainstay of therapy (e.g., omeprazole, lansoprozole, rabeprazole, pantoprazole, and esmomprazole). In advanced cases, promotility therapy (e.g., metoclopramide, bethanechol) and surgery are also considered, albeit controversial (DeVault & Castell, 2005)

GERD is one of the most prevalent gastrointestinal disorders. The American College of Gastroenterology estimates that over 60 million Americans experience symptoms of GERD at least once a month and over 15 million Americans experience symptoms daily (www.acg.gi.org). Estimates of the total cost (both indirect and direct costs) of treating GERD are approximately $9.8 billion, making GERD the most expensive disease of the alimentary tract (Sandler *et al.*, 2002).

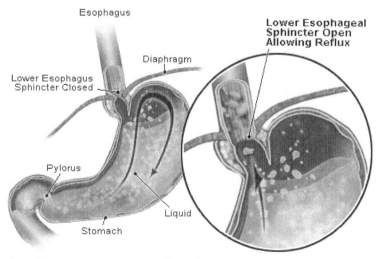

Figure 1: Anatomy of the GI tract and the mechanism of GERD
Source: www.empowereddoctor.com

DIAGNOSIS OF GERD

The most common way GERD is recognized is through a patient's complaint of frequent heartburn, usually occurring after meals and worsening upon lying down. However, this can be unreliable in that other conditions that can mimic GERD have been reported. Diagnostic tests for GERD include x-rays, endoscopy, and esophageal acid testing (DeVault & Castell, 2005). Initially x-ray was the only tool for diagnosing GERD; patients would swallow barium and then esophageal x-rays were taken. This test was limited in that it only showed complications that occasionally arose from GERD, such as ulcers. Patients who did not yet have esophageal lining damage would exhibit normal x-rays. Today, x-rays have been almost completely abandoned as a means of diagnosing GERD.

Endoscopy is another test used to diagnose GERD (DeVault & Castell, 2005). Endoscopy uses a tube with an optical system; this tube is placed down the gastrointestinal tract and the lining of the esophagus, stomach, and duodenum is directly visualized and examined. Endoscopy is beneficial in that it is the only test that can determine the presence of Barrett's esophagus (endoscopy coupled with biopsy) and other complications such as ulcers. However, the esophagus of most patients with GERD will present with a normal endoscopy.

Esophageal acid testing is the preferred method for diagnosing GERD (DeVault & Castell, 2005). There are several methods of esophageal acid testing; in the ambulatory nasoesophageal pH monitoring method a catheter is passed through the nose and into the esophagus. On one end of the catheter is a sensor that measures acid levels, and the other end exits the nasal passage and is attached to a recorder. The catheter monitors the level of acid in the esophagus for a 24-hour period. Each time acid is refluxed back into the esophagus, the catheter will sense the change in pH and send a signal back to the recorder. This also allows for the determining the acid levels in the esophagus with symptoms of GERD, such as heartburn, to confirm the diagnosis of GERD. Ambulatory pH monitoring is helpful in detecting whether respiratory symptoms, including wheezing and coughing, are triggered by reflux.

A new diagnostic test was developed by Sandhill that allows the clinician to observe all reflux episodes ("total reflux monitoring"), not just those that are acid in nature ("pH" or "acid reflux monitoring"). This combines traditional pH with something referred to as "impedance" where the pH catheter is modified by containing a series of conductive rings. During patient monitoring a small current is sent between ring pairs. As the background conductivity changes (such as during swallows or during reflux), or its corresponding reciprocal, impedance, changes, one can separate swallows from true reflux and examine all reflux episodes—those that are non-acid as well as those that are acidic. As such, impedance monitoring measures the rate of fluid movements along the esophagus by detecting changes in resistance to alternating electrical currents (impedance) within the esophagus (Tutuian & Castell, 2006). The appearance of bolus in esophagus is seen as a rapid drop in impedance as the bolus improves the electrical conductivity between the two electrodes (due to increased ionic content of bolus), the impedance will remain low until the bolus passes (Tutuian & Castell, 2006). Combining these diagnostic tests allows for the determination of the quantity and quality of non-acidic or weakly acidic reflux episodes apart from acidic reflux events by determining exactly the exposure time and extent of the refluxed bolus.

Another test used in evaluating GERD is esophageal motility testing (DeVault & Castell, 2005). Motility testing is used to determine how well the muscles of the esophagus [e.g. LES (lower esophageal sphincter)] are working. Similar to esophageal acid testing, a catheter is passed down the nose and into the esophagus. Instead of the catheter having a pH sensor, it has a sensor for pressure and can include impedance sensors as well. The catheter is able to sense the pressure from the contraction of the LES as well as show bolus movement (e.g., the movement of swallowed material during the test). During this evaluation, the pressure during rest, relaxation and contraction of the muscle is recorded and the physician is able to determine whether the patient may also have a motility disorder.

SANDHILL'S DIAGNOSTIC PRODUCTS

Sandhill Scientific is a leader in the field of GI (gastrointestinal) diagnostics; they provide tools to allow physicians to enhance their diagnostic capabilities in both pH monitoring and impedance technology. Sandhill offers a comprehensive series of diagnostic products for esophageal reflux and esophageal motility. Since reflux episodes may or may not occur in the physician's offices or hospital, for monitoring reflux, a small, patient-worn device is used – called the "ZepHr" (Figure 2). This provides both "pH only" and "impedance/pH" monitoring. For general gastrointestinal diagnostics, the company has developed the "InSight" platform. This is a stationary device that can include modularized software packages and electronics to perform a number of diagnostic functions throughout the GI tract, such as esophageal function testing, sphincter of Oddi manometry, and analrectal manometry. While the company manufactures products used for reflux monitoring, it buys various catheter configurations used in GI tract analysis from various third parties.

DEFINING THE PROBLEM AND DECISION CRITERIA

THE PROBLEM

Sandhill Scientific set up an overseas manufacturing plant when the cost of production in the United States became too great. Sandhill manufactures their catheters for the ZepHr product line. The ZepHr catheters are disposable (single-use), therefore they need to be produced in high volume with low cost. The labor rate in the U.S. had increased to $14/hr, and when the impedance technology was first introduced to the company's standard "pH-only" catheter, each catheter was taking as much as approximately 3 man-hours to assemble. With the introduction the new impedance-pH technology, this assembly time would jump to approximately 7.5 man-hours. However, the targeted specification in developing the impedance-pH catheter was to sell the product for under $100.

The first challenge was to design the new line of catheters, which occurred in the company's proto-

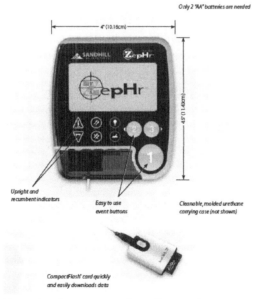

Only 2 "AA" batteries are needed

4" (10.16cm)

4.5" (11.43cm)

Upright and recumbent indicators

Easy to use event buttons

Cleanable, molded urethane carrying case (not shown)

CompactFlash™ card quickly and easily downloads data

Figure 2: ZepHr impedance/pH monitoring system allows for precise, accurate assessment for patients presenting with symptoms such as cough, heartburn, regurgitation and chest pain often are difficult to diagnose using traditional acid (pH) monitoring approaches. Combining impedance and pH, ZepHr enables physicians to reliably distinguish between patients with acid reflux association, nonacid reflux association and symptoms with no reflux association. Comprehensive analysis quantifies all reflux patterns and symptom associations in patients studied on or off acid suppression medication.

type development lab in Denver. As this was being done, a production facility was selected in Prague, Czech Republic. At the time the Czech labor rate was slightly more than US$2.00 per hour. Labor was plentiful as the unemployment rate was approaching 20%, and being in Prague would assist the company's market expansion into the Europe Union. In 2000, Sandhill was selling approximately 42,000 pH-only catheters per year. Launch of the facility was highly successful, however, and within three years the demand for the company's catheters outpaced the production capacity of this initial facility, so the company located additional space and transitioned production to a larger plant in Prague. In 2009, Sandhill sold roughly 60,000 pH-only catheters and 50,000 impedance-pH catheters.

THE DECISION CRITERIA

When determining a suitable location for an overseas plant, Mr. Jory used extensive qualitative and non-qualitative assessments in his due diligence. It was imperative that he had a comprehensive checklist of characteristics that he determined were necessary and important for Sandhill to have a successful plant overseas. To do this, he needed to have a complete understanding of the barriers to entry that were going to present themselves, and what would need to be overcome when doing business in a different country. Since time has a money value and relocating manufacturing would be disruptive, the future benefits must outweigh the projected costs. The company must be evaluated with and without the relocation.

Prague featured high unemployment, an ample pool of skilled workers available at competitive wages, and low turnover. The US dollar was fairly strong, with the exchange rate at 37.65 koruna per US dollar (see Figure 3). Another indicator of the stability or strength of the dollar has been proposed by *The Economist*, albeit a less informal way, to measure the purchasing power parity (PPP) between two currencies (dollar versus the koruna or euro), termed the "Big Mac Index". The Big Mac Index "seeks to make exchange-rate theory a bit more digestible" (Woodall, The Economist 1986). Essentially, the Big Mac Index compares the price of a Big Mac in one country versus another. UBS Wealth Management Research has expanded the idea of the Big Mac Index to include the amount of time that an average worker in a given country must work to earn enough to buy a Big Mac; giving a more realistic view of the purchasing power of the average worker. One method of predicting exchange rate movements is that the rate between two currencies should naturally adjust so that a Big Mac should cost the same in both currencies (this is the PPP or purchasing power parity). The Big Mac PPP exchange rate between two countries is obtained by dividing the price of a Big Mac in one country (in its currency) by the price of a Big Mac in another country (in its currency). This value is then compared with the actual exchange rate; if it is lower, then the first currency is under-valued (according to PPP theory) compared with the second, and conversely, if it is higher, then the first currency is over-valued. *The Economist* publishes these values annually. Figure 4 demonstrates the Big Mac Index for the U.S. versus the Czech

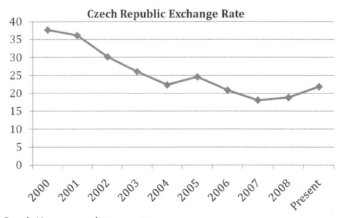

Figure 3: US Dollar to Czech Koruna exchange rate

	Big Mac Price (In local currency)	Big Mac Price (In dollars)	Implied PPP of the dollar	Actual $ exchange rate (04/25/2000 or 01/30/2009	% under (-)/ over(+) valuation against the dollar
United States (2000)	$2.51	2.51	-	-	-
Czech Republic (2000)	Koruna 54.37	1.39	21.7	39.1	-45
United States (2009)	$3.54	3.54	-	-	-
Czech Republic (2009)	Koruna 65.94	3.02	18.6	21.9	-15

Figure 4: Big Mac Index for U.S. and Czech Republic in 2000 and 2009

Republic in both 2000 and 2009. While there were minimal concerns regarding Czech taxation (in that Sandhill's Czech facility would be shipping product back to the U.S. virtually at cost and without profitability), Sandhill did discover, after the fact, that had their production facility been located outside the city limits of Prague they would have probably qualified for a European Union economic incentive that could have offset some of the start-up capitalization costs. Incentives ranged as high as forty cents on the dollar for each dollar invested in the initial start-up.

Many factors must be taken into consideration when moving a production facility from one locale to another. Relocation of a manufacturing facility has many strengths and weaknesses, opportunities and threats. There must absolutely be significant savings in order for a company to move their facilities halfway around the world. With proper research and due diligence, along with well thought out strategic designs, the relocation should be a profitable and will allow for a good return on the investment. Sandhill Scientific was able to establish an extremely successful plant in the Czech Republic as a result of careful and thorough due diligence on the part of Rick Jory. He was able to identify key criteria necessary to Sandhill and developed a successful strategy; he identified as many risks as possible and managed those risks carefully.

CONCLUSION

Rick Jory and the director of manufacturing, Katherine May, were having a discussion about the inflation and increasing labor rates in the Czech Republic, which were causing escalating manufacturing costs for Sandhill Scientific. Mr. Jory charged Ms. May with doing a sensitivity analysis on the impact of exchange rate fluctuations on Sandhill's financials and reporting aback to the upcoming board of directors meeting in a month with a recommendations on how they might the risks, May noted, "I met with an investment banker last week and we outlined three alternatives: (1) do nothing and accept exchange rate induced margin fluctuation; (2) use foreign exchange rate financial hedging instruments such as forward contracts and/or call options to 'lock in' costs[1]; (3) negotiate with customers to share some of the exchange rate variance. We need to do a bit more homework and we'll circle back with the pros and cons of each."

---

1   *Foreign exchange forward contract is a fixed commitment between two parties to exchange one currency for another at a forward or future date. Forward contracts call for delivery on a date beyond the Spot contract settlement, which ordinarily takes place within ten days of the transaction date. In contrast, a foreign exchange option is the right but not the obligation to buy or sell a specified amount of foreign currency on a fixed date and rate of exchange.*

REFERENCES

DeVault, K. R., & Castell, D. O. 2005. Updated guidelines for the diagnosis and treatment of gastroesophageal reflux disease. *American Journal of Gastroenterology*, 100(1): 190-200.

Ferri, F. F. 2009. *Ferri's Clinical Advisor 2010, 1st ed.*: Mosby.

Sandler, R. S., Everhart, J. E., Donowitz, M., Adams, E., Cronin, K., Goodman, C., Gemmen, E., Shah, S., Avdic, A., & Rubin, R. 2002. The burden of selected digestive diseases in the United States. *Gastroenterology*, 122(5): 1500-1511.

Tutuian, R., & Castell, D. O. 2006. Review article: complete gastro-oesophageal reflux monitoring - combined pH and impedance. *Aliment Pharmacology and Therapeutics*, 24 Suppl 2: 27-37.

Lumina Life Sciences: The Challenges of Raising Capital to take to Market a Promising Technology Innovation

MAGDA CHORUZY[1], ANDREW MAXWELL[1], MICHAEL ALVAREZ[2]

[1]*University of Toronto;* [2]*Stanford University*

- **Summary and key issues/decisions:** Lumina Life Sciences is an early stage company that has developed a combination drug-device technology for neurosurgery that improves surgical efficacy and efficiency. Dr. Arjen Bogaards, president of Lumina Life Sciences, has developed a patented technology that uses the fluorescent properties of certain drugs to emphasize the presence of tumor tissues. He has developed an imaging technology that then enables neurosurgeons to remove tumor cells more rapidly and with greater accuracy during the restriction process. At this point the venture needs additional funding in order to proceed, but from an investor standpoint the viability of the company is not clear. Significant concerns remain surrounding the company's ability to secure FDA approval on both the proprietary drugs and imaging technology, prove the efficacy of the solution, and resolve patent issues with the University Health Network (UHN) where the technology was developed. UHN will not agree to conclude an ongoing business relationship unless they can see a business plan that shows that successful market entry is likely. Such a demonstration will be challenging since the market for the solution is small, and it is unclear who benefits or will pay for the deployment of the innovative technology.

- **Companies/institutions:** Lumina Life Sciences, University Health Network in Toronto, Canada

- **Technology**: The Lumina technology is a novel imaging technology to aid surgeons in the removal of brain tumors using a fluorescent targeting drug to identify tumor cells. A targeting dye, TumorTarget®, localizes in the Glioblastoma multiforme (the most aggressive type of primary brain tumor) and causes the tumor to fluoresce. The SurdiGuide® platform, an imaging technology that comprises a visualization software platform and associated hardware technology, then enables the surgeon to visualize the effectiveness of tumor removal in real time during the surgery. This improves the efficiency of the procedure and is thought to improve surgical outcomes.

- **Stage of development at time of issue/decision**: The SurdiGuide® platform has been approved by Health Canada and the FDA for use in human clinical trials. TumorTarget® has been proven, through clinical studies, to provide enhanced fluorescence guided imaging in the brain. The next step is to conduct clinical trials on the drug-device combination, in order to obtain FDA approval, before the product can be marketed.

- **Indication/therapeutic use:** The Lumina technology has been developed to assist neurosurgeons in ensuring that they can more quickly remove all of the tumor cells in a brain tumor. The increased speed of the procedure will reduce the likelihood of complications, while the increased

---

* *This case was prepared as a basis for class discussion rather than to illustrate either effective or ineffective handling of an administrative situation.*

likelihood of removing all of the tumor cells, improves the long-term patient outcome (although the long term prognosis remains poor).

- **Geography**: US, Canada
- **Keywords:** brain tumor, drug-device, approvals, reimbursement, alternate business models, institutional IP policy, critical venture factors

INTRODUCTION

D r. Arjen Bogaards and his management team at Lumina Life Sciences have critical decisions to make in the coming months if they want to bring their novel drug-device technology to commercialization. This drug-device combination technology that provides fluorescence and visualization of brain tumors can potentially bring great benefit to neurosurgeons looking to remove tumors, but the path to commercial success seems even more challenging than the surgical procedure itself. The company needs to show that the technology improves patient outcomes, but the process to do this is both costly and timely, requiring external investment. Shortcuts to obtain such approvals and evidence more quickly may limit the long-term potential of the venture, while an inappropriate choice of entry strategy could limit the potential of ever making this technology commercially available to surgeons.

The company is faced with a number of similar dilemmas that, at first glance, seem to be the classic "chicken and egg situation." The research institution where the technology was developed will not license the technology to the inventor (Dr. Bogaards) without evidence that Lumira has secured a commercial partner. However, without existing legal rights to the pending patent, it will be difficult to convince prospective commercial partners to invest time in understanding the market potential for the product. In addition, although initial research evidence provides evidence of efficacy, continuing trials to confirm this will take time and may dampen investor enthusiasm. Finally, it is not clear that the identified niche market for brain tumor restriction is sufficient to create a new venture, and there are significant uncertainties as to an appropriate business model, which will encourage adoption, especially if there are issues in obtaining an appropriate reimbursement code.

BACKGROUND ON LUMINA LIFE SCIENCES

In early 2007, Dr. Arjen Bogaards, while undertaking research in Dr. Brian Wilson's laboratory in the biophysics and bioimaging division of the University Health Network (UHN) in Toronto, Ontario, Canada, brought together pharmaceutical and medical device technologies to develop a unique imaging technology known as SurdiGuide®. The discovery of the technology led Dr. Bogaards to form Lumina Life Sciences and take on the role of president. With hopes to bring the company to market success, the company was able to take a big step forward when it was granted approval by Health Canada and the FDA for the technology to be used in human clinical studies. This led to the production of four working prototypes, currently utilized in various clinical validation studies in Canada and the United States.

In September 2007, the UHN took responsibility for filing a patent for SurdiGuide® based on institutional Intellectual Property (IP) policies, which state that the institution owns all IP associated with technologies that are developed within their laboratories. At the same time, UHN accepts and assumes responsibility for maximizing the likelihood of such technologies reaching the commercial marketplace. This will both provide a net benefit to society, and generate licensing revenues for the institution. As such, before UHN will even enter license discussions with Lumina, the company will have to both prepare a compelling business plan that answers each of these questions (summarized at the end) and identify a commercialization partner. In addition, Lumina will have to show evidence to UHN that the revenue they will earn from this approach is adequate, and that the development of a spinout company represents a better opportunity than licensing the technology directly to an existing company.

EFFICACY ISSUES

The next steps for Lumina Life Sciences are technology and efficacy validation. Lumina Life Sciences has four clinical trials proceeding, two in the US and two in Canada. However, these are small scale clinical research studies designed to obtain some initial insights into technology effectiveness rather than randomized trials seeking to establish unequivocal statistical significance. Images from these trials are provided in Appendix E. Upon completion, in about 18 months, further studies will still be required before efficacy claims about the product can be made. This is partly due to the need to carry out the trials under controlled circumstances, and also the requirement to measure patient survival rates over longer time frames.

Lumina intends to apply for 510(K) regulatory approval medical device which requires that they demonstrate substantial equivalence to a similar product already marketed known as a predicate (Appendix A). While the 510(K) approval route is significantly less time and resources, it is important to note that product claims must also be equivalent to the predicate device. That is, superiority versus competitors cannot be claimed. As a result, it is difficult to make the case to health insurance companies that the use of Lumina technology should be considered for reimbursement, as there would be no additional benefit provided. If improvements in efficacy or efficiency can be shown with prospective, randomized multi-center clinical trials, then this situation could change. The ability to show the level of efficacy required to obtain funding, however, cannot be completed until funding is in place. Given that the company lacks financial resources and its ability to attract and raise capital in the future is uncertain, funding for the development of the technology represents a major challenge for Lumina.

The SurdiGuide® platform however is only one component of the Lumina technology; the platform is dependent on a tumor targeting dye, TumorTarget®. TumorTarget® was developed by an independent company and has been proven, through clinical studies, to provide enhanced florescence for guided imaging in the brain. Together, the TumorTarget® drug, which has the ability to dye the tumor, and the SurdiGuide® technology to visualize the presence of the dye in tumor tissues, make up the Lumina technology. However, neither TumorTarget® nor SurdiGuide® has been granted FDA approval at this stage, and they do not even have their patents issued. Dr. Bogaards is aware that it can take a long time and cost a considerable amount of money to secure both patents and obtain FDA approval.

POTENTIAL BUSINESS MODELS

The issue of payment for the use of the Lumina Life Science technology is multi-faceted, both in terms of who pays, and where the benefit arises. If the company cannot justify reimbursement for the product, because it is launched as an equivalent solution to other technologies, then the company needs to be clear on who pays. The most obvious person to pay is the patient, whose likelihood of survival may be increased. However, as the longterm prognosis is not good and the costs of the procedure high, this is only likely in a small number of cases. The insurer may be persuaded to pay as the use of Lumina technology likely reduces the likelihood of repeat procedures being required, and thus the overall cost of providing care. However, for this to be the case, the improved efficacy of the treatment would have to be demonstrated.

Others can also benefit from this technology. As the technology was designed to improve the efficacy of the surgeon, the surgeon might be willing to bear the cost of a technology that can improve outcomes, but this is somewhat unlikely, unless there are other compelling factors and/or external pressures. Furthermore, the hospital itself stands to benefit by being able to accommodate more of these procedures in the same theatre due to improvements in operating efficiencies; by enhancing outcomes, the hospital may also be able to attract additional patients and revenues.

The issues of payment for the use of the Lumina technologies are closely tied to the issue of finding the most appropriate business model for this technology. In the first case, there are a relatively small number of such procedures completed annually (Appendix D). Thus there would be a limited number of facilities that would regard this technology as a necessary cost. Total costs for the system include the technology, service and software licensing fees, and per-use costs associated with the TumorTarget® dye. Upfront investment in the technology (about $60,000) may make it of limited interest to most hospitals, as they may be concerned that it would not be used enough to justify the expense relative to marginal benefits of the system. For this reason Lumina has proposed a market and sales strategy that entails providing the SurdiGuide® technology platform at a very low cost (or possibly free) and then charging for the TumorTarget® drug as well as a pay per use fee. Despite the fact that this model has been used by other technologies, and seems to address some of the concerns of hospital administrators, it is challenging in many ways. For starters, it is difficult to forecast revenues for the company if the number of procedures at an installation is not known in advance. In addition, the deployment of Lumina technology solution across the country represents a large negative cash flow, and this represents a challenge for a company at this stage.

LUMINA'S MARKET OPPORTUNITY

The National Cancer Institute reports that the incidence rate of brain cancer is 6.4 per 100,000 men and women per year (SEER Cancer Statistics Review, 1975-2006) (refer to Appendix D for a list of the most common tumors in humans). In 2008 there were approximately 21,800 new cases of brain cancers diagnosed in the United States of America (NCI, 2008). This represents a 13% increase from 2007 with the trend likely to continue to grow. Nearly 60% of brain cancers are malignant Glioblastoma Multiforme (GBM), representing roughly 13,000 patients per year and the prognosis for a patient is dismal. Upon diagnosis of GBM, a patient will survive for an average of 3 months if no therapy is initiated. Assuming the patient is given standard of care therapy, prognosis still remains grim with only 50% of patients living after one year, 25% of patients living after two years, and less than 10% of patients alive after five years (Stein, 2008).

Treatment for GBM is initiated most often by attempting surgical resection, followed by chemotherapy and/or radiotherapy (British Journal of Neurosurgery, 2008). The main goal of surgery is to achieve the safest level of complete gross total resection (GTR) of the GBM without damaging healthy brain tissues. This is a challenging task even to the most skilled neurosurgeons as the tumor margins are poorly defined and difficult to visualize making it difficult to achieve GTR.

It can take many years of experience to distinguish between subtle differences in tumor and normal tissues, and even experienced surgeons vary in the completeness of tumor resection, highlighting the opportunity to improve standard of care. At present, a common tool used in surgical removal of GBM is Image Guided Surgery (IGS). This technology has a major drawback when used for GTR as static brain images are obtained pre-operatively, and the static picture is not a useful tool to navigate the dynamic environment of the brain as brain shifts occur throughout the surgical procedure as tumor is slowly removed. Intra-operative MRI surgery is another standard of care surgical procedure that allows multiple MRI images to be taken before, during, and after surgery allowing an increased chance to achieve GTR, however this is not really real time and MRI machines are expensive and in high demand in hospitals.

Currently the Lumina technology is the only that offers a quantitative approach to fluorescence GBM surgical removal in a clinical setting. The technology in the SurdiGuide® is unique in that it is able to quantify the completeness of the tumor removal. There are however a number of companies that have similar technologies that can be used outside the operating room (see Appendix E), as well as operating room imaging technologies that might be modified for this application. While this represents a competitive threat, as it is possible that these companies could find a way around the patent, it is also possible that these organizations could be interested in licensing the technology. However, as there is no company operating directly in this space, it is not clear whether or not any of these competitors would be better able or inclined to capture the specific market that Lumina Life Sciences seeks to enter itself. In addition, it seems likely that a non-traditional business model may be required, where instead of selling the technology outright, some form of charge per use, or embedded costs in the drug could be the business model.

LUMINA'S TECHNOLOGY

Lumina's unique technology uses the fluorescent properties of certain drugs, when present in tumors, to assist neurosurgeons in removing tumors more rapidly and with greater accuracy. (Pictures of the patent pending SurdiGuide® technology can be found in Appendix C, with images from the device shown in Appendix D). The targeting drug TumorTarget® localizes in the Glioblastoma multiforme and dyes the tumor causing it to fluoresce. This fluorescence can be visualized and measured by the SurgiGuide® platform, enabling the surgeon to visualize the tumor in real-time during the entire surgical procedure. This allows the neurosurgeon to operate with better accuracy and increase the likelihood of achieving the safest level of complete GTR.

Lumina technology is already proving effective in the four clinical studies being undertaken, but the approvals process is long, and the technology may require several rounds of development before the product can be commercialized. The company has collaborated with Kangaroo Design, a specialist medical device designer in Toronto, to develop a second generation of the device and develop a prototype for manufacturing. As more trials are designed and executed, particularly for the 510(K) approval, the company will have to establish new contracts with the manufacturer of TumorTarget®. If the 510(K) is granted, Lumina will need to evaluate prospective relationships with current and potential drug suppliers and other technology companies with which they might be able to co-operate profitably.

MANAGEMENT

Lumina Life Sciences Inc. is owned by Dr. Arjen Bogaards (PhD, MBA), with a board of experienced advisors (see Appendix F) who provide in-kind resources and advice that address each of the critical areas of business. The board members are part of a strategic alliance that has been established to enable the company to succeed. These individuals are responsible for providing input in areas such as prototype development, industrial design and manufacturing, business development, strategic business consulting, marketing and sales, regulatory affairs and quality assurance, scientific advisory board, and medical advisory board.

While Lumina presently functions as a virtual company, with no office and no full time personnel, it is clear that in the future the company will have to operate from its own facility or those of one of its partners. Currently Dr. Arjen Bogaards is the only dedicated employee, making it challenging for him to address each of the many critical business and technological development issues simultaneously. In addition, the inability to sign a license deal with UHN in advance of attracting finance is restricting the options open to Dr. Bogaards. It is therefore up to Dr. Bogaards to develop a plan that will be able to attract the required stakeholders. If you were providing Arjen some advice, how would you suggest he proceed to make Lumina more attractive to potential investors, in order to best commercialize the technology? How would you suggest he addresses the question raised by the technology transfer office of UHN, about the advantages and disadvantages of commercializing this technology through a new venture in contrast to collaborating with an existing company that already provides operating room technology?

APPENDICES

APPENDIX A: SUMMARY EXPLANATION OF 501(K) FDA APPROVALS

MEDICAL DEVICE APPROVAL IN THE US

In the US, medical devices are subject to controls detailed in the *Federal Food Drug and Cosmetic Act*. These controls detail the baseline requirements for all medical devices, and include conditions for marketing, proper labeling, and monitoring of performance once the device is on the market. Lumina must have 510(k) approval met in order to entice an in-licensing company or sign a licensing agreement with UHN.

The US Food and Drug Agency (FDA) designates three classifications for medical devices, according to which manufactures must obtain specific approvals to market their device. Class I devices are routine devices which are generally exempt from premarket notification 510(k) (e.g., Band-Aids); Class II devices require 510(k) pre-marketing notification (e.g., intubation tubes); and Class III devices are more complex and require pre-market approval (e.g., cardiovascular stents). Generally, the 510(k) approval allows a company to market their device based on proof that it is substantially equivalent to another device already on the market known as a predicate. Substantive equivalence to a predicate is determined based on many factors, including but not limited to: intended use, design, energy used or delivered, materials, chemical composition, manufacturing process, performance, safety, effectiveness, labeling, biocompatibility, and standards. For small companies, the FDA has a number of programs to make the application process simpler and less costly.

APPENDIX B: TOP TEN CANCER SITES 2009 IN THE UNITED STATES

Leading Sites of New Cancer Cases and Deaths – 2009 Estimates

Estimated New Cases*		Estimated Deaths	
Male	**Female**	**Male**	**Female**
Prostate 192,280 (25%)	Breast 192,370 (27%)	Lung & bronchus 88,900 (30%)	Lung & bronchus 70,490 (26%)
Lung & bronchus 116,090 (15%)	Lung & bronchus 103,350 (14%)	Prostate 27,360 (9%)	Breast 40,170 (15%)
Colon & rectum 75,590 (10%)	Colon & rectum 71,380 (10%)	Colon & rectum 25,240 (9%)	Colon & rectum 24,680 (9%)
Urinary bladder 52,810 (7%)	Uterine corpus 42,160 (6%)	Pancreas 18,030 (6%)	Pancreas 17,210 (6%)
Melanoma of the skin 39,080 (5%)	Non-Hodgkin lymphoma 29,990 (4%)	Leukemia 12,590 (4%)	Ovary 14,600 (5%)
Non-Hodgkin lymphoma 35,990 (5%)	Melanoma of the skin 29,640 (4%)	Liver & intrahepatic bile duct 12,090 (4%)	Non-Hodgkin lymphoma 9,670 (4%)
Kidney & renal pelvis 35,430 (5%)	Thyroid 27,200 (4%)	Esophagus 11,490 (4%)	Leukemia 9,280 (3%)
Leukemia 25,630 (3%)	Kidney & renal pelvis 22,330 (3%)	Urinary bladder 10,180 (3%)	Uterine corpus 7,780 (3%)
Oral cavity & pharynx 25,240 (3%)	Ovary 21,550 (3%)	Non-Hodgkin lymphoma 9,830 (3%)	Liver & intrahepatic bile duct 6,070 (2%)
Pancreas 21,050 (3%)	Pancreas 21,420 (3%)	Kidney & renal pelvis 8,160 (3%)	Brain & other nervous system 5,590 (2%)
All sites 766,130 (100%)	All sites 713,220 (100%)	All sites 292,540 (100%)	All sites 269,800 (100%)

*Excludes basal and squamous cell skin cancers and in situ carcinoma except urinary bladder.

©2009, American Cancer Society, Inc., Surveillance and Health Policy Research

Source: American Cancer Society, 2009

APPENDIX C: SURGIGUIDE® FLUORESCENCE IMAGING SYSTEM

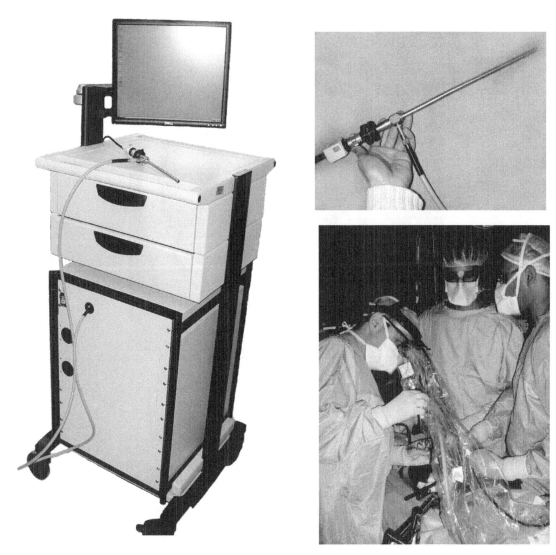

Upper left: the proprietary SurgiGuide® fluorescence imaging system, upper right: the camera head and laparoscope, lower right: the imaging system in use during radical prostatectomy.
Source: Lumina

APPENDIX D: IMAGING AND CLINICAL EVIDENCE FROM APPLICATION OF LUMINA TECHNOLOGY
Source: Lumina

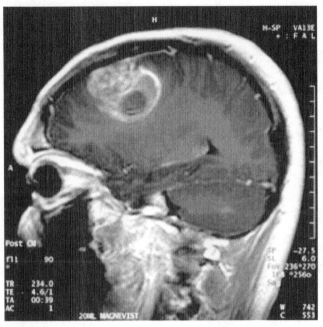

Figure 1: Brain image, showing tumor

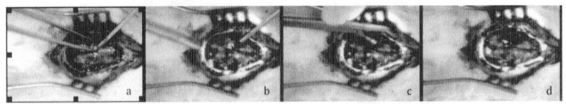

Figure 2: Demonstration of fluorescence guided brain tumor resection. In (a) we show the white light image of the anterior brain in a preclinical model. The forceps point to the surgical cavity after resection of a VX2 brain tumor under white light. In (b) the red PpIX fluorescence in surgical cavity is identified, while in (c) we show the resection of red fluorescing tissues until no fluorescence is detected in the cavity as shown in (d).

Figure 3: Demonstration of image quantification in patients with prostate cancer. a) A white light image of the prostate capsule with forceps around a nodule. (b) The raw unprocessed raw fluorescence image showing small amounts of blue reflectance, green autofluorescence of the prostate capsule and bright red fluorescence of the nodule. c) The same fluorescence image now made quantitative through image processing according to method Q1. As can be observed most of the anatomical/structural information is lost. To alleviate this problem this image is thresholded (d) and overlaid on the raw fluorescence image so that the resulting (e) image contains both structural/anatomical information as well as functional quantitative information.

APPENDIX E: POTENTIAL STRATEGIC ALLIANCE PARTNERS

Possible license partner	Details
Novadaq (TSX: NDQ)	Develops medical imaging systems and real-time image guided procedures for the operating room. Novadaq's intellectual property focuses on the visualization of blood vessels, nerves, lymphatics, and other tissue during a variety of open and minimally invasive surgical procedures. FDA approved products include Spy® Imaging System for cardiac surgery and Pinpoint™ Autofluorescence Endoscopic Imaging System –assist with thoracic surgery. Novadaq has an experienced direct Sales and Clinical Education team in US and is looking for new technology platforms. In May 2007, they purchased Xillix business for $1million
Leica Microsystems GmbH	One of the world's leading manufacturers of microscopes and related scientific equipment for research in industries including healthcare. Leica manufactures a wide range of neurosurgical microscopes. SurgiGuide® would enhance Leica's portfolio making the company and its products more attractive to its neurosurgeon customer base.
Karl Storz	A leader in minimally invasive technologies enforced by rigid or flexible endoscopes used in intracranial surgery. The endoscopes facilitate the neurosurgeon's view into the brain more so than their current microscopic products, allowing them to carry out more extensive and precise surgery. Recipient of 2006 Market Penetration Leadership Award for the company's remarkable growth in the US endoscopic industry. Karl Storz has built a reputation for itself as one of the most responsive and customer-centric companies in terms of the service, technology, and training it offers. The Karl Storz reputation for precision, clarity, ergonomics, and versatility will increase acceptance and adoption rate of SurgiGuide® by the neurosurgeons. Karl Storz's brand name and customer relationship is an added benefit for the commercialization of SurgiGuide®.
General Electric Healthcare (NYSE: GE)	General Electric (GE) Healthcare is a $17 billion unit of GE and in 2008 the division received the "Pioneer in Technology" award from The International Brain Mapping & Intraoperative Surgical Planning Society. GE Healthcare offers a broad range of market-leading intra-operative imaging products such as C-arm. GE Healthcare also offers a broad range of diagnostic and imaging technologies including CT, MRI, and PET. GE Healthcare is continuously seeking to be the leader in the state-of-the-art imaging platforms. It is a market leader in quantifiable fluorescence technology. With a large sales force and markets in more than 100 countries, GE offers a large network for global launch of SurgiGuide®.

APPENDIX F: BOARD OF ADVISORS

PROTOTYPE DEVELOPMENT

Robert Weersink, Ph.D., Scientific Associate, Laboratory of Applied Biophotonics (LAB), Toronto. Since 2002, Robert has been the director of operations of the Photonics Research Ontario - UHN Biophotonics Facility. His primary activities are working with industrial, clinical and academic collaborators on biophotonics projects. These have included photodynamic therapy instrumentation and planning, the use of nanotechnology for disease diagnosis, optical glucose monitoring, and ex corporeal blood treatment using UV light.

INDUSTRIAL DESIGN AND MANUFACTURING

Lahav Gil, CEO and Director of Design, Kangaroo Design Inc. Lahav Gil is a senior industrial designer and product development expert with 22 years of experience in commercialization of emerging technologies and actively participated in the design and launch of numerous high technology products resulting in more than $200 million in sales. Lahav is the CEO and founder of Kangaroo Design Inc., a Toronto based product development firm offering services to the medical devices/diagnostic industries, operating in Canada, Europe, Israel and the Far East.

BUSINESS DEVELOPMENT

Joe Miller, Business Development Officer, Research Business Development Office, UHN, Toronto. Joseph joined the RBDO in July 2006, Prior to joining UHN, he was a licensing associate in the technology transfer office at the Wistar Institute in Philadelphia. Before Wistar, he helped to create and develop Shamrock Structures (a structural proteomics start-up out of Chicago). Joe was also involved with the clinical management department at Advanced Life Sciences Inc and holds a BSc in science and business from the University of Notre Dame.

STRATEGIC BUSINESS CONSULTING

Andrew Maxwell MBA P.Eng, Chief Innovation Officer at Kangaroo Design. Andrew started his career working for two technology multinationals, in Europe and North America and left to found two environmental technology companies with his customers. Since them he went on to create a wireless, medical device and web company. He subsequently joined the U of Toronto's Innovation Foundation and founded its technology incubator. He teaches at Universities of Toronto and Waterloo, and is pursuing a Ph.D. in the area of new venture creation at the University of Waterloo.

MARKETING AND SALES

Mathijs Gajentaan, Marketing Manager at VisualSonics Inc, Toronto. Mathijs had over 10 years international marketing experience in developing, managing and executing marketing plans & budgets. He is now marketing manager for VisualSonics a medical device company that specializes in high resolution ultrasound machines for the pre-clinical market. Previously Mathijs is a strategic and creative thinker and an entrepreneur with a hands-on mentality.

REGULATORY AFFAIRS AND QUALITY ASSURANCE

Daryl Wisdahl, Senior Consultant at the Emergo Group, Inc. Daryl Wisdahl brings a wealth of experience in medical device regulations in the Canadian, American and European marketplaces. Prior to his work at Emergo Daryl worked with Xillix Technologies, a company commercializing fluorescence imaging products for cancer detection. As such Daryl has relevant expertise to the business proposed here. Daryl held positions as director at BC Medical Technology Industry Association and director of RA at VSM MedTech.

SCIENTIFIC ADVISORY BOARD

Prof. Brian Wilson, Head Division of BioPhysics and Imaging, Ontario Cancer Institute, Toronto. Prof Wilson is deemed one of the world's authorities in biophysics applied to photodynamics and fluorescence diagnostics. The focus of the research the division of biophysics and imaging is the development and application of new therapeutic and diagnostic techniques based on the use of lasers and other optical technologies.

Thomas S. Mang, Ph.D., Ass. Professor at University of Buffalo, Dept. of oral and maxillofacial surgery. Dr. Tom Mang previously held a position as director of PDT marketing for Diomed, a world leader in diode lasers for photodynamic therapy (PDT). Mang has a wealth of experience with years of clinical involvement at PDT centers of excellence, holding positions as PDT director at the Roswell Park Cancer Institute and department of radiation oncology at Buffalo General Hospital.

MEDICAL ADVISORY BOARD

Paul J. Muller, MD, MSc, FRCSC, Professor Department of NeuroSurgery at University of Toronto. Dr. Muller entered the Neurosurgical Training Program at the University of Toronto in 1970. He received his FRCSC in Neurosurgery in 1975. Dr. Muller then undertook graduate studies in the Institute of Medical Sciences at The University of Toronto, and received his Masters on the topic of experimental neuro-oncology. He is currently professor in the department of surgery at the University of Toronto. His main clinical and research interests are in the field of photodynamic therapy and fluorescence-guided surgery for patients with malignant gliomas.

Victor Yang, MD, PhD, Resident at the Dept of NeuroSurgery, University of Toronto. Victor has been involved in the development of earlier generations of the fluorescence imaging system presented here. Victor received a Honors BASc in Engineering Science Biomedical option from the University of Toronto in 1997. He then went on to complete a Masters in Electrical and Computer Engineering in 1998.

FOUNDER AND MANAGING DIRECTOR

Arjen Bogaards. Over the last 10 years Arjen has been involved in "Bench-top to Bedside" research specializing in the development of drug-device combination products and its evaluation in clinical trials. Arjen held academic and industry positions and build up a broad network of scientists, clinicians, medical device and pharmaceutical companies. Prior to Lumina, Arjen acted as chief technology officer for Steba Biotech N.V. where Arjen was involved with commercializing a novel treatment for prostate cancer based on administration of a drug that requires activation by laser light. As CTO, Arjen was responsible for the development of prototype medical devices and their manufacture. As a scientific associate at the University Health Network in Toronto, is author of over 10 peer reviewed scientific articles and continues to be involved in the researching applications of light sensitive drugs.

REFERENCES

American Cancer Society Inc. *Cancer Facts and Figures 2005*, Atlanta, Georgia 2005

Anik, I., Cabuk, B., Ceylan S., Koc, K. 2008. Fluorescein sodium-guided surgery in glioblastoma multiforme: a prospective evaluation. *British Journal of Neurosurgery*, 22(1):99-103.

Crous, A.T., Drouin, R., Fortin D., Gadji, M., Klonisch, T., Krcek, J., Mai, S., Torchia, M. 2009. EGF receptor inhibitors in the treatment of glioblastoma multiform: Old clinical allies and newly emerging therapeutic concepts. *European Journal of Pharmacology*, 625 (1-3): 23-30.

National Cancer Institute, 2008. *Brain Tumour.* www.cancer.gov/cancertopics/types/brainSEER Cancer Statistics Review, 1975-2006, National Cancer Institute. Bethesda, MD, http://seer.cancer.gov/csr/1975_2006/, based on November 2008 SEER data submission. Accessed February 14, 2010.

Stein, Rob. 2008. Malignant Gliomas Affect About 10,000 Americans Annually. *Washington Post*, May 20. Accessed 10 February 2010.

RESOURCES

Resources

BIOTECHNOLOGY INDUSTRY

Biospace
http://www.biospace.com/

Biotechnology Industry Organization
http://www.bio.org/

Biotechnology Institute
http://biotechinstitute.org/index.php

Ernst & Young Life Sciences Reports
http://www.ey.com/US/en/Industries/Life-Sciences

Fierce Biotech
http://www.fiercebiotech.com/

Genetic Engineering & Biotechnology News
http://www.genengnews.com/

Journal of Commercial Biotechnology
http://www.palgrave-journals.com/jcb/index.html

Nature Biotechnology
http://www.nature.com/bioent/index.html

Pharmaceutical Research and Manufacturers of America (PhRMA)
http://www.phrma.org/

US FDA (Food and Drug Administration)
http://www.fda.gov/

BIOENTREPRENERSHIP

BlueMineGroup: Maximizing the Value of Innovation
http://www.blueminegroup.com/

Invention 2 Venture Workshops in Technology Entrepreneurship
http://www.invention2venture.org/

Kauffman Foundation
http://www.kauffman.org/

Society of Physician Entrepreneurs
http://www.sopenet.org/

The Entrepreneur's Guide to a Biotech Startup
http://www.evelexa.com/index.cfm

United States Association for Small Business and Entrepreneurship
http://usasbe.org/

FINANCING

Angel Capital Association
http://www.angelcapitalassociation.org/

AUTM (The Association of Technology Transfer Managers)
http://www.autm.net/home.htm

National Venture Capital Association
http://www.nvca.org/

OnBioVC (Bioscience Venture Capital Data)
http://onbiovc.com/

PricewaterhouseCoopers MoneyTree
https://www.pwcmoneytree.com/MTPublic/ns/index.jsp

Small Business Innovation and Research Grant Program
http://www.sbir.gov/

VC Experts: Private and Venture Capital Expertise
http://vcexperts.com/vce/library/encyclopedia/glossary.asp

Related titles from Logos Press

http://www.logos-press.com

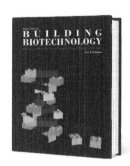

Building Biotechnology

Scientists know science; businesspeople know business. This book explains both.

Hardcover ISBN: 978-09734676-5-9
Softcover ISBN: 978-09734676-6-6

Best Practices in Biotechnology Education

22 International Best Practices in K-12, College, Certificate, Master's, Doctoral, MBA, Distance Education Programs and Student Groups

ISBN: 978-0-9734676-7-3

Best Practices in Biotechnology Business Development

Valuation, Licensing, Cash Flow, Pharmacoeconomics, Market Selection, Communication, and Intellectual Property

ISBN: 978-09734676-0-4

CPSIA information can be obtained at www.ICGtesting.com
Printed in the USA
LVOW021232281212

313547LV00002B/4/P